Latest Edition

KT-556-319

Exercise Biochemistry

Vassilis Mougios, PhD

Associate Professor,
University of Thessaloniki, Greece

Human Kinetics

Library of Congress Cataloging-in-Publication Data

Mougios, Vassilis, 1958-
    Exercise biochemistry / Vassilis Mougios.
        p. ; cm.
    Includes bibliographical references and index.
    ISBN-13: 978-0-7360-5638-0 (hard cover)
    ISBN-10: 0-7360-5638-6 (hard cover)
    1. Exercise--Physiological aspects.   2. Muscle--Metabolism.
3. Biochemistry.   I. Title.
    [DNLM: 1. Exercise--Physiology.   2. Movement--physiology.
3. Biochemistry.   4. Muscles--metabolism   WE 103 M9245e
2006]
    QP301.M693 2006
    612'.044--dc22

                                        2006007134

ISBN-10: 0-7360-5638-6
ISBN-13: 978-0-7360-5638-0

Copyright © 2006 by Vassilis Mougios

Acquisitions Editor: Michael S. Bahrke, PhD; **Developmental Editor:** Renee Thomas Pyrtel; **Assistant Editor:** Kevin Matz; **Copyeditor:** Joyce Sexton; **Proofreader:** Pamela Johnson; **Indexer:** Betty Frizzéll; **Permission Manager:** Dalene Reeder; **Graphic Designer:** Bob Reuther; **Graphic Artist:** Dawn Sills; **Photo Manager:** Sarah Ritz; **Cover Designer:** Bob Reuther; **Photographer (interior):** Sarah Ritz, unless otherwise noted. Photo on page 89 © Human Kinetics. Photos on pages 102 and 103 © Getty Images; **Art Manager:** Kelly Hendren; **Illustrator:** Paris Parissis, Keri Evans (figures 3.10, 8.5, and III.1); **Printer:** Versa Press

Printed in the United States of America   10   9   8   7   6   5   4   3   2   1

Human Kinetics
Web site: www.HumanKinetics.com

United States: Human Kinetics
P.O. Box 5076
Champaign, IL 61825-5076
800-747-4457
e-mail: humank@hkusa.com

Canada: Human Kinetics
475 Devonshire Road Unit 100
Windsor, ON N8Y 2L5
800-465-7301 (in Canada only)
e-mail: orders@hkcanada.com

Europe: Human Kinetics
107 Bradford Road
Stanningley
Leeds LS28 6AT, United Kingdom
+44 (0) 113 255 5665
e-mail: hk@hkeurope.com

Australia: Human Kinetics
57A Price Avenue
Lower Mitcham, South Australia 5062
08 8277 1555
e-mail: liaw@hkaustralia.com

New Zealand: Human Kinetics
Division of Sports Distributors NZ Ltd.
P.O. Box 300 226 Albany
North Shore City
Auckland
0064 9 448 1207
e-mail: info@humankinetics.co.nz

To the insecurity that makes us better

Contents

Preface **xiii** • A Guided Tour for the Student **xv** • Acknowledgments **xvii**

PART I BIOCHEMISTRY BASICS 1

Chapter 1 Introduction.........................3

1.1	Chemical Elements	3
1.2	Chemical Bonds	4
1.3	Molecules	5
1.4	Ions	6
1.5	Polarity Influences Miscibility	6
1.6	Solutions	7
1.7	Chemical Reactions and Equilibrium	7
1.8	pH	8
1.9	Acid–Base Interconversions	9
1.10	Classes of Biological Substances	10
1.11	Cell Structure	11
	Problems and Critical Thinking Questions	12

Chapter 2 Metabolism 13

2.1	Free-Energy Changes Earmark Metabolic Reactions	14
2.2	Determinants of Free-Energy Change	15
2.3	ATP, the Energy Currency of Cells	16
2.4	Phases of Metabolism	17
2.5	Oxidation–Reduction Reactions	18
2.6	Overview of Catabolism	21
	Problems and Critical Thinking Questions	24

Chapter 3 Proteins 25

3.1	Amino Acids	25
3.2	The Peptide Bond	28
3.3	Primary Structure of Proteins	29
3.4	Secondary Structure	30
3.5	Tertiary Structure	31
3.6	Denaturation	32
3.7	Quaternary Structure	33
3.8	Protein Function	34

3.9   Oxygen Carriers 35
3.10   Myoglobin 35
3.11   Hemoglobin 36
3.12   The Wondrous Properties of Hemoglobin 37
3.13   Enzymes 39
3.14   The Active Site 39
3.15   Enzymes Affect the Rate But Not the Direction of Reactions 40
3.16   Factors Affecting the Rate of Enzyme Reactions 42
       Problems and Critical Thinking Questions 43

Chapter 4   Nucleic Acids and Gene Expression 45
       4.1   Introducing Nucleic Acids 45
       4.2   Flow of Genetic Information 46
       4.3   Deoxyribonucleotides, the Building Blocks of DNA 46
       4.4   Primary Structure of DNA 48
       4.5   The Double Helix of DNA 48
       4.6   The Genome of Living Organisms 49
       4.7   DNA Replication 50
       4.8   Mutations 53
       4.9   RNA 55
       4.10   Transcription 56
       4.11   Genes and Gene Expression 57
       4.12   Messenger RNA 58
       4.13   Translation 59
       4.14   The Genetic Code 60
       4.15   Transfer RNA 61
       4.16   Translation Continued 62
       4.17   In the Beginning, RNA? 64
       Problems and Critical Thinking Questions 66

Chapter 5   Carbohydrates and Lipids 67
       5.1   Carbohydrates 67
       5.2   Monosaccharides 68
       5.3   Oligosaccharides 69
       5.4   Polysaccharides 70
       5.5   Lipids 72
       5.6   Fatty Acids 73
       5.7   Triacylglycerols 74
       5.8   Phospholipids 75
       5.9   Steroids 77
       5.10   Cell Membranes 77
       Problems and Critical Thinking Questions 79

PART I   SUMMARY 81

PART II BIOCHEMISTRY OF THE NEURAL AND MUSCULAR PROCESSES OF MOVEMENT 85

Chapter 6   Neural Control of Movement 87

6.1   Nerve Signals Are Transmitted in Two Ways 88
6.2   The Resting Potential 89
6.3   The Action Potential 91
6.4   Propagation of an Action Potential 93
6.5   Transmission of a Nerve Impulse From One Neuron to Another 94
6.6   Birth of a Nerve Impulse 96
6.7   The Neuromuscular Junction 98
6.8   A Lethal Arsenal at the Service of Research 101
        Problems and Critical Thinking Questions 103

Chapter 7   Muscle Contraction 105

7.1   Structure of a Muscle Cell 105
7.2   The Sliding-Filament Theory 107
7.3   The Wondrous Properties of Myosin 108
7.4   Structure of Myosin 108
7.5   Actin 109
7.6   Sarcomere Architecture 110
7.7   Mechanism of Force Generation 111
7.8   Myosin Isoforms and Muscle Fiber Types 113
7.9   Control of Muscle Contraction 114
7.10 Excitation–Contraction Coupling 114
        Problems and Critical Thinking Questions 117

PART II   SUMMARY 119

PART III EXERCISE METABOLISM 121

III.1   Principles of Exercise Metabolism 121
III.2   Exercise Parameters 122
III.3   Experimental Models Used to Study Exercise Metabolism 122
III.4   Five Means of Metabolic Control in Exercise 124
III.5   Four Classes of Energy Sources in Exercise 126

Chapter 8   Compounds of High Phosphoryl Transfer Potential 127

8.1   The ATP-ADP Cycle 127
8.2   The ATP-ADP Cycle in Exercise 128
8.3   Creatine Phosphate 129
8.4   Window Into the Sarcoplasm 131
8.5   Loss of AMP by Deamination 133
        Problems and Critical Thinking Questions 135

Chapter 9   Carbohydrate Metabolism in Exercise **137**

9.1   Glycogen Metabolism 138
9.2   Exercise Speeds Up Glycogenolysis in Muscle 143
9.3   The Cyclic-AMP Cascade 144
9.4   Recapping the Effect of Exercise on Muscle Glycogen Metabolism 147
9.5   Glycolysis 148
9.6   Exercise Speeds Up Glycolysis in Muscle 150
9.7   Pyruvate Oxidation 152
9.8   Exercise Speeds Up Pyruvate Oxidation in Muscle 154
9.9   The Citric Acid Cycle 155
9.10 Exercise Speeds Up the Citric Acid Cycle in Muscle 156
9.11 The Electron Transport Chain 157
9.12 Oxidative Phosphorylation 158
9.13 Energy Yield of the Electron Transport Chain .159
9.14 Energy Yield of Carbohydrate Oxidation 160
9.15 Exercise Speeds Up Oxidative Phosphorylation in Muscle 161
9.16 Lactate Production in Muscle During Exercise 162
9.17 Features of the Anaerobic Carbohydrate Catabolism 164
9.18 Utilizing Lactate 164
9.19 Gluconeogenesis 165
9.20 Exercise Speeds Up Gluconeogenesis in the Liver 168
9.21 The Cori Cycle 170
9.22 Exercise Speeds Up Glycogenolysis in the Liver 170
9.23 Control of the Plasma Glucose Concentration in Exercise 173
9.24 Blood Lactate Accumulation 176
9.25 Blood Lactate Decline 177
9.26 "Thresholds" 178
       Problems and Critical Thinking Questions 179

Chapter 10 Lipid Metabolism in Exercise **181**

10.1   Triacylglycerol Metabolism in Adipose Tissue 181
10.2   Exercise Speeds Up Lipolysis 185
10.3   Fate of the Lipolytic Products During Exercise 188
10.4   Fatty Acid Degradation 190
10.5   Energy Yield of Fatty Acid Oxidation 193
10.6   Fatty Acid Synthesis 193
10.7   Exercise Speeds Up Fatty Acid Oxidation in Muscle 196
10.8   Changes in the Plasma Fatty Acid Concentration and Profile During Exercise .197
10.9   Interconversion of Lipids and Carbohydrates .199
10.10 Plasma Lipoproteins 200
10.11 A Lipoprotein Odyssey 201
10.12 Effects of Exercise on Plasma Triacylglycerols 205
10.13 Effects of Exercise on Plasma Cholesterol 207
10.14 Exercise Increases Ketone Body Formation 209
        Problems and Critical Thinking Questions 211

Chapter 11 Protein Metabolism in Exercise **213**

   11.1 Protein Metabolism 213
   11.2 Effect of Exercise on Protein Metabolism 214
   11.3 Amino Acid Metabolism in Muscle During Exercise 215
   11.4 Amino Acid Metabolism in the Liver During Exercise 219
   11.5 The Urea Cycle 220
   11.6 Amino Acid Synthesis 222
   11.7 Plasma Amino Acid, Ammonia, and Urea Concentrations During Exercise 223
   11.8 Contribution of Proteins to the Energy Expenditure of Exercise 223
   11.9 Effects of Training on Protein Metabolism 224
         Problems and Critical Thinking Questions 225

Chapter 12 Effects of Exercise on Gene Expression **227**

   12.1 Stages in the Control of Gene Expression 228
   12.2 Which Stages in the Control of Gene Expression Does Exercise Affect? 229
   12.3 Kinetics of a Gene Product After Exercise 230
   12.4 Exercise-Induced Changes That May Modify Gene Expression 231
   12.5 Mechanisms of Exercise-Induced Muscle Hypertrophy 232
   12.6 Mechanisms of Exercise-Induced Mitochondrial Biogenesis 235
         Problems and Critical Thinking Questions 236

Chapter 13 Integration of Exercise Metabolism **237**

   13.1 Interconnections of Metabolic Pathways 237
   13.2 Energy Systems 239
   13.3 Energy Sources in Exercise 239
   13.4 Choice of Energy Sources During Exercise 241
   13.5 Effect of Exercise Intensity on the Choice of Energy Sources 241
   13.6 Effect of Exercise Duration on the Choice of Energy Sources 243
   13.7 Interaction of Duration and Intensity:
         Energy Sources in Running and Swimming 244
   13.8 Effect of the Exercise Program on the Choice of Energy Sources 246
   13.9 Effect of Heredity on the Choice of Energy Sources in Exercise 247
   13.10 Conversions of Muscle Fiber Types 248
   13.11 Effect of Nutrition on the Choice of Energy Sources During Exercise 249
   13.12 Adaptations of the Proportion of Energy Sources
         During Exercise to Endurance Training 250
   13.13 How Does Endurance Training Modify
         the Proportion of Energy Sources During Exercise? 251
   13.14 Adaptations of Energy Metabolism to Anaerobic Training 252
   13.15 Effect of Age on the Choice of Energy Sources During Exercise 253
   13.16 Do Sex and Ambient Temperature Affect
         the Choice of Energy Sources During Exercise? 254
   13.17 The Proportion of Fuels Can Be Measured Bloodlessly 254
   13.18 Hormonal Effects on Exercise Metabolism 254
   13.19 Fatigue 257
   13.20 Central Fatigue 258

13.21 Peripheral Fatigue 259
13.22 Restoration of the Energy State After Exercise 260
13.23 Metabolic Changes in Detraining 264
       Problems and Critical Thinking Questions 265

PART III   SUMMARY 267

PART IV   BIOCHEMICAL ASSESSMENT
          OF EXERCISING PERSONS 273
       IV.1 The Blood 274
       IV.2 Aims and Scope of the Biochemical Assessment 274
       IV.3 The Reference Interval 275
       IV.4 Classes of Biochemical Parameters 276

Chapter 14 Iron Status 277
       14.1 Hemoglobin 278
       14.2 Hematologic Parameters 278
       14.3 Does Sports Anemia Exist? 279
       14.4 Iron 280
       14.5 Total Iron-Binding Capacity 280
       14.6 Transferrin Saturation 280
       14.7 Soluble Transferrin Receptor 281
       14.8 Ferritin 281
       14.9 Iron Deficiency 282
             Problems and Critical Thinking Questions 283

Chapter 15 Metabolites 285
       15.1 Lactate 285
       15.2 Estimating the Anaerobic Lactic Capacity 286
       15.3 Programming Training 287
       15.4 Estimating Aerobic Endurance 288
       15.5 Glucose 289
       15.6 Triacylglycerols 289
       15.7 Cholesterol 290
       15.8 HDL Cholesterol 290
       15.9 LDL Cholesterol 291
       15.10 Recapping Cholesterol 292
       15.11 Glycerol 292
       15.12 Urea 292
       15.13 Ammonia 293
       15.14 Creatinine 293
             Problems and Critical Thinking Questions 294

Chapter 16   Enzymes and Hormones **295**

       16.1   Enzymes 295
       16.2   Creatine Kinase 296
       16.3   Aminotransferases....................................297
       16.4   Steroid Hormones 297
       16.5   Cortisol...300
       16.6   Testosterone..300
       16.7   Overtraining..301
       16.8   Epilogue..302
              Problems and Critical Thinking Questions...............303

PART IV   SUMMARY **304**

Answers to Problems and Critical Thinking Questions   **305**

Glossary   **309**

Suggested Readings   **316**

References   **319**

Index   **320**

About the Author   **332**

Preface

Exercise Biochemistry examines how exercise affects the functioning of human and animal organisms at the molecular level. The main tool for such an endeavor can be none other than the principles of basic biochemistry, which occupy part I of the book. These are then applied to the state of excitation of the nervous and muscular systems to produce muscle contraction; part II deals with this. How exercise modulates metabolism forms the core of the book, part III. Finally, part IV describes the use of simple biochemical tests to assess an exercising person's health and performance.

The discipline of exercise biochemistry is a rather young child of a rather young mother. The mother (biochemistry) is a little over one century old. During this period, she has taken huge steps (even leaps) toward understanding the phenomenon of life in its most intricate details. As a result, she has offered invaluable services to the welfare of human beings. All health sciences today depend on biochemical findings for their own success. Purely biochemical topics occupy a considerable part of textbooks on physiology, pathology, microbiology, pharmacology, and so on. Bright minds from the areas of other health sciences and chemistry have become biochemistry converts. And biochemistry (having no Nobel prize established for herself) has repeatedly looted the Nobel prizes in chemistry and physiology or medicine.

As for the child, it possesses all the freshness, grace, and promise hidden in youth. Exercise biochemistry aspires to answer the countless "hows" and "whys" born of observation and experimentation on physical activity, through the delicate, precise, and demanding language of molecular interactions. It is a language that takes human knowledge—itself a product of immensely complex molecular interactions—through one of its hardest ordeals. As such, exercise biochemistry serves as an abutment to work physiology, sport medicine, sport nutrition, and other branches of the health sciences focusing on exercise.

The main purpose of this book is to serve as a self-inclusive source of teaching material for undergraduate courses in exercise biochemistry. Given the introduction of more and more courses of this kind in universities around the world and the indispensability of this discipline for an understanding of how exercise changes bodily functions, I believe the time is ripe for exercise biochemistry to have its own, stand-alone textbooks. Furthermore, it is my firm belief that the publication of such textbooks will encourage more institutions to introduce exercise biochemistry courses in their curricula. Having the necessary elements of basic biochemistry along with extensive coverage of exercise biochemistry topics in a single volume should facilitate the work of both instructors and students.

Exercise Biochemistry may also prove useful to graduate students in sport science who did not have the chance to be formally introduced to the discipline during their undergraduate years. Additionally, it can supplement exercise physiology textbooks by covering the molecular basis of physiological processes described therein. The book is also addressed to physical education and sport professionals with an interest in how the human body operates during and after exercise. Finally, the book is addressed to those health scientists who have been impressed or wish to be impressed by the transformations brought about in human metabolism by physical activity.

There is a lot of beauty in modern biochemistry and exercise biochemistry. This beauty stems primarily from the revelation and apprehension of the ingenious ways in which animals have been solving the problems of self-conservation, reproduction, and response to the demands of movement on planet Earth for some hundred million years now, no one knows through how many failed efforts. But there is also beauty in more tangible things such as countless drawings and microscopic images (quite a few of which I present), as well as similarities with works of art like the one on the following page.

Although books have been traditionally viewed as monologues of the authors, I have tried, to the best of my abilities, to give this one the character of a conversation with the reader (dare I say, a long chat about exercise biochemistry). Therefore, I have interspersed the text with questions that sprang up in my head when I was a student (or later), questions that my students ask frequently during my teaching, and questions that I ask to tease them.

I have written this book with the understanding that many of its readers may not feel comfortable with the laws of chemistry and science in general. Therefore, I have used simple language, albeit without compromising scientific accuracy. Whether I have managed to maintain this delicate balance is up to the reader to decide. Likewise, it's up to the reader to decide if this work has succeeded in its ultimate goal: to arm the future sport scientist or physical education instructor with knowledge that will serve his or her professional, scientific, and personal needs, as well as protect him or her from the ignorance, smatter, and misinformation that persist in many areas of exercise and sport. In the end, all of us hopeless hunters of the elusive trophy of excellence will find consolation in the words that Peter Ustinov put in the mouth of Mr. Smith (the Devil) in one of his exquisite dialogues with the Old Man (God):

. . . there is nothing in all of your Creation as sterile, as lifeless, as overwhelmingly negative as perfection.

▶ Interior of an ancient Greek pot of the sixth century BCE filled with scenes of physical activity. In the perimeter, 17 young men dressed in women's gowns are dancing in worship of Hercules. In the middle, the demigod wrestles with Triton, a sea monster, to force him to tell where the golden apples of the Hesperides are kept.

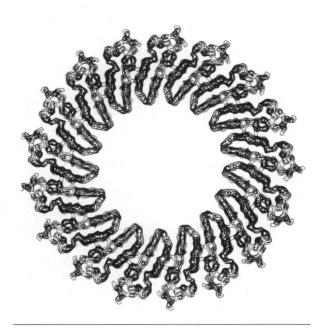

▶ Electron density map of a protein disc from tobacco mosaic virus. Protein molecules in a helical array surround the genetic material of the virus, which attacks the leaves of the tobacco plant. Seventeen such molecules form a full circle. The similarity with the previous image lies not only in the number but also in the angle and shape of the protein molecules, which look like people holding hands in a circular dance.

Courtesy of Dr. Aaron Klug.

A Guided Tour for the Student

The book is divided into 16 chapters organized in four parts. Part I contains basic biochemical information, necessary for the unhindered reading of subsequent parts. I begin with an introduction to elementary bits of chemistry and biology (chapter 1). This is followed by an overture to metabolism through the presentation of the general principles governing the exchange of mass and energy in living organisms, along with the molecules starring in this exchange (chapter 2). The next three chapters acquaint the reader with the four major classes of biological molecules. Chapter 3 belongs to proteins, participants in almost every biochemical process. We examine the proteins in terms of both structure and function, with special reference to proteins involved in oxygen transport and catalysis of reactions. Chapter 4 deals with nucleic acids and the flow of genetic information from DNA to RNA to protein. Finally, chapter 5 describes the structure of carbohydrates and lipids, leaving their metabolism to part III.

In part II we explore the biochemical basis of the neural (chapter 6) and muscular processes (chapter 7) that enable, for example, our lungs to fill with air, our heart to beat, and our eyes to move from left to right and back again as we read these lines. This is an amazing sequence of delicate, precise, and coordinated processes endowing us with the synonym of life, movement.

How does the metabolism of humans and animals change with exercise? Part III answers this question. Its first four chapters examine the effects of exercise on the metabolism of the four classes of compounds supplying energy, that is, compounds of high phosphoryl transfer potential (chapter 8), carbohydrates (chapter 9), lipids (chapter 10), and proteins (chapter 11). Chapter 12 describes how exercise alters the expression of our genes, a prerequisite for adaptations to training. Finally, chapter 13 integrates all previous chapters of part III by examining the interaction and interdependence of energy sources during exercise, the adaptations to the various types of training, and matters related to the interruption of exercise and training.

Part IV shows how biochemical tests can aid an athlete or, generally, an exercising person. The three chapters of this part examine two dozen parameters providing useful information on health and performance. I divide these into parameters indicating the iron status (chapter 14), metabolites (chapter 15), enzymes, and hormones (the latter two in chapter 16). Each parameter has its own value and makes a unique contribution to forming the picture of an exercising person's health and performance.

In the page margins you will find explanatory notes, bits of nonbiochemical information aiding in understanding the text, brief descriptions of complicated processes such as biochemical pathways, and trivia. Naturally, margins are for your own notes too. Finally, the margins contain three kinds of icons that mark the presentation of issues related to disease, sport nutrition, and doping.

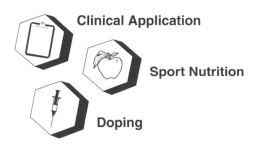

Each chapter closes with problems and critical thinking questions, and each part closes with a summary. At the end of the book, you will find references, suggested readings, answers to the problems and questions, a glossary, and an index. Glossary terms are in color and boldface in the text; other important terms and key concepts are emphasized with italics.

Acknowledgments

A number of people contributed to bringing this book to completion. Paris Parissis created most of the art in a professional and conscientious manner. My wife Maria and daughter Niki created a warm and supporting environment at home, where I did most of my writing. I am grateful to my associates Anatoli Petridou and Michalis Nikolaidis for carefully reviewing the original manuscript. Anatoli, in addition, oversaw the creation of many drawings and saved me valuable time by taking over laborious teaching and research tasks. Mike Bahrke, my acquisitions editor, was the one who, back in 2001, inspired in me the idea of writing an international book and supported me all the way through the submission of the manuscript. My developmental editor, Renee Thomas Pyrtel, assisted me in greatly improving the manuscript. Joyce Sexton did fine copyediting work. Dalene Reeder managed the permissions, Kevin Matz assisted throughout the publication process, Dawn Sills skillfully laid out the book, and Pamela Johnson carefully proofed it. Finally, I owe many thanks to the students who attended my classes and worked in the lab during the past 17 years. Through their love, inquisitiveness, encouragement, and demand, they became the catalysts (biochemically speaking, the enzymes) of this book.

Biochemistry Basics

Organic chemistry is the chemistry of carbon compounds. Biochemistry is the study of carbon compounds that crawl.

—Mike Adams

Biochemistry is one of the sciences dealing with the phenomenon of life. Along with biology, medicine, pharmacology, and other related sciences, it examines living organisms and their components with the aim of deciphering their structure and function. Humans exploit the knowledge produced by biochemical research to improve the quality of life and the environment.

What distinguishes biochemistry from other cognate sciences is that it moves at the deepest possible level, the fundamental level of life's expression: the level of atoms, molecules, and their interactions. The major tool for such a science can be none other than the laws of chemistry. That is why it is called biochemistry, that is, the chemistry of life.

Part I of the book you are holding offers an acquaintance with the building blocks of living organisms, the principles governing their operation, and certain basic expressions of life, such as the catalysis of biochemical reactions and the flow of genetic information. In a nutshell, part I offers biochemical knowledge that is a prerequisite for entry into the world of exercise biochemistry in parts II, III, and IV.

▶ The introductory figure of part I (facing page) is a bridge between the world of mathematics and the natural environment. Named a Mandelbrot set, it is one of the magnificent creations of fractal geometry, a not-so-well-known branch of geometry, which aspires to describe natural shapes and forms with unprecedented accuracy and grace. This particular figure—a product of complex mathematical algorithms—resembles one of the stages of gene expression: the exodus of mRNA (the snakelike shapes in the center) from the nucleus (the black orb occupying the bottom) to the cytosol and the membranes of the endoplasmic reticulum. See chapter 4 for a full description of the process.

CHAPTER **1**

Introduction

In this introductory chapter of part I we examine elements that are indispensable for the unhindered reading of the remainder of the book. In particular, the chapter contains elementary *concepts* of chemistry, a presentation of the classes of biological substances, and a brief description of the cell and its components.

1.1 Chemical Elements

The matter that surrounds us, and the matter that we are made of, is composed of *chemical elements* such as hydrogen, oxygen, and carbon. The smallest unit of each element that maintains its properties is the *atom.* Every atom consists of the *nucleus* and the surrounding *electrons.* The nucleus contains *protons,* which carry a positive electric charge, and *neutrons,* which are neutral. Electrons carry a negative charge of equal absolute value to that of protons. Thus, an atom—in which the number of electrons equals the number of protons—is electrically neutral.

Electrons move around a nucleus not in definite orbits, but within *atomic orbitals.* Each atomic orbital is described by a complex mathematical equation from which chemists can calculate the probability of finding an electron in a specific position relative to the nucleus. Schematically, the position of electrons is depicted by *electron clouds,* which are denser where there is a high probability of finding electrons than where there is a low probability of finding electrons.

Ninety-two elements exist in nature. In living organisms, however, we find just about 29. Six of them occupy the overwhelming majority of the mass of biological substances. These elements are hydrogen (symbolized as H), carbon (C), nitrogen (N), oxygen (O), phosphorus (P), and sulfur (S). Table 1.1 presents some features of these elements.

▶ Table 1.1 The Most Abundant Elements in Living Organisms

Name	Symbol	Atomic number	Atomic mass (Da)[a]	Bonds with other atoms
Hydrogen	H	1	1	1
Carbon	C	6	12	4
Nitrogen	N	7	14	3
Oxygen	O	8	16	2
Phosphorus	P	15	31	5
Sulfur	S	16	32	2

[a]Rounded off to the nearest integer and expressed as daltons.

Atomic Number

The atomic number of an element is the number of protons in its nucleus. For the elements in table 1.1, it ranges from 1 to 16. In contrast to the number of neutrons in the nucleus, the atomic number is characteristic of and unique to every element. This means that two atoms of the same element have to have the same number of protons, but their number of neutrons may differ. If this is the case, the atoms are called *isotopes*. For example, while the vast majority of carbon atoms in nature have six protons and six neutrons (these atoms are symbolized as ^{12}C), a small percentage (1.1%) possess seven neutrons (these atoms are therefore symbolized as ^{13}C).

Atomic Mass

Atomic mass (also known as *atomic weight*) of an element is the mass of one of its atoms. The unit of atomic mass is the *dalton*. It is symbolized as Da and defined as one-twelfth of the mass of a ^{12}C atom. One dalton is an inconceivably small mass, just $1.66 \cdot 10^{-24}$ g.

Rounded up to the nearest integer, the atomic mass coincides with the sum of protons and neutrons in the nucleus of the predominant isotope of each element, for the relatively light elements. Thus, the atomic mass of hydrogen is approximately 1 Da, and its main isotope has only one proton in its nucleus, while the atomic mass of phosphorus is almost 31 Da, and its predominant isotope has 15 protons and 16 neutrons.

Their mass having been described, it is worth completing the picture of atoms with reference to their size. The size, too, is infinitesimal. As a unit of measure we use the *angstrom* (Å), which equals 10^{-10} m. The atomic diameter of the six most abundant elements in living organisms ranges from 0.7 to 2.2 Å.

Before we proceed, and since we have already considered some units, it is useful to remember that in order to express multiples and submultiples of units, we frequently add prefixes to their symbols; some prefixes are presented in table 1.2.

▶ **Table 1.2** The Most Common Prefixes of Units

Symbol	Name	Equivalent to
M	mega	10^6
k	kilo	10^3
d	deci	10^{-1}
c	centi	10^{-2}
m	milli	10^{-3}
μ	micro	10^{-6}
n	nano	10^{-9}
p	pico	10^{-12}

1.2   Chemical Bonds

Atoms form chemical bonds with atoms of the same element or different elements. A chemical bond requires at least two electrons, which in most cases are contributed mutually by the atoms participating in the bond. According to the current scientific view, the movement of these electrons is constrained. Thus, whereas before bond formation it is described by isolated atomic orbitals, after bond formation it obeys the equations of new orbitals, termed *molecular orbitals*.

The bond formed when two atoms share electrons is called *covalent*. If each atom contributes one electron, a single bond is formed, symbolized as a thin line between the atoms. Two atoms may be linked by a double or even a triple bond. These are formed by two or three electron pairs and are symbolized as a double or triple line, respectively.

The number of bonds that an atom can form is dictated by the distribution of electrons in its atomic orbitals. Knowing this number is essential for the correct construction of molecular formulas, as we will see later. The numbers of covalent bonds formed by the six most abundant elements in living organisms are shown in the last column of table 1.1. Thus, the atoms of the elements that compose biological molecules can be joined covalently to one, two, three, or four atoms. (Note that although table 1.1 shows that

P can form five bonds, two of them are directed toward one atom as a double bond. Therefore, P bonds with four atoms. See figure 2.3 for examples.)

1.3   Molecules

Atoms are joined by covalent bonds to form molecules. For example, two hydrogen atoms connected by a single bond form a hydrogen molecule. If the molecules of a substance are composed of atoms belonging to different elements, then the substance is called a *compound*. Water, consisting of two hydrogens linked to an oxygen, is the most abundant compound in our bodies.

Molecular Formula

What a compound is made of is depicted by its molecular formula, which contains the symbols of the elements present in the compound along with the numbers of their atoms as subscripts if they exceed 1. Thus, the molecular formula of water is H_2O. From a molecular formula we can calculate the *molecular mass,* that is, the sum of the atomic masses of the elements constituting the molecule (naturally, first we have to multiply the atomic mass of every element by the number of its atoms in the compound). Like atomic mass, molecular mass is measured in daltons. If, alternatively, we express it in grams, we get one *mole* of the compound, which is symbolized as mol. Thus, the molecular mass of water is 18 Da ($1 \cdot 2 + 16$), and 1 mol of it is 18 g.

Constitutional Formula

In addition to the kind and number of atoms in the molecule of a compound chemists are interested in the way the atoms are connected. Such information is provided by constitutional formulas, which we can construct by knowing the number of bonds that each atom can form. This is where the last column of table 1.1 comes into play. The rule is that each atom has to be surrounded by as many bonds as that column dictates. Verify the rule by examining the compound of figure 1.1a, which is the amino acid alanine. (Amino acids are dealt with in chapter 3.)

Constitutional formulas can be detailed (showing all bonds among atoms) or abbreviated. For example, we can simplify the formula of alanine by substituting —COOH for the group of atoms at the right-hand side of the molecule, known in organic chemistry as the *carboxyl group* (figure 1.1b). Likewise, we can substitute —NH_2 for the group of atoms at the left-hand side, known as the *amino group,* and —CH_3 for the group of atoms at the top, known as the *methyl group.* Two compounds may have the same molecular formula but different constitutional formulas because of different configurations of the same atoms. Such compounds are called *isomeric.*

Two compounds differ in *configuration* if one cannot be converted to the other without the breaking and reformation of certain covalent bonds. If, on the contrary, one can be converted to the other by the mere twisting of part of it, we say that they differ in *conformation.*

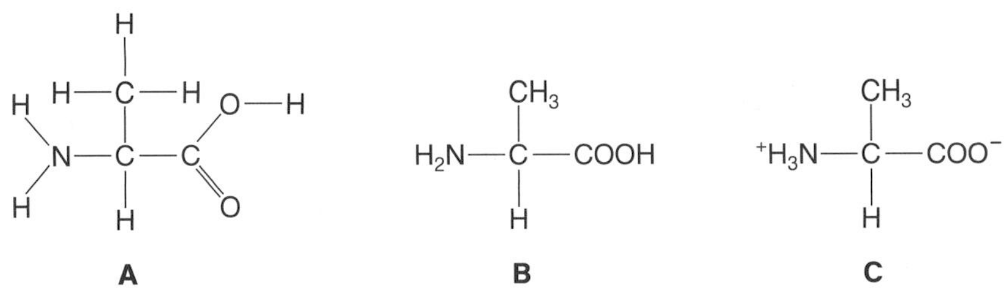

▶ **Figure 1.1**   Constitutional formulas. Chemical compounds arise by the linking of atoms of different elements with covalent bonds. Pictured here are constitutional formulas of a relatively simple biological compound, alanine. All bonds are shown in *a,* whereas for the sake of brevity only the bonds around the central carbon atom are shown in *b.* This form of representation (omitting the most common bonds) is more usual. In aqueous solutions such as biological fluids, alanine is ionized *(c).*

1.4   Ions

Molecules are electrically neutral, as they are composed of neutral atoms. However, most compounds in biological fluids are in the form of ions; that is, they carry electric charges. This happens because some atoms are more stable if they have more or fewer electrons than the number tallying with their protons.

A molecule in a biological fluid can be easily converted to an ion through the exchange of one or more hydrogen ions (H^+) with its surroundings. H^+ is nothing more than a proton; thus, it is extremely mobile. Where does H^+ come from? Water itself dissociates to a small degree to H^+ and hydroxyl ion (OH^-), thus supplying material for the formation of ions.

Because of their structure, some groups, such as the carboxyl group, have the tendency to release a proton, thus acquiring a negative charge ($-COO^-$). In contrast, other groups, like the amino group, have the tendency to attract a proton, thus acquiring a positive charge ($-NH_3^+$). An example of an ion (in fact, an ion bearing two charges) is presented in figure 1.1c. Note that an ion such as that of alanine may be neutral as a whole if the positive charges equal the negative ones. Positively charged ions are termed *cations;* negatively charged ions are *anions;* and ions bearing both kinds of charge are *zwitterions* (*zwitter* is German for "both at the same time").

Most ions have their electrons in pairs, but some do not. These are called *radicals* (or free radicals, a redundant term), and we usually denote their unpaired electron by a dot in addition to the charge symbol. For example, the superoxide radical, produced by the addition of an electron to an oxygen molecule, is symbolized as $O_2^{\cdot-}$. Other radicals, such as the hydroxyl radical ($HO^{\cdot}$, not to be confused with the hydroxyl ion), carry no charge and are not ions. As we will see in section 9.11, radicals are produced naturally in the body and increase during exercise.

1.5   Polarity Influences Miscibility

Although the positive and negative charges are equal in the neutral molecules of any compound, they may not be evenly distributed. The reason is that the nuclei of some atoms (notably N and O) attract bonding electrons more strongly than the nuclei of other atoms do. Thus, atoms of the former kind acquire a *partial negative charge* (symbolized as δ^- and pronounced delta minus), whereas atoms of the latter kind acquire a *partial positive charge* (δ^+). In such molecules, we can discern a negative and a positive electric pole, and we call the compound **polar** (figure 1.2). On the other hand, if charges are evenly distributed within the molecules of a compound, we call it **nonpolar.**

The *polarity* of a substance (that is, whether and to what degree it is polar) affects one of its important physical properties: the *miscibility* (that is, the ability to mix) with other substances. Here's how:

▶ **Figure 1.2** Polar compound. An uneven distribution of charges in the molecule of a compound results in the appearance of two poles having partial positive and negative charges.

- Polar substances tend to mix with polar substances.
- Nonpolar substances tend to mix with nonpolar substances.
- A polar and a nonpolar substance do not mix readily.

These interactions verify the saying "Birds of a feather flock together."

1.6   Solutions

When the mixing of two or more substances results in a homogeneous mixture, that is, a mixture having the same composition all over its mass, we call it a solution. In a solution, we usually distinguish the *solvent,* that is, the substance present in the highest proportion, from the *solute(s),* that is, the dissolved substance(s). We define the **concentration** of a solute as the amount of it that is contained in a certain amount of solution or solvent. A common unit of concentration is mole per liter of solution (mol/L, or mol · L^{-1}, or M). This is referred to as molar concentration.

The term concentration is often used loosely for mixtures that do not qualify as solutions (they are not homogeneous), such as many biological fluids. The concentrations of substances dissolved in biological fluids are relatively low. The highest ones are in the order of 10^{-2} mol · L^{-1}, whereas the lowest ones are as low as 10^{-12} mol · L^{-1}.

The solvent in biological systems is water. It is a polar compound (figure 1.3) and, as such, it mixes readily with other polar compounds, which are thus called **hydrophilic** (meaning "water loving" in Greek). Sugar is an example of a hydrophilic compound. In contrast, nonpolar compounds do not mix readily with water and are thus called **hydrophobic** ("water fearing"). Oils are examples of hydrophobic compounds: Their droplets are repelled by water and tend to aggregate and form drops. Naturally, nonpolar compounds dissolve in nonpolar solvents. A common nonpolar solvent is the spot remover used to take away oily stains from clothes.

Polar is hydrophilic; nonpolar is hydrophobic.

▶ **Figure 1.3**   The polarity of water. The molecule of water is polar for two reasons. First, its two bonds are not aligned; rather, they form a 105° angle. Second, the oxygen nucleus attracts the bonding electrons more strongly than the hydrogen nuclei do. Thus, oxygen carries a partial negative charge, whereas the area between the hydrogens presents a partial positive charge.

1.7   Chemical Reactions and Equilibrium

Often the compounds present in a mixture do not stay inert but interact to form new compounds. These interactions are called chemical reactions. A chemical reaction differs from a physical process in that in the latter, no new compounds are produced. For example, the dissolution of sugar in a cup of coffee is a physical process, since no new compounds are formed (sugar molecules just go from being embedded in a crystal to being surrounded by water molecules). In contrast, the burning of a piece of wood involves several chemical reactions (one of which is the conversion of cellulose to carbon dioxide and water).

Is life the outcome of chemical reactions or physical processes? Both! Thousands of reactions take place within organisms to enable them to produce energy and build their components. On the other hand, thousands of physical processes, such as the dissolution or diffusion of an ion into a biological fluid or the binding together of two molecules, let biological molecules interact and convey messages. In fact, most biological processes involve both chemical and physical interactions in the aqueous environment of biological fluids.

The substances participating in a chemical reaction are termed *reactants,* while the substances produced are the *products.* To represent a reaction chemists write its equation, which includes two sides usually divided by a one-way arrow or two opposite arrows. The following is an example of a chemical equation.

$$H_2N-\overset{\overset{\textstyle O}{\|}}{C}-NH_2 + H_2O \rightleftharpoons 2\,NH_3 + CO_2 \qquad\qquad \text{(equation 1.1)}$$

Equation 1.1 represents the **hydrolysis,** that is, the breakdown by water, of urea—a simple biological compound—to ammonia and carbon dioxide. The bidirectional arrows signify that the reaction is *reversible,* and this is the rule for chemical reactions. Which way a reaction will go depends on energy factors that we will explore in section 2.1. If a reaction is left to proceed far enough, it reaches a state in which no change in the concentration of any of the participating substances is detected. This is called the equilibrium and is characterized by the **equilibrium constant**, K_{eq}, which is the ratio of the molar concentrations of the products to the molar concentrations of the reactants. For equation 1.1,

> The formula or the name of a substance enclosed in brackets denotes its concentration. For example, NH_3 denotes the concentration of ammonia. $[NH_3]$ or [ammonia] denotes the concentration of ammonia.

$$K_{eq} = \frac{[NH_3]^2 \, [CO_2]}{[H_2NCONH_2] \, [H_2O]}$$

(equation 1.2)

If a substance participates in a reaction by a number of molecules that is different from 1 (as is the case with ammonia in our example), then we raise its concentration to this number in the K_{eq} expression—thus, $[NH_3]^2$ instead of $[NH_3]$.

The value of K_{eq} depends on the nature of the reactants and products, the temperature, the pressure, and, occasionally, the presence of other substances in the reaction medium. Certain reactions go almost completely (*quantitatively*, as we say) in one direction under certain circumstances and are characterized as *irreversible*. In these cases, we are allowed to use a one-way arrow.

> Unlike chemical reactions, nuclear reactions involve the vanishing of existing atoms and the formation of new ones.

Chemical reactions are governed by the *principle of mass conservation,* which dictates that atoms neither form nor vanish (they are only rearranged). Because of this, the two sides of a chemical equation must have the same kind and number of atoms. To ensure the principle of mass conservation, one may have to add numbers in front of some reactants or products. In the case of equation 1.1, we had to insert the number 2 in front of ammonia.

Reactions involving ions are additionally governed by the *principle of charge conservation,* which requires that the algebraic sum of charges be equal on the two sides of the equation. To achieve this, one may have to add one or more H^+ to one of the two sides.

A *balanced* chemical equation is one that complies with the two principles just laid out. Compared to an unbalanced equation, it has the advantage of offering quantitative as well as qualitative information on the reaction it represents. In other words, it shows not only which compounds participate in the reaction and what the products are, but also what the proportions of molecules and moles are. Equation 1.1 informs us that one mole(cule) of urea and one mole(cule) of water produce two mole(cule)s of ammonia and one mole(cule) of carbon dioxide. All chemical equations in this book are balanced.

1.8   pH

The ease with which protons are detached from chemical compounds or added to them (as we saw happening during the formation of ions) endows them with an important role in chemical processes. Because $[H^+]$ is usually very low (for example, $10^{-7} \, mol \cdot L^{-1}$ in pure water), chemists have established a convenient index of it, pH. pH is defined as the negative decimal logarithm of the molar concentration of protons.

$$pH = -\log_{10} [H^+], \text{ or } [H^+] = 10^{-pH}$$

(equation 1.3)

pH is a dimensionless quantity (it has no units). The pH of pure water is 7 because of the dissociation of a minuscule fraction of its molecules to H^+ and OH^-. When the pH of a solution is 7, the solution is *neutral* (figure 1.4). If an acid (defined here as a proton donor) is added, then the $[H^+]$ increases and the pH drops (because of the negative sign on the right-hand side in equation 1.3).

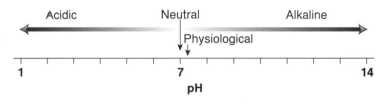

▶ **Figure 1.4**   The pH scale. pH is an index of a solution's acidity and plays an important role in biochemical processes. Shown here is its most useful range (⁻ to 14).

The solution then becomes *acidic* (examples of acidic solutions are lemon juice and vinegar). If, on the other hand, a base (defined here as a proton acceptor) is added to a neutral solution, then the [H⁺] decreases and the pH rises. The solution then becomes *alkaline* or *basic* (examples of alkaline solutions are whitewash and soapy water). The pH of most biological fluids is nearly neutral and is called physiological. As we will see in subsequent chapters, this pH can change with exercise. The pH of the blood is 7.3 to 7.4.

Do not confuse the electric neutrality of a molecule with the pH neutrality of a solution; the two concepts are not related.

1.9 Acid–Base Interconversions

If a compound in aqueous solution can exchange protons with its environment, its form is not fixed but depends on the pH of the solution. Such a compound binds protons when the pH decreases (because the [H⁺] increases) and loses protons when the pH increases (because the [H⁺] decreases). Let's explore these transitions by using alanine as an example.

The predominant form of alanine in a neutral solution is the one shown in figure 1.1c. If an acid (such as HCl) is added, its H⁺ will tend to associate with the negatively charged carboxyl group of alanine and neutralize it.

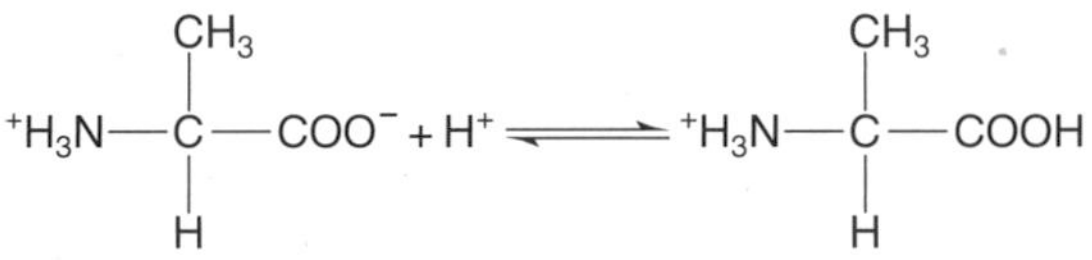

$$^+H_3N-\underset{\underset{H}{|}}{\overset{\overset{CH_3}{|}}{C}}-COO^- + H^+ \rightleftharpoons {}^+H_3N-\underset{\underset{H}{|}}{\overset{\overset{CH_3}{|}}{C}}-COOH \qquad \text{(equation 1.4)}$$

or $A + H^+ \rightleftharpoons AH^+$ if we substitute A for alanine. We refer to the K_{eq} of the reverse reaction (that is, the dissociation of AH⁺ to A and H⁺) as the *dissociation constant, K*.

$$K = \frac{[A][H^+]}{[AH^+]} \qquad \text{(equation 1.5)}$$

We further define pK as –log K. The pK of equation 1.4 is 2.3. We can introduce pK and pH into equation 1.5 by taking the logarithms of its two sides.

$$\log K = \log \frac{[A]}{[AH^+]} + \log [H^+] \qquad \text{(equation 1.6)}$$

Remember that log (x · y) = log x + log y.

Therefore,

$$-\log [H^+] = -\log K + \log \frac{[A]}{[AH^+]} \qquad \text{(equation 1.7)}$$

Therefore,

$$pH = pK + \log \frac{[A]}{[AH^+]} \qquad \text{(equation 1.8)}$$

Two compounds differing by one H⁺ are known as conjugate acid (the protonated one) and conjugate base.

which is known as the *Henderson-Hasselbalch equation*. This equation establishes a relationship between pH, pK, and the concentration ratio of a base to its conjugate acid, which permits the calculation of any of the three when the other two are known.

An interesting relationship arises when pH = pK. According to the Henderson-Hasselbalch equation,

$$\log \frac{[A]}{[AH^+]} = 0$$

therefore,

$$\frac{[A]}{[AH^+]} = 1$$

or

$$[A] = [AH^+]$$

Thus, *equal concentrations of a conjugate acid and base are present when pH = pK.* In the case of alanine, half of it carries an ionized and half of it carries an un-ionized carboxyl group at pH 2.3. It also follows from the Henderson-Hasselbalch equation that *the conjugate acid predominates when pH < pK, whereas the conjugate base predominates when pH > pK.* This is why the carboxyl group of alanine is deprotonated in a neutral solution.

Let's consider now what happens when we add a base (such as NaOH) to a neutral alanine solution. By analogy to the previous discussion, the OH$^-$ of the base will tend to remove H$^+$ from the positively charged amino group and neutralize it.

$$^+H_3N-\overset{\displaystyle CH_3}{\underset{\displaystyle H}{\overset{|}{\underset{|}{C}}}}-COO^- + OH^- \rightleftharpoons H_2N-\overset{\displaystyle CH_3}{\underset{\displaystyle H}{\overset{|}{\underset{|}{C}}}}-COO^- + H_2O \qquad \text{(equation 1.9)}$$

Equation 1.9 is characterized by a different pK equaling 9.9. Thus, at pH 9.9, half of alanine will carry an ionized amino group and half of it will carry an un-ionized amino group, while the amino group will be protonated in a neutral solution.

Throughout this book, constitutional formulas of ionizable compounds will depict the predominant form at physiological pH. Additionally, the names of acids will reflect the fact that they are ionized at physiological pH. Thus, we will refer to phosphate rather than phosphoric acid, aspartate rather than aspartic acid, palmitate rather than palmitic acid, lactate rather than lactic acid, and so on.

Because changes in the pH of a solution affect the form of the ionizable solutes, which in turn affects the interactions and reactions among them, cells and multicellular organisms have devised ways to maintain the pH of their fluids constant. For example, as mentioned in the previous section, the pH of the blood remains between 7.3 and 7.4. To protect the pH of their fluids against fluctuations caused by the production of acids or bases, living organisms have compounds of high **buffering capacity,** defined as the amount of strong acid or base needed to change the pH by one unit. These compounds usually come in pairs of conjugate acid and base, which are interconverted upon absorbing H$^+$ or OH$^-$ (as in equations 1.4 and 1.9), thus preventing H$^+$ and OH$^-$ from changing the pH.

A *buffer system* consisting of a conjugate acid–base pair in solution is not efficient at every pH value. Rather, its buffering capacity is maximal at pH = pK, where the concentrations of the conjugate acid and base are equal. The buffering capacity gradually decreases as one draws away from the pK. We will consider three important buffer systems, based on proteins, bicarbonate, and phosphate, later in this book (sections 3.5, 3.12, and 8.4, respectively).

1.10   Classes of Biological Substances

Studying biochemistry is facilitated by dividing biological substances into classes. More versatile and more interesting are the organic compounds, those based on carbon. There are four major classes of biological organic compounds: **proteins, nucleic acids, carbohydrates,** and **lipids.**

These classes differ greatly in structure and function. Most of their molecules are big and are characterized as *macromolecules*. How big are these macromolecules? They can contain millions of atoms! To study them would be despairingly difficult did they not consist of smaller units, identical or similar to each other. These building blocks are easier to study; they are called *monomers* (meaning "single parts" in Greek), whereas the macromolecules resulting from joining monomers together are called *polymers* ("multiple parts").

Apart from organic substances, living organisms contain inorganic substances. Water alone composes over half of their mass. In addition, a multitude of inorganic ions are dissolved in biological fluids or bound to organic compounds. Most abundant among these ions are sodium (Na^+), potassium (K^+), chloride (Cl^-), calcium (Ca^{2+}), magnesium (Mg^{2+}), hydrogen carbonate or bicarbonate (HCO_3^-), and hydrogen phosphate (HPO_4^{2-}).

1.11 Cell Structure

The cell is the building block of living organisms. Primitive organisms such as bacteria are unicellular, whereas organisms that appeared later in the course of the evolution of life are multicellular. The human body consists of approximately 10^{14} (100 trillion!) cells.

The size of cells varies. A bacterial cell is about 1 μm in diameter, whereas the cells of multicellular organisms are usually 10 to 100 times larger. An average human cell is about 25 μm in diameter.

All cells of multicellular organisms are not the same. Cells in the skin are different from those in muscle, which are different from those in the brain. Almost 200 cell types have been identified in the human body.

Despite their great diversity—both from species to species and within the same organism—cells have many features in common (figure 1.5). To begin with their boundaries, all are enclosed in a membrane called the **plasma membrane.** The plasma membrane is only 60 to 100 Å thick and consists of lipids and proteins, to which carbohydrates may be attached.

The interior of a cell is called the **cytoplasm.** Its main component is water, which surrounds a multitude of molecules, ions, and molecular complexes. The cytoplasm is fairly uniform in the simplest of cells, the **prokaryotic cells.** Prokaryotes were the first form of life to appear on Earth some 3.5 billion years ago and comprise two main groups, *bacteria* and *archaea*. On the other hand, multicellular organisms (but also many unicellular organisms such as fungi), which appeared on Earth during the past two billion years, consist of cells that are more complex. These are called **eukaryotic cells.**

Eukaryotic cells have internal compartments, or organelles, bounded by membranes similar to the plasma membrane. The largest, densest, and most conspicuous intracellular organelle is the **nucleus.** The nucleus contains the main genetic material of the cell, which is a nucleic acid known by the initials DNA (short for deoxyribonucleic acid).

The nucleus is surrounded by an extensive network of tubules and flattened sacs, the **endoplasmic reticulum.** The endoplasmic reticulum is the site of synthesis, storage, and transport of substances to other parts of the cell or outside the cell. It ends in a similar, though more distinct, system of flattened sacs, stacked one on top of each other and called the *Golgi apparatus,* or *Golgi complex.* This is where proteins to be secreted outside the cell are packaged and, sometimes, processed.

The cytoplasm of eukaryotic cells is filled with oval organelles named **mitochondria** (singular: mitochondrion). Mitochondria are the power plants of the cell: They are the main sites where nutrients are burned by oxygen and energy is produced for cellular functions. Other organelles include *lysosomes* and *peroxisomes,* specialized sites where cellular components are broken down. What is left of the cytoplasm outside all intracellular organelles is the **cytosol.**

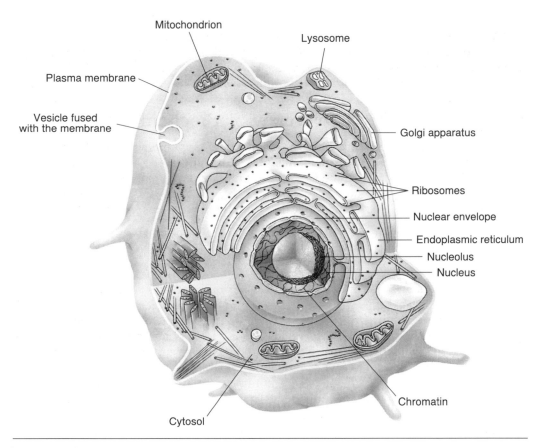

Mitochondrion
Lysosome
Plasma membrane
Vesicle fused
with the membrane
Golgi apparatus
Ribosomes
Nuclear envelope
Endoplasmic reticulum
Nucleolus
Nucleus
Chromatin
Cytosol

▶ **Figure 1.5** Animal cell. A typical animal cell contains a nucleus and nucleolus. The nucleus harbors most of the cell's genetic material. Other compartments include the endoplasmic reticulum, where many substances are made; the Golgi apparatus, which packages proteins for export; hundreds or thousands of mitochondria, where most of our energy is produced; and lysosomes, which are demolition centers. All these organelles bathe in the cytosol and are segregated from it by membranes. The entire cell is wrapped in the plasma membrane. Other components shown here will be presented in upcoming sections.

Reprinted, by permission, from G. Zubay, 1993, *Biochemistry*, 3rd ed. (New York: McGraw-Hill Companies), 4.

Problems and Critical Thinking Questions

1. Are detergents hydrophilic or hydrophobic compounds? Explain.

2. Is the evaporation of a fluid a physical process or a chemical reaction? What about the rusting of a metal?

3. The following equation depicts a lipolysis. Balance it!

$$
\begin{array}{ccc}
H_2COCOR & & H_2COH \\
| & & | \\
HCOCOR + H_2O \longrightarrow & HCOH + RCO_2^- \\
| & & | \\
H_2COCOR & & H_2COH
\end{array}
$$

(R represents a so-called aliphatic chain.)

4. The pH of the cytosol in a muscle cell may drop by nearly one unit after hard exercise, as we will see in ensuing chapters. Suppose, for the sake of simplicity, that it drops from 7 to 6. What happens to the $[H^+]$?

   a. It halves.

   b. It becomes 10 times lower.

   c. It doubles.

   d. It becomes 10 times higher.

CHAPTER 2

Metabolism

The origin of the term metabolism (the Greek word *metabolé* meaning "change") is indicative of what it denotes: **Metabolism** is the sum of the chemical reactions occurring in a living organism. More than 1,000 reactions take place in even the simplest of organisms. However, the picture of metabolism is not a picture of chaos but instead of an organized and coordinated hive (figure 2.1). This facilitates the study of metabolism immensely. Some of its features are as follows.

- Although the number of metabolic reactions is huge, many are similar in terms of reactants and chemical mechanisms.

- Reactions can be placed in a row such that the product of the first is a reactant in the second, the product of the second is a reactant in the third, and so on (figure 2.2). Reaction sequences of this kind are called **metabolic pathways.**

- A relatively small number of compounds play a central role in the metabolism of all living organisms.

- The velocity of metabolic reactions is not constant. Instead, it is controlled by a multitude of biochemical factors. Through this multitude, some common ways of controlling metabolism emerge, several of which we will discuss in part III.

▶ **Figure 2.1** Complexity and order. Silkscreen entitled *Metabolism 90M* by Matsuo Kato depicts, in an artistic manner, the complexity and at the same time the order of the chemistry of living organisms.

Courtesy of Matsuo Kato.

Figure 2.2 Metabolic relay. Reactions form metabolic pathways if they are linked in such a way that the product of one is the reactant of another.

In this chapter, we become acquainted with some principles of metabolism. This will enable us to easily follow the processes described in the subsequent chapters.

2.1 Free-Energy Changes Earmark Metabolic Reactions

Let's begin the presentation of the principles governing metabolism by picking up what we left pending in section 1.7, that is, by exploring the factors that determine which way a reaction will go. These factors are the subject of **chemical thermodynamics**, a branch of chemistry and physics dealing with energy changes in chemical systems. Its application to living organisms constitutes the field of **bioenergetics.**

When a chemical reaction takes place in a system such as the cell, the energy of the system usually changes because the products have a different energy from that of the reactants. In biological systems operating at constant temperature and pressure, energy change is described by *enthalpy change,* symbolized as ΔH (delta H). Another useful thermodynamic parameter is *entropy change,* or ΔS, entropy being a measure of disorder in a system. The two terms are combined to produce a third term, **free-energy change**, or ΔG, as follows.

$$\Delta G = \Delta H - T \Delta S \qquad \text{(equation 2.1)}$$

where T is the absolute temperature.

Of the terms just presented, ΔG is the one most useful to biochemists because it provides a single, clear criterion as to which way a reaction—or physical process for that matter—will go. In a reaction, ΔG is the free energy of the products minus the free energy of the reactants. Three possibilities exist:

- $\Delta G < 0$. This means that the free energy of the products is lower than the free energy of the reactants. The reaction can proceed by itself (spontaneously, as we say); therefore, it is favored. It is said to be **exergonic.**

- $\Delta G > 0$. The free energy of the products is higher than the free energy of the reactants. The reaction cannot proceed spontaneously; therefore, it is not favored. It is said to be **endergonic.** It is understood, however, that if a reaction is not favored in one direction, it is favored in the reverse direction. Reverse reactions have opposite ΔG, since the reactants of one are the products of the other.

- $\Delta G = 0$. There is no free-energy change. The reaction is in equilibrium. Note that equilibrium is dynamic, not static. That is, the reaction does not stop, but while a certain amount of reactants is converted into products, an equal amount of products is converted back into reactants.

The ΔG of a reaction can be negative through a proper combination of values of its components, ΔH and ΔS. Thus, ΔG will be negative if

ΔH is negative and ΔS is positive,

both ΔH and ΔS are negative but the absolute value of ΔH is higher than that of $T \Delta S$, or

both ΔH and ΔS are positive but the absolute value of ΔH is lower than that of $T \Delta S$.

However, there is no way in which ΔG can be negative if ΔH is positive and ΔS is negative.

ΔG serves another function in addition to showing whether a reaction is favored. In exergonic reactions, ΔG is the part of the overall energy change that can produce

Δ (Greek delta) before the symbol of a function denotes change, as Δ is the first letter of the Greek word for difference *(diaphorá)*. See section 3.15 for more uses of Δ.

T is measured in Kelvin degrees (K) and equals the Celsius degrees plus 273. T is always positive.

useful work (defined in physics as the product of force by transposition). In fact, the term exergonic is derived from the Greek word for work *(érgon)*. Work production is vital to a multitude of biological functions such as muscle contraction and transport of solutes. Thus, ΔG *represents the most valuable form of energy in biochemical processes.*

The rest of the energy change in an exergonic reaction is released as heat. Although useful in sustaining body temperature, heat cannot produce work unless there is a change in temperature or pressure. This is, for example, how internal combustion engines (those that move trains, cars, ships, and airplanes) work.

2.2 Determinants of Free-Energy Change

What determines the value of a reaction's ΔG? Let's consider the reaction

A $\rightleftharpoons$ B   **(equation 2.2)**

in which A and B are not necessarily single substances; rather, they symbolize the reactants and products in a collective fashion. ΔG depends on the nature and concentrations of both reactants and products. The ΔG of reaction 2.2 (equation 2.2) is given by the equation

$$\Delta G = \Delta G^\circ + R T \ln \frac{[B]}{[A]}$$ **(equation 2.3)**

where ΔG° is the **standard free-energy change**, R is the gas constant $(1.987 \text{ cal} \cdot \text{mol}^{-1} \cdot \text{K}^{-1})$, and ln is the natural logarithm. ΔG° is defined as the free-energy change of the reaction when the concentration of every participating substance in solution is $1 \text{ mol} \cdot \text{L}^{-1}$ (standard state). In this case, the concentration ratio in equation 2.3 becomes 1, whose natural logarithm is 0; therefore, $\Delta G = \Delta G^\circ$. Being based on a standard state, ΔG° is independent of concentrations and permits a "fair" comparison of different reactions.

A relationship between ΔG° and K_{eq}—the equilibrium constant introduced in section 1.7—can be established from equation 2.3. Since, at equilibrium, $\Delta G = 0$ and K_{eq} = [B]/[A], it follows that

$0 = \Delta G^\circ + R T \ln K_{eq}$, or $\Delta G^\circ = -R T \ln K_{eq}$ **(equation 2.4)**

Equation 2.4 shows that the higher the K_{eq}, the more negative the ΔG°. In other words, the farther a reaction is shifted toward the products at equilibrium (resulting in a high [B]/[A]), the more negative the free-energy change in the standard state will be. Note, however, that ΔG can be quite different from ΔG° depending on the actual concentrations of reactants and products. For example, the ΔG of a reaction having a positive ΔG° can be negative if [B] is much lower than [A] (see equation 2.3 and remember that numbers smaller than 1 have negative logarithms). In general, a reaction can be favored (shifted to the right) if the concentrations of the reactants are much higher than the concentrations of the products.

For biochemical reactions, which usually take place in aqueous solutions of nearly neutral pH, we use a modified ΔG° denoted by $\Delta G^{\circ\prime}$ and defined as the standard free-energy change at pH 7. I will be reporting this function of metabolic reactions in the remainder of the book.

Free-energy changes are measured in energy units. The most commonly used energy unit in biochemistry is the *kilocalorie* (kcal). One kilocalorie is the amount of energy required to raise the temperature of 1 kg water by one degree—in particular, from 14.5° to 15.5° C. The kilocalorie is connected to the *joule* (J), the unit of energy in the Système Internationale, through the equation 1 kcal = 4,184 J (1 kcal = 4.184 kJ).

2.3 ATP, the Energy Currency of Cells

Living organisms are in a state of continuous exchange of mass and energy with their environment. As far as energy is concerned, plants obtain it from sunlight and harness it through photosynthesis, whereas animals and humans obtain it by burning foodstuffs. Part of the energy released by these processes is captured by **a̲denosine tri̲phosphate**, or **ATP**.

We call ATP an energy currency because this is the compound primarily used by cells in their energy transactions. As we will see later in detail, cells exploit the energy from sunlight or foodstuffs to synthesize ATP and spend energy by breaking down ATP. Thus, living organisms use ATP the way we use money in everyday life (we work to earn it, and we spend it to meet our needs).

ATP (figure 2.3) is a complex molecule consisting of three discrete units. The first one is **adenine**, a nitrogenous base that we will examine in a systematic fashion in chapter 4. The second unit is **β-D-ribose**, or simply ribose, a carbohydrate with five carbon atoms that we will examine in a systematic fashion in chapter 5. Finally, ATP contains three phosphoryl groups connected by two **phosphoanhydride linkages.**

The high energy content of ATP resides in its phosphoanhydride linkages: Their hydrolysis releases a high amount of energy. In biological systems, ATP can be hydrolyzed at either phosphoanhydride linkage but not at both simultaneously. Hydrolysis at the outermost linkage—the one between the β and γ phosphoryl groups—is more common and takes place according to the equation

$$ATP + H_2O \rightleftharpoons ADP + P_i + H^+ \quad \Delta G^{\circ\prime} = -7.3 \text{ kcal} \cdot \text{mol}^{-1}$$

(equation 2.5)

in which **ADP** is **a̲denosine di̲phosphate** (figure 2.3) and P_i is **inorganic phosphate** (figure 2.4). Note that the energy data next to equation 2.5 indicate that the standard free-energy change during the hydrolysis of 1 mol ATP is –7.3 kcal.

Hydrolysis at the other phosphoanhydride linkage—the one between the α and β phosphoryl groups—follows the equation

$$ATP + H_2O \rightleftharpoons AMP + PP_i + H^+ \quad \Delta G^{\circ\prime} = -10.9 \text{ kcal} \cdot \text{mol}^{-1}$$

(equation 2.6)

in which **AMP** is **a̲denosine mono̲phosphate** (figure 2.3) and **PP_i** is **inorganic**

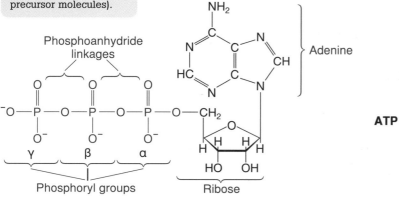

ATP

ADP

AMP

▶ **Figure 2.3** ATP, ADP, and AMP. ATP stars in the energy transactions taking place in biological systems. ATP is converted to ADP or AMP while releasing energy. At physiological pH, the phosphoryl groups are ionized and the three compounds have charges of –4, –3, and –2, respectively. The phosphorus atoms and phosphoryl groups are marked α (alpha), β (beta), and γ (gamma) beginning on the side of ribose. The ribose ring is perpendicular to the plane of the paper, with the bond that is drawn thicker projecting toward you (it is, nevertheless, a single bond). Four of the five atoms in the ribose ring are carbons that for the sake of simplicity are not shown.

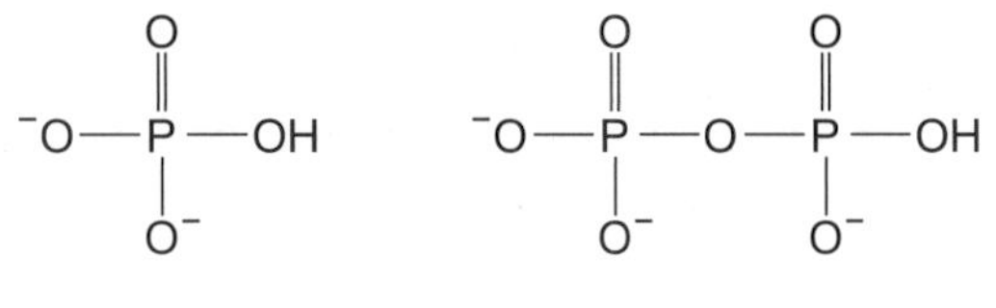

▶ **Figure 2.4** P_i and PP_i. Inorganic phosphate (P_i) and inorganic pyrophosphate (PP_i) are produced during the hydrolysis of ATP to ADP and AMP, respectively. The ionic forms shown here are the ones prevailing at physiological pH. The term inorganic is used to stress that these phosphates are not parts of organic molecules, as the phosphoryl groups in ATP, ADP, AMP, and many other biological compounds are.

pyrophosphate (figure 2.4). PP_i can be further hydrolyzed to two P_i, thus liberating additional energy.

> AMP is also called **adenylate.**

$$PP_i + H_2O \rightleftharpoons 2\,P_i + H^+ \qquad \Delta G^{\circ\prime} = -4.6\ \text{kcal} \cdot \text{mol}^{-1} \qquad \textbf{(equation 2.7)}$$

Another way of looking at this is to add equations 2.6 and 2.7.

$$ATP + 2\,H_2O \rightleftharpoons AMP + 2\,P_i + 2\,H^+ \qquad \Delta G^{\circ\prime} = -15.5\ \text{kcal} \cdot \text{mol}^{-1} \qquad \textbf{(equation 2.8)}$$

The $\Delta G^{\circ\prime}$ of this combined reaction, which is calculated as the sum of the $\Delta G^{\circ\prime}$ of reactions 2.6 and 2.7 (equations 2.6 and 2.7), shows that it is extremely favored thermodynamically.

ATP is often called a *high-energy compound;* by analogy, the phosphoanhydride linkages are called *high-energy bonds,* and they are symbolized as $\sim P$. For the sake of accuracy, one has to note that the *hydrolysis* of these linkages yields high amounts of energy. An equally accurate way to phrase this is to say that ATP possesses a *high phosphoryl transfer potential*. This term will become clearer in chapter 8.

2.4 Phases of Metabolism

We divide metabolism into two phases, **catabolism** and **anabolism.**

Catabolism

Catabolism includes degradation processes, that is, series of reactions by which biological molecules are broken down into smaller molecules. These processes have a double utility:

- They produce raw materials for the synthesis of our macromolecules (we will define this process as anabolism later).
- They release energy, part of which is used in the synthesis of ATP.

To understand the second utility of catabolism better, let's assume that the breakdown of a large molecule M to n smaller molecules m has a $\Delta G^{\circ\prime}$ of –10 kcal.

$$M \rightleftharpoons n\,m \qquad \Delta G^{\circ\prime} = -10\ \text{kcal} \cdot \text{mol}^{-1} \qquad \textbf{(equation 2.9)}$$

Now let's write the equation of ATP synthesis from ADP and P_i by reversing equation 2.5.

$$ADP + P_i + H^+ \rightleftharpoons ATP + H_2O \qquad \Delta G^{\circ\prime} = 7.3\ \text{kcal} \cdot \text{mol}^{-1} \qquad \textbf{(equation 2.10)}$$

This synthesis is highly endergonic, therefore not favored. However, it can proceed if it is chemically coupled to exergonic reaction 2.9 (equation 2.9) by some mechanism. Addition of the two equations yields

$$M + ADP + P_i + H^+ \rightleftharpoons n\,m + ATP + H_2O \qquad \Delta G^{\circ\prime} = -2.7\ \text{kcal} \cdot \text{mol}^{-1} \qquad \textbf{(equation 2.11)}$$

The negative $\Delta G^{\circ\prime}$ value of the combined reaction shows that it is favored. This, in principle, is how catabolic reactions fuel the synthesis of ATP.

Anabolism

In contrast to catabolism, anabolism includes biosynthetic processes, in which cells form molecules from smaller units. Cells need anabolism to grow and divide, to replace molecules that wear out, and to create energy depots.

Biosynthetic reactions utilize intermediate products of catabolism as starting materials and are endergonic, therefore not favored. However, they can proceed if they are coupled to an exergonic reaction such as ATP hydrolysis. Suppose we wish to synthesize the large molecule M.

$$n\,m \rightleftharpoons M \qquad\qquad \Delta G^{\circ\prime} = 10\ \text{kcal} \cdot (\text{mol M})^{-1} \qquad \textbf{(equation 2.12)}$$

If this synthesis is coupled to reaction 2.5 (equation 2.5), the following reaction will ensue.

$$n\,m + ATP + H_2O \rightleftharpoons M + ADP + P_i + H^+ \qquad \Delta G^{\circ\prime} = 2.7\ \text{kcal} \cdot (\text{mol M})^{-1} \qquad \textbf{(equation 2.13)}$$

The $\Delta G^{\circ\prime}$ of this reaction is positive; therefore, the synthesis of M is still not favored. This means that the hydrolysis of one ATP is not sufficient to make the synthesis of one M feasible. However, the hydrolysis of two ATP will do, since it liberates twice as much energy, forcing the total $\Delta G^{\circ\prime}$ to plummet below zero [10 + 2 · (−7.3) = −4.6].

$$n\,m + 2\,ATP + 2\,H_2O \rightleftharpoons M + 2\,ADP + 2\,P_i + 2\,H^+ \quad \Delta G^{\circ\prime} = -4.6\ \text{kcal} \cdot (\text{mol M})^{-1} \quad \textbf{(equation 2.14)}$$

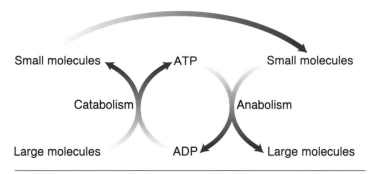

▶ **Figure 2.5** ATP-ADP cycling. Catabolism fuels the synthesis of ATP, which is degraded to ADP while fueling anabolism.

The previous examples illustrate the central concept of metabolism: *Catabolism yields energy for the synthesis of our energy currency, ATP, which in turn is spent in biological functions requiring an input of energy.* One of these is anabolism (the others will be presented in chapter 8). Thus, anabolism depends on catabolism in terms of both raw materials—referred to as **metabolites**—and energy (figure 2.5). Moreover, the last example shows that a *biosynthetic process is more expensive than the reverse degradation process is lucrative.* This "extravagance" of ATP ensures that anabolism as well as catabolism is thermodynamically favored.

2.5   Oxidation–Reduction Reactions

Metabolic reactions often involve **oxidations** and **reductions** of biological substances. We say that a substance is oxidized when it loses one or more electrons. Conversely, it is reduced when it accepts one or more electrons. However, the transfer of electrons is not always evident in a reaction. Instead, it may be more helpful to look at the transfer of hydrogen or oxygen atoms to decide whether a substance is oxidized or reduced. When a compound loses hydrogen (not hydrogen ion), it is oxidized; when it accepts hydrogen, it is reduced. The opposite is the case for oxygen, when the addition is an oxidation and the removal is a reduction.

Since no atoms are created or lost in a reaction (principle of mass conservation, section 1.7), when a substance is oxidized, another is necessarily reduced and vice versa. That is why we speak of oxidation–reduction reactions.

Catabolic processes usually include oxidations of metabolites by the removal of H. Such oxidations are specifically called *dehydrogenations.* Conversely, anabolic processes usually include reductions of metabolites by the addition of H *(hydrogenations)*. Where do the H that are removed go, and where do the H that are added come from? These hydrogen transactions are performed by specialized compounds, the most common of which is **nicotinamide adenine dinucleotide**, or **NAD**.

NAD

NAD (figure 2.6) consists of an AMP unit and a similar unit containing *nicotinamide* instead of adenine. Nicotinamide is a form of the vitamin *niacin*. The two units of NAD are connected by a phosphoanhydride linkage identical to the ones in ATP.

NAD exists in two forms, one oxidized and one reduced. In the oxidized form, the six-membered nicotinamide ring—which is the reactive part of the whole molecule—bears one positive charge. For this reason, we symbolize this form as **NAD$^+$** (although the molecule has a net negative charge overall because of the two phosphoryl groups). When a metabolite participating in an oxidation–reduction reaction along with NAD$^+$ is oxidized, it loses a hydrogen, which is detached with both electrons of the bond connecting it to the rest of the metabolite molecule. Thus, H$^-$ (called a *hydride ion*) is removed. H$^-$ is then transferred to the nicotinamide ring, thus neutralizing it and converting NAD$^+$ to **NADH**. This is the reduced form of NAD.

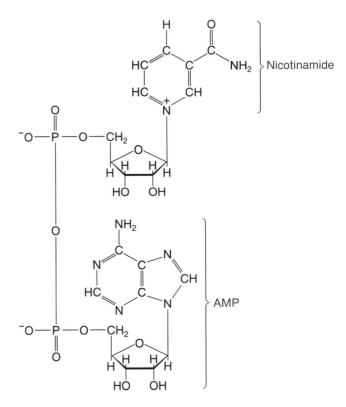

▶ **Figure 2.6** NAD$^+$. Nicotinamide adenine dinucleotide (NAD) is a major hydrogen acceptor in biological oxidations. Its oxidized form (NAD$^+$) is presented here.

(equation 2.15)

NAD$^+$ NADH

R is the rest of the NAD molecule as shown in figure 2.6.

NADP

Some oxidation–reduction reactions employ a compound very similar to NAD, **nicotinamide adenine dinucleotide phosphate**, or **NADP**. NADP differs from NAD in having an additional phosphoryl group attached to the ribose of the AMP unit (figure 2.7). Like NAD, NADP comes in two interconvertible forms—**NADP$^+$** and **NADPH**. The latter is the main hydride donor in reactions involving reductions of metabolites in anabolic pathways.

FAD

Another H acceptor in metabolite oxidations is **flavin adenine dinucleotide**, or **FAD**. FAD (figure 2.8) consists of an AMP unit and a *flavin mononucleotide* (abbreviated FMN) unit, the two linked by a phosphoanhydride linkage. FMN is derived from *vitamin B$_2$*, also called *riboflavin*.

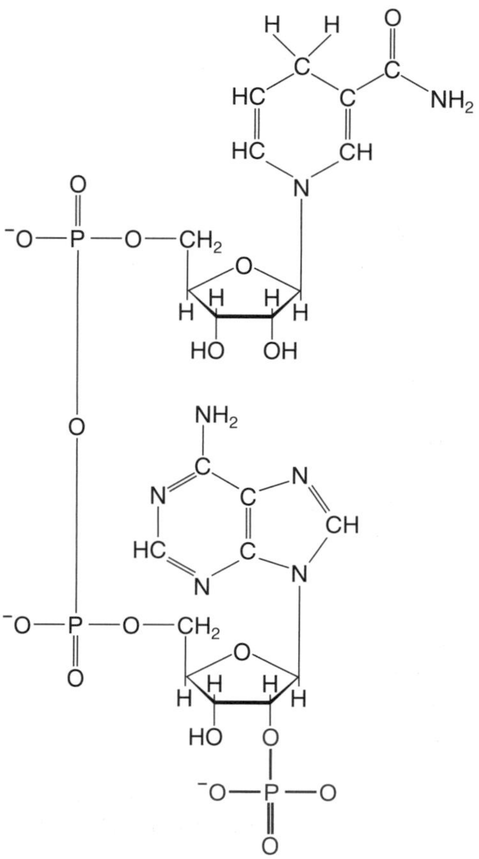

▶ **Figure 2.7** NADPH. Nicotinamide adenine dinucleotide phosphate (NADP) differs from NAD by one phosphoryl group only. This addition earmarks the reduced form (NADPH) for participation in biosynthetic pathways.

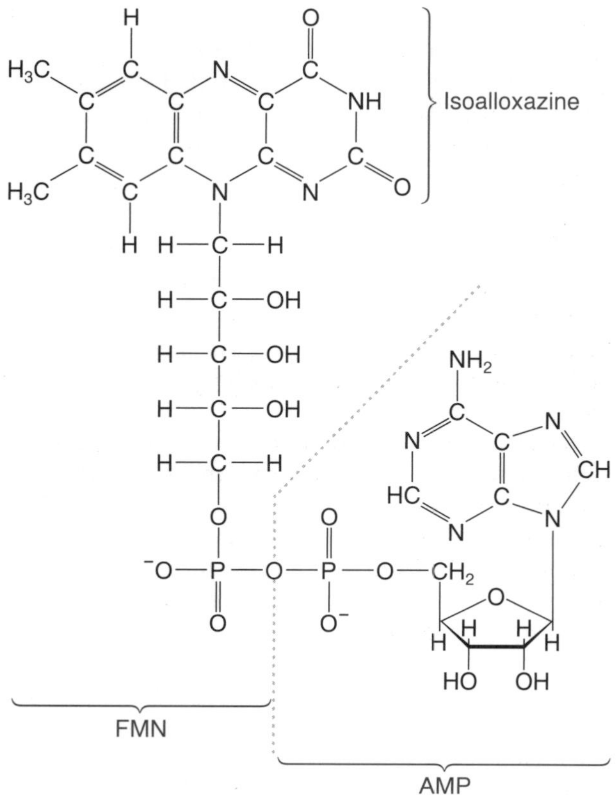

▶ **Figure 2.8** FAD. Flavin adenine dinucleotide (FAD, oxidized form) participates in biological oxidations as a hydrogen acceptor.

By analogy to nicotinamide in NAD, the *isoalloxazine* unit of FMN is the reactive part in FAD. Isoalloxazine accepts two H (to be precise, one H^+ and one H^-) from the compound being oxidized, thus converting FAD to its reduced form, **FADH$_2$**.

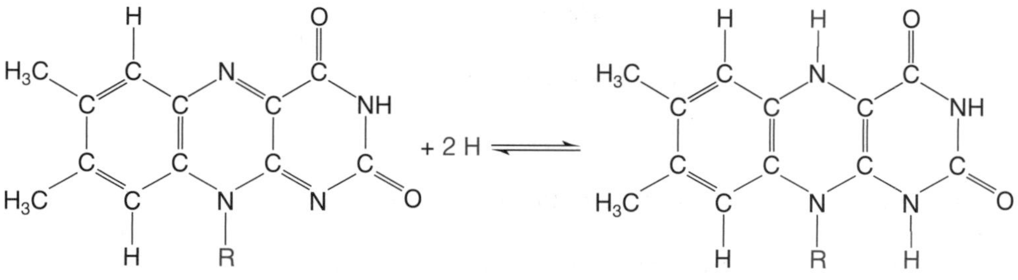

FAD   **FADH$_2$**

(equation 2.16)

R is the rest of the FAD molecule as shown in figure 2.8.

By accepting H from metabolites, NAD^+, $NADP^+$, and FAD serve as *oxidizing agents*, or *oxidants*. Conversely, by donating H to metabolites, NADH, NADPH, and FADH$_2$ serve as *reducing agents* (note the bidirectional arrows in reactions 2.15 and 2.16 [equations 2.15 and 2.16]). Additionally, NADH and FADH$_2$ donate H (to be precise, electrons)

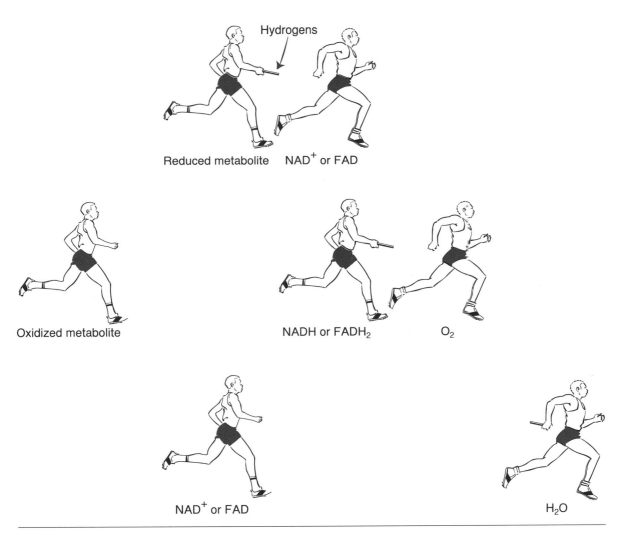

▶ **Figure 2.9** Oxidation–reduction relay. Metabolites are oxidized by donating their hydrogens to ntermediate oxidants such as NAD^+ and FAD. The NADH and $FADH_2$ thus formed pass the hydrogens to oxygen, the final ox dant, and turn it into water.

to molecular oxygen. In fact, the transfer of electrons from NADH and $FADH_2$ to O_2 marks the end of the oxidation of metabolites within cells, as we will see in the next section briefly and in chapter 9 in detail. This renders oxygen the ultimate oxidant in the body, whereas NAD^+ and FAD act as intermediates in this oxidation–reduction relay (figure 2.9).

2.6   Overview of Catabolism

Of the two phases of metabolism, this book emphasizes catabolism because physical activity requires increased amounts of ATP, which is synthesized during catabolic processes. For this reason I will close the present chapter with a bird's-eye view of the processes producing the vast majority of ATP in animal cells, processes that will be the focus of part III.

As mentioned in section 2.3, animals obtain their energy by burning foodstuffs. The components of food that yield energy belong to three of the four broad classes of biological compounds presented in section 1.10. These are—in order of abundance in the

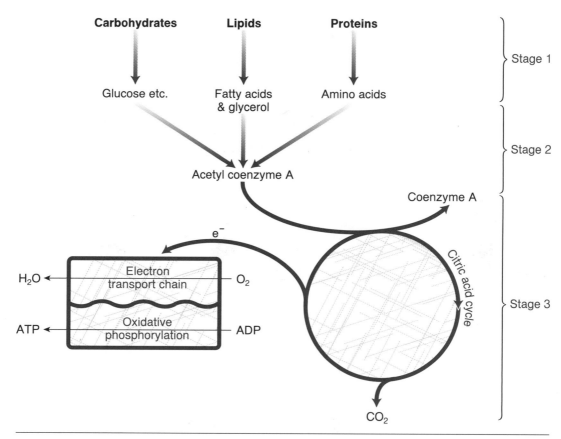

▶ **Figure 2.10** Stages of catabolism. Animals and humans extract energy from foodstuffs in three stages. In stage 1, large molecules are broken down into smaller molecules. In stage 2, these are converted to acetyl coenzyme A (except in the case of a number of amino acids that follow a different course). In stage 3, the acetyl group of acetyl coenzyme A is funneled to the citric acid cycle—leaving coenzyme A behind—and is converted to carbon dioxide. The electrons produced in the cycle are channeled to the electron transport chain, and the energy released feeds the synthesis of ATP through oxidative phosphorylation.

The citric acid cycle is so named for historical reasons; because it was discovered long before authors agreed to call acids in biological fluids by their ionic names (see section 1.9), we do not call it the citrate cycle.

usual human diet—carbohydrates, lipids, and proteins. Nucleic acids are not thought of as energy sources, as they are present in small quantities in foodstuffs.

We can divide the course of extracting energy from carbohydrates, lipids, and proteins into three stages (figure 2.10). In stage 1, the macromolecules in food or similar macromolecules in our cells are broken down to their building blocks. Carbohydrates are converted to **glucose** and related compounds, lipids are converted to **fatty acids** and **glycerol,** and proteins are converted to **amino acids.** These reactions take place in the digestive tract, being part of the process of digestion, or in the cytosol of most cells. No ATP is synthesized in stage 1; rather, stage 1 is a preparatory step for ATP synthesis in subsequent stages.

Stages 2 and 3 take place solely inside cells. In stage 2, the products of stage 1 are degraded to a few simpler units lying at the core of metabolism. Glucose is catabolized through the pathway of **glycolysis,** which accommodates glycerol from the breakdown of lipids as well. Fatty acids are subjected to β **oxidation,** while amino acids follow individual routes. Most of the compounds entering stage 2 end up in a group of two carbons, the **acetyl group,** which is part of **acetyl coenzyme A.** A small amount of ATP is produced in stage 2.

In stage 3, the acetyl group is oxidized to carbon dioxide through the **citric acid cycle.** The electrons released by this oxidation are taken up by NAD^+ and FAD initially. Next, the electrons are transferred to oxygen, which reaches the cells from the lungs through the bloodstream. Oxygen is thus reduced to water in a series of exergonic reactions forming the **electron transport chain.** Coupled to this process is **oxidative phosphorylation,** in which part of the energy of the electron transport chain is harnessed to synthesize most of a cell's ATP. Because carbohydrates, lipids, and proteins are finally burned by O_2 to CO_2 to produce energy, they are referred to as fuels for cells, and therefore for the processes of the human body, and therefore for exercise.

> **Phosphorylation** is the addition of a phosphoryl group to a compound. Thus, ATP is produced by phosphorylation of ADP.

Tools in the Study of Metabolism

How do biochemists and exercise biochemists study human and animal metabolism? Most laboratory techniques used are invasive: The body is pierced or cut open, and a sample, such as blood, muscle, or liver, is removed in order for the biological substances in it to be measured (see, however, sections 8.4 and 13.17 for two noninvasive techniques). Muscle is obviously a particularly popular starting material for studying exercise metabolism. Studying human muscle metabolism has been facilitated in recent decades by the advent of the needle biopsy, whereby small samples (in the area of 10-100 mg) are removed from a volunteer's muscle with minimal damage to the muscle. Needle biopsy has been complemented by very sensitive analytical methods that enable the measurement of very low amounts of biological substances.

One of the oldest methods used in biochemical research is **spectrophotometry.** Spectrophotometry is based on the principle that many compounds in solution absorb light in proportion to their concentration. Light absorbance is responsible for the color of things: The more light an object absorbs, the darker its color is. Thus, much in the way that you can judge how concentrated or dilute orange juice is by looking at it through a glass, biochemists can deduce how much of the compound of interest (be it a protein, nucleic acid, carbohydrate, lipid, or other metabolite) is in a sample by accurately measuring its absorbance in a *spectrophotometer.* In fact, we can apply the method even to colorless compounds if we make them react in a specific manner to produce colored compounds. In a similar technique, rather than measuring how much light a compound absorbs, we measure how much light a compound emits at a particular wavelength when it returns to its basal energy state (the *ground state*) after having been excited by light of a different wavelength. Since this phenomenon is called *fluorescence,* the name of the method is **fluorometry.**

Much of what we know about metabolism comes from experiments in which we administer substances carrying an unusual isotope and let them metabolize, so that we can trace which substances the isotope is passed on to. This is much like planting a radio emitter in a car in order to know where it goes. The isotope can be either radioactive (such as ^{14}C) or nonradioactive (stable, as we call it, such as ^{13}C). The two kinds of isotopes are measured by different devices called *radioactivity counters* and *mass spectrometers,* respectively.

Biological substances are separated from complex mixtures, identified, and, if desirable, isolated in pure form by a variety of techniques falling under two broad categories, chromatography and electrophoresis. In **chromatography,** compounds in a liquid or gaseous phase are forced by pressure or capillary action to migrate through a solid or liquid medium. Because they differ in chemical structure, they migrate at different speeds and are thus separated from each other. Think of a swim race, in which athletes who started together finish separately because they are just different. The same principle applies to **electrophoresis** except that the driving force of the separation is electricity. Consequently, only ions can be separated by this method. Electrophoresis is usually carried out in gels and is used for protein and nucleic acid separations.

Problems and Critical Thinking Questions

1. What are the links between the phases of metabolism?

2. Which one of the following is true? The higher the equilibrium constant of a reaction,

   a. the more positive the free-energy change.

   b. the more positive the standard free-energy change.

   c. the more negative the free-energy change.

   d. the more negative the standard free-energy change.

3. Gluconeogenesis is an anabolic pathway consisting of the synthesis of glucose from pyruvate (to be discussed in section 9.19). The $\Delta G°'$ of this conversion is 35 kcal $\cdot$ mol^{-1}. How many mol ATP must be hydrolyzed to make the synthesis of 1 mol glucose favorable?

4. When a substance burns, is it oxidized or reduced? Explain.

CHAPTER 3

Proteins

Proteins are the first class of biological macromolecules that we are going to explore. Their name is derived from the Greek word *proteíos* and the ending *-in*. *Proteíos* means "first in rank" and denotes the primary role of these compounds in biological processes. As for the ending, it characterizes a category of organic compounds named amines and reflects the fact that protein molecules contain amino groups. In fact, the nitrogen of amino groups constitutes a substantial part of the protein mass (16% on average).

We will examine proteins (as all classes of biological macromolecules) from two sides, structure and function. We will start with structure, from building blocks to the full, three-dimensional shape of proteins, and then we will consider some of their diverse functions.

3.1 Amino Acids

Amino acids are the building blocks of proteins. Biological fluids contain a multitude of amino acids, but those that compose proteins are limited to 20. To be exact, one may find additional amino acids in a protein molecule, but these arise by modification of some of the 20 fundamental amino acids after a protein has been synthesized. We will encounter examples of modified amino acids in sections 3.5 and 9.2.

All but one of the 20 amino acids of proteins can be depicted by the same general formula (figure 3.1). They are made up of a carbon atom to which four different groups are attached: a carboxyl group, an amino group, a hydrogen, and a variable group. The latter determines the identity of an amino acid; it is also called a *side chain* for a reason that will become clear when we examine how a protein is assembled. Amino acids of this kind are called *α-amino acids* because the amino group is connected to the carbon next to the carboxyl group; this was denoted α carbon in older nomenclature.

Amino acids are ionized in aqueous solutions. The predominant form at physiological pH is negatively charged at the carboxyl group, which has a pK (see section 1.9) of around 2, and positively charged at the amino group, which has a pK between 9 and 11. Because the two charges cancel each other out, an amino acid

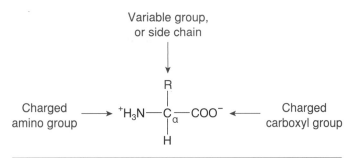

▶ **Figure 3.1** General formula of an amino acid. An α-amino acid is shown in its zwitterionic form, which predominates in most biological fluids.

zwitterion is electrically neutral unless the side chain is charged. Another detail regarding the structure of amino acids is that they can exist in two isomeric forms, one of which is the mirror image of the other. We discriminate these forms by using the prefixes D and L. Amino acids in proteins are of the L form, and all formulas in this book depict L-amino acids (according to a convention called the Fischer projections).

Table 3.1 presents the names of the 20 amino acids of proteins. Every amino acid has a three-letter abbreviation, which in most cases coincides with the three first letters of its name. For greater brevity, biochemists use one-letter symbols of amino acids.

As already mentioned, amino acids differ with respect to the side chain only. Side chains vary in size, charge, polarity, and chemical reactivity. You can review their versatility in figure 3.2, which presents the constitutional formulas of all 20 amino acids of proteins. Amino acids can be as small as glycine (number 1) or as large as tryptophan (number 9). They can have no net charge (numbers 1 to 13, 16, and 17), negative charge (numbers 14 and 15), or positive charge (numbers 18 and 19) at physiological pH. Histidine (number 20) is present in two forms (positive charge or no charge on the side chain) in considerable amounts, as the pK of the side chain is close to the physiological pH. Proline (number 6) is the one amino acid not obeying the general formula of figure 3.1, since its side chain is connected to the amino nitrogen. Finally, there are

▶ **Table 3.1** Amino Acid Names and Abbreviations

Name	Three-letter abbreviation	One-letter symbol
Alanine	Ala	A
Arginine	Arg	R
Asparagine	Asn	N
Aspartate	Asp	D
Cysteine	Cys	C
Glutamate	Glu	E
Glutamine	Gln	Q
Glycine	Gly	G
Histidine	His	H
Isoleucine	Ile	I
Leucine	Leu	L
Lysine	Lys	K
Methionine	Met	M
Phenylalanine	Phe	F
Proline	Pro	P
Serine	Ser	S
Threonine	Thr	T
Tryptophan	Trp	W
Tyrosine	Tyr	Y
Valine	Val	V

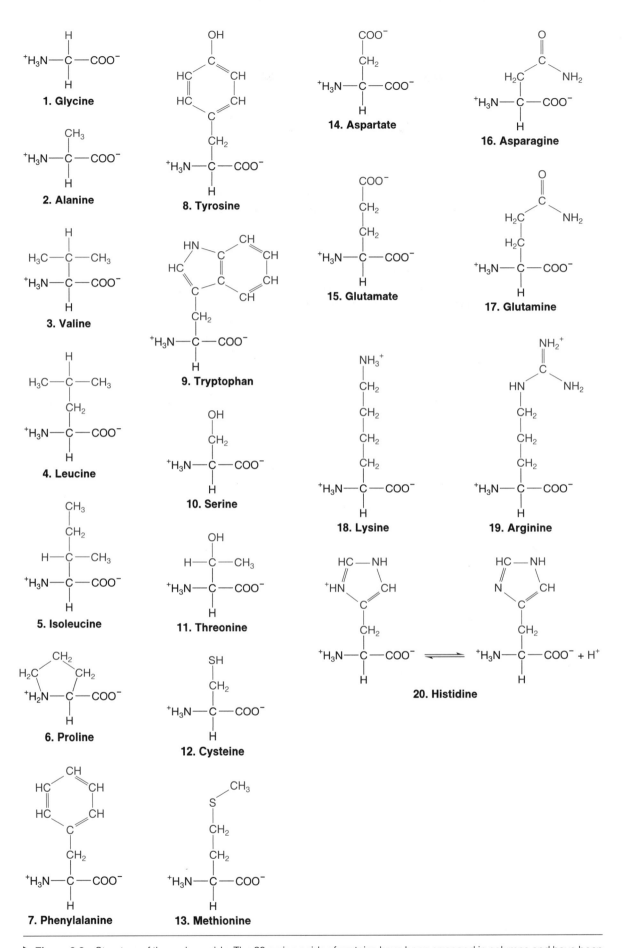

▶ **Figure 3.2** Structure of the amino acids. The 20 amino acids of proteins have been arranged in columns and have been grouped according to the similarity of their side chains.

amino acids like alanine (number 2), with hydrophobic side chains, and amino acids like serine (number 10), with hydrophilic, though uncharged, side chains.

3.2   The Peptide Bond

Amino acids form proteins by being joined covalently in a row by the tens or hundreds or even thousands. Short stretches of amino acids are called **peptides,** a term derived from the Greek verb *pépto,* meaning "to digest." Indeed, as we will see in section 11.1, proteins are digested to small amino acid sequences, which are therefore called peptides. Depending on the number of amino acids constituting them, peptides may be specified as dipeptides, tripeptides, and so on, while proteins are often referred to as **polypeptide chains.** The part of an amino acid participating in a peptide is called a *residue.*

The bonds linking amino acids are formed between the amino group of one amino acid and the carboxyl group of another. To see how this is done, let's consider two amino acids, phenylalanine and aspartate. The nitrogen of aspartate can be linked to the carboxyl carbon of phenylalanine by a bond called amide bond in organic chemistry and **peptide bond** in biochemistry (figure 3.3).

Before we proceed, let me clarify that the reaction of figure 3.3 is endergonic and thus not favored. (In contrast, the reverse reaction, that is, the hydrolysis of the peptide bond, is favored.) I have presented it only to describe the peptide bond. In reality, peptide bond formation for protein synthesis takes place in a more complicated way that is described in the next chapter.

The dipeptide of figure 3.3 has a free (unbound) amino group belonging to phenylalanine and a free carboxyl group belonging to aspartate. It is called phenylalanyl aspartate and is abbreviated as Phe-Asp. If the two amino acids were linked in the reverse way, the resulting dipeptide, Asp-Phe, would have a free amino group belonging to aspartate and a free carboxyl group belonging to phenylalanine. It would thus be a different compound (an isomer of Phe-Asp).

To avoid confusion, biochemists have agreed to write and name peptides—and proteins—from the free amino group to the free carboxyl group. The residue bearing the free amino group is termed the *amino terminal* or *N-terminal residue,* whereas the residue bearing the free carboxyl group is termed the *carboxyl terminal* or *C-terminal residue.* Thus, the conventional direction of a polypeptide chain is N → C.

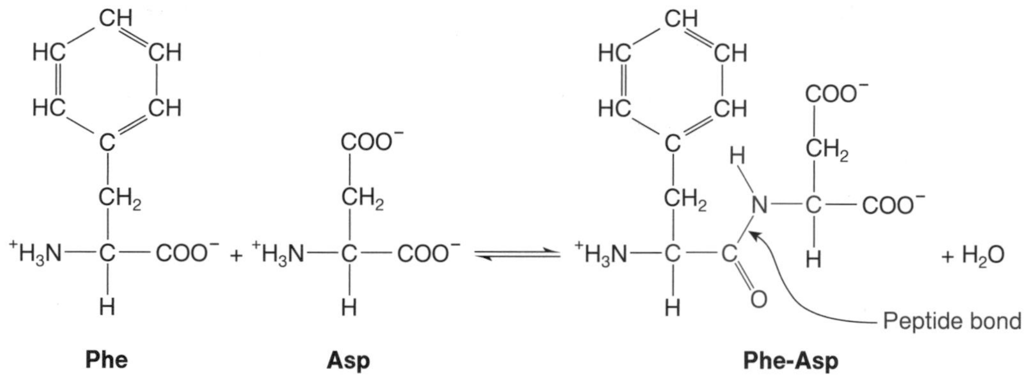

The dipeptide Asp–Phe, with the addition of a methyl group to the free carboxyl group of Phe, is the artificial sweetener *aspartame,* which is added to diet beverages and other low-calorie products. Although not devoid of calories, aspartame is 180 to 200 times as sweet as sugar and can replace it in much smaller quantities.

▶ **Figure 3.3**   Peptide bond. Two amino acids can be joined by a peptide bond, forming a dipeptide and water. The six atoms in color are coplanar (they lie on the same plane); of these, H, N, C, and O form a *peptide unit.*

3.3 Primary Structure of Proteins

Proteins are intricate three-dimensional entities in which we can discern four structure levels of increasing complexity: primary structure, secondary structure, tertiary structure, and quaternary structure. The simplest of these, primary structure, is defined as *the amino acid sequence in a protein.*

The general formula of the primary structure of a protein is shown in figure 3.4a. The "backbone" of the polypeptide chain is formed by alternating atoms of nitrogen, α carbon, and formerly carboxyl, presently amide carbon. The variable groups of the amino acids protrude from the backbone, and that is why they have been named side chains.

The primary structure of a protein can be likened to a human chain (figure 3.4b) in which each human is an amino acid and his or her hands are the amino group and the carboxyl group. The person on one end of the chain has a free right hand (say, the free amino group), and the person on the other end has a free left hand (the free carboxyl group). Everybody else in the chain has a right hand linked to somebody else's left hand and a left hand linked to somebody else's right hand (the peptide bonds). Finally, the heads and legs, just like the side chains and hydrogens, take no part in the linking.

▶ **Figure 3.4** Primary structure. *(a)* A protein is a chain of amino acids linked by peptide bonds. The broken lines mark the boundaries of each residue. The symbols R_1-R_n represent the side chains. Note the distinct ends of the polypeptide chain. *(b)* A protein is like a human chain. See text for parallels.

Protein molecules are usually assembled from hundreds to thousands of amino acids. Thus, the repertoire of the 20 different amino acids—which can be joined in any possible sequence—permits the existence of an inconceivably large number of proteins. Their real number, of course, is finite, though impressive. For example, humans are thought to have 100,000 to 200,000 different proteins.

What determines the primary structure of a protein? We will explore this question in detail in the next chapter. For now, we can say that the amino acid sequence in a protein is dictated by the genetic information contained, in an encoded form, in what we call the gene.

Let me now reverse the question: What does the primary structure of a protein determine? The answer is simple: It determines the next, more complex levels of structure (figure 3.5). Until a couple of decades ago, biochemists were adamant about this. They stressed that the shape (and hence function) of a protein was determined exclusively by its primary structure—in particular, the kind of the side chains of its residues. However, recent findings suggest that at least some proteins are aided by other proteins (aptly named *chaperones*) in assuming their definite shape. Thus, it is safer to say that the primary structure is an important (probably, the most important) determinant of the shape of a protein.

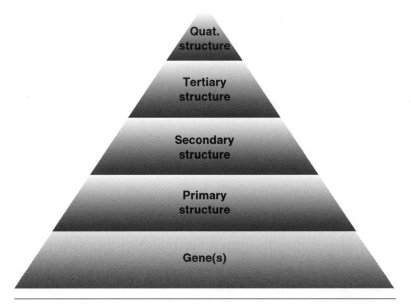

▶ **Figure 3.5** Information buildup. The primary structure of a protein is the basis for the higher levels of structure (secondary, tertiary, and quaternary structures). The primary structure itself is encoded by the gene(s).

3.4   Secondary Structure

A polypeptide chain does not extend in a straight line, as figure 3.4 may suggest. Instead, the chain folds in various ways, giving rise to a multitude of three-dimensional structures. *The conformation of a short stretch of the polypeptide chain is the secondary structure.*

The main force responsible for secondary structure in a protein is the **hydrogen bond.** This is a primarily noncovalent—though with some covalent character—bond appearing between a hydrogen atom with partial positive charge and an atom with partial negative charge, usually oxygen or nitrogen (figure 3.6). A hydrogen bond is 10 to 100 times weaker than a covalent bond. However, there are many H atoms with δ^+ and many O and N atoms with δ^- in a protein, giving rise to many hydrogen bonds. The sum of the resulting forces is sufficient to stabilize the secondary structure.

The secondary structure can be irregular (although strictly defined) or regular, with repeating forms. In the latter case, we speak of periodic secondary conformations. The two most widely known and most frequently encountered of these are the α *helix* and the β *pleated sheet*, or simply β *sheet*. In the helix, the backbone of the polypeptide chain forms the core of a right-handed spiral having a pitch of 5.4 Å and encompassing 3.6 residues per turn (figure 3.7*a*), while their side chains point outward. The helix is stabilized by hydrogen bonds between H and O atoms of peptide units in adjacent turns.

In the β sheet, two or more segments of a polypeptide chain (called β *strands*) are almost fully stretched, and they undulate in such a way that the planes of every other peptide unit are parallel (figure 3.7*b*). The side chains of adjacent residues in a β strand

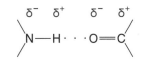

▶ **Figure 3.6** Hydrogen bond. The hydrogen in a peptide unit, bearing a δ^+ because of its covalent bonding with N, can by attracted by the oxygen in another peptide unit, bearing a δ^-. This is but one of several kinds of hydrogen bond that stabilize protein structure.

We are not going in circles, we are going upwards. The path is a spiral; we have already climbed many steps.—Herman Hesse, *Siddhartha*

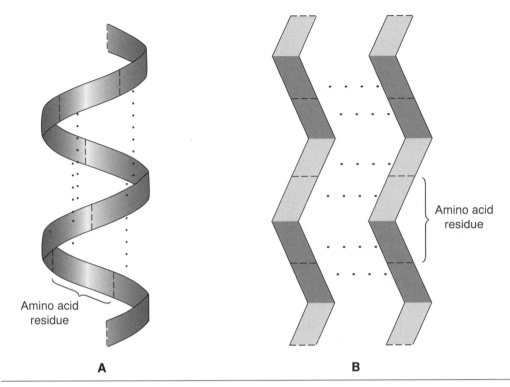

Amino acid
residue

Amino acid
residue

A   **B**

▶ **Figure 3.7** Secondary structure. Two of the most common conformations of polypeptide chains are the α helix *(a)* and the β pleated sheet *(b)*. The chains are presented here as ribbons. Broken lines mark the boundaries of the amino acid residues and the position of the peptide bonds. Colored dots represent hydrogen bonds. The α helix is a right-handed spiral held in place by hydrogen bonds almost parallel to its axis. The β sheet is an undulating structure. Two or more β strands may approach in parallel and form a sheet stabilized by hydrogen bonds perpendicular to the axes of the strands.

point in opposite directions. β Strands are held parallel by hydrogen bonds between H and O atoms of peptide units in neighboring strands, thus forming the β sheet.

An amino acid occupies 3.5 Å along a β strand as opposed to only 1.5 Å along an α helix.

3.5   Tertiary Structure

The conformation of an entire polypeptide chain is its tertiary structure. The chain is flexible and can fold to bring together amino acids that lie far apart along the primary structure. Like the secondary structure, the tertiary structure is stabilized by hydrogen bonding. In addition, it employs four other kinds of interactions, the *electrostatic bond,* the *van der Waals interactions,* the *disulfide bond,* and the *hydrophobic interactions.*

Electrostatic Bond

The electrostatic, or ionic, bond is formed between positively and negatively charged groups and is due to the attraction of opposite charges (figure 3.8). Positively charged groups in a protein are the terminal amino group and the side chains of lysine, arginine, and (possibly) histidine. On the other hand, negatively charged groups are the terminal carboxyl group and the side chains of aspartate and glutamate. Electrostatic bonds within or between molecules in biological fluids are weaker than hydrogen bonds.

It is interesting to note that all positively charged groups in a protein can donate H^+, whereas all negatively charged groups in a protein can accept H^+. This endows proteins with considerable buffering capacity (section 1.9) and an important role in maintaining the pH of biological fluids.

▶ **Figure 3.8** Electrostatic, or ionic, bond.

Van der Waals Interactions

The van der Waals interactions are nonspecific, noncovalent attractions that develop between atoms as a result of the movement of their electrons in atomic orbitals. Because of this, there are momentary asymmetries in the distribution of charge inside an atom, which induce similar asymmetries in neighboring atoms. Thus, attractive electrostatic forces appear. Van der Waals interactions are weaker than electrostatic bonds.

Disulfide Bond

The disulfide bond (figure 3.9) is a covalent bond formed between two residues of cysteine, an amino acid containing sulfur in its side chain (number 12 in figure 3.2). Two S atoms, one from each side chain, shed their H and are linked directly. This is a case of amino acid modification after protein synthesis, as mentioned in section 3.1. The disulfide bond is stronger than the peptide bond. Thus, when a protein is broken down to its constituent amino acids, the two cysteine residues remain united, forming a *cystine* molecule.

▶ **Figure 3.9** Disulfide bond.

Hydrophobic Interactions

Finally, hydrophobic side chains such as those of alanine, valine, and leucine may come together and recede inside a protein molecule. These hydrophobic interactions are not so much the result of mutual attraction as of repulsion by the surrounding water. We first encountered this tendency when we discussed hydrophobic substances (section 1.6). The clustering of hydrophobic groups limits their surface of contact with water. This compact conformation is more stable than a conformation in which every hydrophobic group would be exposed to water. Hydrophobic interactions are the weakest of all the interactions that we have considered.

Note that, with the exception of the disulfide bond, the interactions presented are not confined to protein molecules. Thus, hydrogen bonds, electrostatic bonds, van der Waals interactions, and hydrophobic interactions develop within all kinds of biological molecules and among biological molecules. In fact, *it is these kinds of rather weak interactions among molecules in biological fluids that enable biochemical processes to occur.*

Figure 3.10 is an example of the tertiary structure of a protein. This protein features three α helices and six β strands, four of which form two β sheets (one on the right and one on the left). These periodic secondary structures are joined by short irregular stretches of amino acid residues. The protein contains eight cysteine residues linked in pairs by four disulfide bonds.

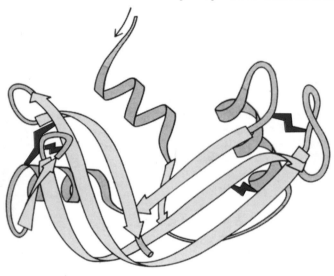

▶ **Figure 3.10**   Tertiary structure. Ribonuclease is a relatively small protein combining α helices and β pleated sheets in its structure. α Helices are depicted as spiraling ribbons, while β pleated sheets are depicted as flat ribbons with arrowheads showing the conventional direction of the polypeptide chain (N → C). The thin arrow at the top indicates the amino end. Four disulfide bonds (two on the left and two on the right, all shown as colored lightning bolts) stabilize the tertiary structure. Try to follow the polypeptide chain from one end to the other. In case you are wondering what ribonuclease does, it is an enzyme, that is, a protein accelerating chemical reactions (see section 3.13). In particular, ribonuclease accelerates the breakdown of a nucleic acid (RNA).

Courtesy of Jane S. Richardson.

3.6   Denaturation

Being stabilized by all the bonds mentioned in the previous section, each protein acquires a definite tertiary structure. This structure determines the interaction of the protein with other molecules and therefore its function. The tertiary structure is sensitive, since it is the result of mostly weak bonds. Things like high temperature and a highly acidic or alkaline environment can easily disrupt the delicate structure of a protein. High temperatures increase the thermal movement of the segments of the polypeptide chain as well as the surrounding molecules, resulting in alteration of the chain's structure. Extreme pH values, on the other hand, cause the addition or removal of hydrogens to or from chemical groups having the tendency to exchange H^+. This elicits changes in hydrogen and electrostatic bonding, which affect protein structure. In all these cases, we speak of protein **denaturation.** Denaturation is usually permanent (irreversible).

We encounter many cases of protein denaturation in food processing. The egg white and egg yolk solidify during cooking because of their high protein content and the collapse of protein structure with heating. Yogurt, on the other hand, offers an example of denaturation because of a drastic drop in pH: Milk proteins lose their solubility and settle when microorganisms degrade the carbohydrate of milk to lactate. In fact, as we will see in part III, a similar metabolic pathway (the anaerobic breakdown of carbohydrates) reduces the pH and contributes to the emergence of fatigue in an intensely exercising muscle through milder—and reversible—alterations in muscle proteins.

3.7   Quaternary Structure

One would expect tertiary structure (which shows how the entire polypeptide chain folds) to be the highest level of organization in a protein. However, many proteins consist of not just one but two or more polypeptide chains, which are called **subunits** or simply chains. Proteins of this kind have one additional level of organization, quaternary structure, defined as *the conformation of the entire subunit complex.*

Quaternary structure is as accurate as tertiary structure: The position of each subunit relative to the others is strictly defined. The subunits are held in place by the same bonds as those that stabilize the tertiary structure. A classical example of a protein having subunits (therefore, a quaternary structure) is hemoglobin, which we will examine in terms of both structure and function in sections 3.11 and 3.12. The quaternary structure of another protein is shown in figure 3.11.

> Tertiary and quaternary structures are stabilized by hydrogen bonds, electrostatic bonds, van der Waals interactions, disulfide bonds, and hydrophobic interactions.

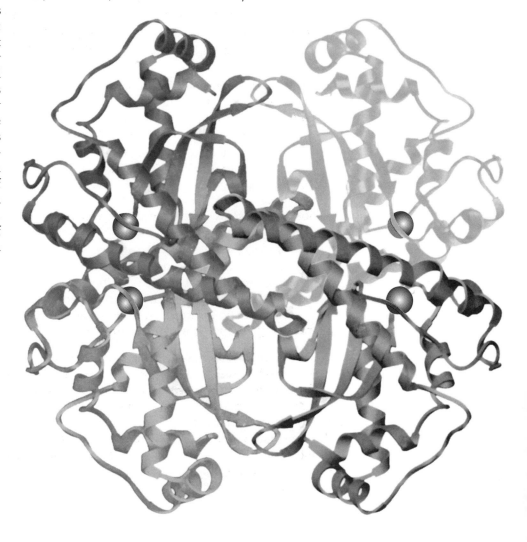

▶ **Figure 3.11** Quaternary structure. Superoxide dismutase is an enzyme with antioxidant activity; that is, it prevents the superoxide radical (section 1.4) from acting as an oxidant. The protein consists of four identical polypeptide chains shown in different shades for clarity. Can you discern α helices and β sheets in each subunit? The four colored orbs—one in each subunit—are manganese ions (Mn^{2+}) required for dismutase activity.

Image by M. Pique, G. Borgstahl, and J. Tainer.

3.8   Protein Function

Proteins have a vast repertoire of functions. This endows them with the ability to participate in any biochemical process, for example, reactions; movement of substances, cells, or whole organisms; and transmission of information. To facilitate the study of such diverse roles, we assign proteins to categories such as the following.

- *Catalytic proteins.* A large group of proteins catalyze (accelerate) biochemical reactions. These proteins are called **enzymes** and will be the topic of subsequent sections.

- *Transport proteins.* Many substances are transported in and out of cells, in and out of intracellular organelles, and from one site in the body to another with the help of specific proteins. For example, oxygen from the air is carried to our cells by hemoglobin, the major protein of the blood.

- *Storage proteins.* Other proteins store cellular components. Ferritin belongs to this category; it stores iron in the body (see section 14.8).

- *Motile proteins.* Movement in living organisms is due to the interaction of organized protein assemblies. Muscle contraction, for example, is feasible thanks to the movement of myosin relative to actin, as we will see in part II.

- *Structural proteins.* Proteins offer mechanical support to cells and tissues. The cytoskeleton, which gives cells a particular shape—otherwise they would be amorphous sacs of cytoplasm—consists of proteins. The elasticity of the connective tissue is due to the remarkable properties of proteins like collagen and elastin.

- *Defensive proteins.* This category contains the *antibodies,* which recognize and neutralize foreign substances and cells very selectively, thus protecting us from invaders.

- *Messenger proteins.* Multicellular organisms require communication among their cells to ensure concerted action. Many of the compounds carrying messages from one cell to another are proteins. This category includes hormones such as insulin and neurotransmitters such as endorphins.

- *Receptors.* Molecular messengers—whether protein in nature or not—afford cell-to-cell communication not directly but indirectly, that is, by binding to proteins that recognize them. These proteins are called **receptors**, an example being the insulin receptor. The binding of a messenger to its receptor changes the structure of the latter, thus altering its biological activity. There is at least one receptor for every messenger.

- *Other regulatory proteins.* The proteins of most of the previous categories control some biological processes and could thus be characterized as regulatory in one sense or another. There are additional regulatory proteins that do not fit into any of these categories. These include, for example, the proteins that control the flow of genetic information by binding to nucleic acids. Biochemical research is revealing a rapidly growing number of regulatory roles for proteins.

I have to stress that the classification just presented is not meant to confine proteins within mutually exclusive categories but rather to demonstrate the inexhaustible variety of functions that proteins perform. Many proteins do not limit themselves to one function. For example, myosin is a motile protein and an enzyme at the same time, as we will see in section 7.3.

In the ensuing sections of this chapter, we turn our attention to some proteins of special interest. We will examine hemoglobin and myoglobin as representatives of transport and storage proteins. Our interest in them lies in the fact that they bind the oxygen that we need to live and exercise. In addition, we will examine the enzymes, since almost none of the reactions described in the book can take place to an appreciable extent unless an enzyme is present.

3.9   Oxygen Carriers

Animals and many microorganisms cannot live without oxygen; thus, they are **aerobic** organisms. Oxygen is vital to them because it serves as the ultimate oxidant in the electron transport chain, which provides the energy needed to synthesize the vast majority of their ATP (section 2.6). The first aerobic bacteria appeared on Earth approximately in the middle of the present age, some two billion years ago. Their emergence became possible thanks to the liberation of oxygen as a product of **photosynthesis** in the primeval atmosphere. Photosynthesis had enabled even more primitive bacteria to harness solar energy and synthesize glucose from carbon dioxide in the air and water.

$$6\ CO_2 + 6\ H_2O \rightleftharpoons C_6H_{12}O_6 + 6\ O_2 \qquad\qquad \textbf{(equation 3.1)}$$

Aerobic organisms prevailed on Earth thanks to the high oxidizing potential of O_2 and the large amounts of energy released by the burning of fuel molecules. The entry of O_2 into microorganisms was no problem, since it could diffuse easily from the surrounding air or water—where it was dissolved—into the cells by crossing their membranes. Things, however, were not quite so simple with multicellular organisms, as only the cells at their surface made contact with air or water. Thus, the imperative arose to carry O_2 to every internal cell.

Unfortunately, the solubility of O_2 in water is so low that its mere transport through the body fluids could not cover the needs of active animal cells. To meet these needs—a prerequisite for the development of large and strong animals—proteins of high oxygen-carrying and -storing capacity appeared and improved in the course of evolution. We will explore these proteins in the next three sections, starting with the simplest one.

3.10   Myoglobin

Myoglobin is a relatively small protein. The human form contains 153 amino acid residues and has a molecular mass of 17 kDa. It is located in muscle and serves to receive oxygen from the blood, store it, and transport it inside muscle cells. This helps muscle cope with sudden increases in oxygen demand during exercise. A human skeletal muscle contains around 18 g of myoglobin per kilogram.

The highest concentrations of myoglobin are found in the muscles of sea mammals like whales, which need large reservoirs of O_2 for each dive.

The ability to bind oxygen resides not in the polypeptide chain of myoglobin per se but in a nonprotein compound attached to the polypeptide chain. This is not uncommon: Many proteins owe their function to relatively small organic molecules that are tightly bound to the polypeptide chain. These are collectively known as *prosthetic groups*. The prosthetic group of myoglobin is **heme**, a complex compound with one ion of iron (Fe^{2+}) in the middle. Heme is red and is responsible for the color of the blood, muscles, and other organs (in the following section we will see that hemoglobin contains heme as well).

The tertiary structure of myoglobin can be seen at high resolution in figure 3.12. Its molecule is compact, 45 by 35 by 25 Å in size. Most of the polypeptide chain is α helices. Heme is located in a cavity of the molecule and is held in place through bonds with the side chains of residues from adjacent α helices. One of these residues, histidine, binds iron almost perpendicularly to the heme plane. Iron binds O_2 on the other side of the heme plane. The oxygenated form of myoglobin is called *oxymyoglobin.*

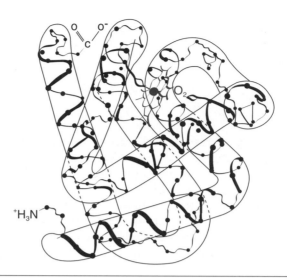

▶ **Figure 3.12** Myoglobin. A continuous line of variable thickness (to produce the sense of depth) depicts the backbone of the polypeptide chain in this classical representation of myoglobin. The dots along the backbone are the α carbons of the amino acid residues. The thin line surrounding the spiraling chain helps one follow it from the amino end to the carboxyl end. Most of the chain is folded into α helices. The colored disc is heme, and the bulge in its center is iron, which binds O_2. The seemingly empty space among the α helices is in fact occupied by amino acid side chains, only two of which are shown. These are the tiny pentagons flanking heme, and they belong to histidine residues. One (to the left) binds iron, while the other contributes to making iron relatively inaccessible, thus preventing it from being oxidized by O_2.

Reprinted from *The Proteins*, vol. 2, H. Neurath (ed.), R.E. Dickerson, X-ray analysis and protein structure, pg. 634, Copyright 1964, with permission from Elsevier.

If only Fe^{2+} binds O_2, what is the rest of myoglobin doing? The answer is that neither Fe^{2+} nor heme alone can hold O_2 because O_2 is so reactive that it oxidizes Fe^{2+} to Fe^{3+}, rendering it incapable of attracting O_2. However, the polypeptide chain offers a protective environment that inhibits iron oxidation and enables the binding of oxygen. Thus, the entire myoglobin molecule is perfectly suited for this biological function.

3.11   Hemoglobin

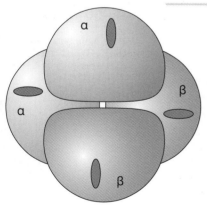

▶ **Figure 3.13** Hemoglobin. The oxygen carrier in the blood is a tetramer having the subunit composition $\alpha_2\beta_2$. Each hemoglobin subunit is in touch with all three other subunits. The colored discs, one in each subunit, represent heme.

Hemoglobin is the most abundant protein of the **erythrocytes** (or *red blood cells*) and the blood as a whole. A typical blood hemoglobin concentration is 14 g · dL^{-1} (see more on blood hemoglobin concentration in section 14.1). Hemoglobin binds atmospheric oxygen in the capillaries surrounding the alveoli of the lungs, thus becoming *oxyhemoglobin*. The erythrocytes then transport oxyhemoglobin to all tissues. Oxygen dissociates from hemoglobin in the capillaries of the tissues, mainly because the oxygen concentration there is lower than in the alveoli, and diffuses into the cells. The dissociation of O_2 from hemoglobin is facilitated by the presence of CO_2 and H^+ in the capillaries, as we will see shortly.

Hemoglobin binds O_2 through heme, exactly as myoglobin does. However, hemoglobin enjoys a higher level of organization and biological function. It consists of not just one but four polypeptide chains called *globins*. In adults, two globins are of the α type and two are of the β type (figure 3.13). The tertiary structure of each globin is very similar to that of myoglobin, although the primary structures differ markedly. Globins contain α helices having almost the same arrangement as in myoglobin. Additionally, every globin contains one heme. As a result, there are four oxygen-binding sites in hemoglobin.

The four chains of hemoglobin are held together by noncovalent bonds and form a tetrahedral structure displaying impressive spatial economy. The resulting complex is nearly globular, having a diameter of 55 Å. The name of hemoglobin—and, by extension, myoglobin—reflects this globular shape. (As for *hemo* and *myo*, they derive from the Greek words *héma*, meaning "blood," and *mýs*, meaning "muscle.") The heme molecules are located on the surface of the complex.

3.12   The Wondrous Properties of Hemoglobin

The complexity of structure is reflected in the function of hemoglobin. The protein is not just an oxygen carrier but an accurate sensor and regulator of the blood environment. In this section, we will explore those properties of hemoglobin that justify this claim and fall within the scope of the book.

Cooperativity

Let's begin with the primary role of hemoglobin, oxygen binding and transport. The binding of oxygen by hemoglobin is *cooperative*. This means that the **affinity** (that is, the binding strength) of the protein for O_2 is not constant but relates to the number of O_2 molecules already bound. Thus, if we started adding O_2 to hemoglobin by increasing the $[O_2]$ in the surrounding fluid, every molecule would bind more tightly than the previous one (as if every molecule paved the way for the next one). In fact, the fourth O_2 molecule binds over 20 times as tightly as the first one! Conversely, if we started unloading oxyhemoglobin by decreasing the $[O_2]$, the last O_2 would be removed much more easily than the first one (again, as if every molecule paved the way for the next one).

The importance of cooperative binding becomes evident if we consider that the $[O_2]$ is high in the lungs but low in the rest of the body. Thus, thanks to cooperativity, we achieve a higher loading of hemoglobin with O_2 in the lungs and a more efficient unloading in the other organs. In the end, *cooperativity increases oxygen delivery to the cells.*

How can the binding of oxygen to one globin affect the binding to an adjacent globin? The binding of O_2 causes a translocation of Fe^{2+} by a mere 0.4 Å relative to the heme plane, which triggers a series of translocations of amino acid residues inside the globin molecule. This conformational change modifies the interface with the other globins and is thus transmitted to them, facilitating O_2 binding. This results in an increasingly tighter binding of subsequent O_2 molecules.

The delivery of oxygen to the muscles in particular is facilitated by an additional factor: Compared to hemoglobin, myoglobin has a higher affinity for O_2 under physiological conditions. Thus, myoglobin binds and sequesters the O_2 molecules that diffuse into the muscle cells after being detached from oxyhemoglobin in the capillaries.

Apart from oxygen, hemoglobin carries carbon dioxide produced by the burning of fuel molecules in the cells. However, most (about three-quarters) of CO_2 is carried in the bloodstream as soluble bicarbonate formed inside the erythrocytes according to the equations

$$CO_2 + H_2O \rightleftharpoons H_2CO_3 \rightleftharpoons HCO_3^- + H^+ \qquad \textbf{(equation 3.2)}$$

The compound in the middle is carbonic acid. The reactions are reversed in the alveolar capillaries because of the low CO_2 content of the atmospheric air. Thus, CO_2 passes to the exhaled air. These interconversions work as an important buffer system, the *bicarbonate system,* which cooperates with proteins and other solutes to maintain the pH of the blood.

Hemoglobin plays an additional role in binding and transporting part of the H^+ produced in equation 3.2 and from the anaerobic breakdown of carbohydrates during hard exercise. Removal of H^+ is essential for protecting the blood from a dangerous drop in pH (known as *metabolic acidosis*).

The Bohr Effect

The binding of CO_2, H^+, or both to hemoglobin reduces its affinity for O_2; conversely, the binding of O_2 reduces the affinity of hemoglobin for CO_2 and H^+ (O_2, CO_2, and H^+ bind to different sites on hemoglobin). These interactions, known as the Bohr effect (after Christian Bohr, who discovered them in 1904), are due to intramolecular rearrangements similar to the ones described earlier when explaining the mechanism of cooperativity. The Bohr effect is of great physiological importance: CO_2 and H^+ abound in the capillaries of active organs such as exercising muscles. This facilitates the detachment of O_2 from oxyhemoglobin and O_2 entry to the cells that need it. Hemoglobin takes up CO_2 and H^+ and travels back to the alveoli of the lungs. There, the abundance of O_2 forces hemoglobin to release CO_2 and H^+.

Allostery

When the binding of a molecule to a protein affects the binding of another molecule to a different site in the protein, we speak of **allostery** (meaning "other site" in Greek). We encountered two cases of allostery in hemoglobin:

- Any bound O_2 molecule(s) favor(s) the binding of additional O_2 molecules until hemoglobin is fully saturated.
- The binding of CO_2, H^+, or both favors the dissociation of O_2, and vice versa.

The importance of CO_2 for the delivery of O_2 to the tissues is evident in the case of *hyperventilation:* If a person breathes too fast, he or she removes an unnaturally high amount of CO_2 from the blood, resulting in inadequate delivery of O_2 to cells, including brain cells. Consequently, the person feels dizzy and, in extreme cases, faints. Hyperventilation is corrected easily by breathing into a bag; this returns CO_2 from the exhaled air to the blood.

Hemoglobin and Aerobic Capacity

As the transporter of oxygen to our cells, hemoglobin is intimately connected with aerobic capacity, that is, the maximal amount of oxygen delivered to the tissues during maximal exertion. This is expressed as **maximal oxygen uptake** and symbolized as **$\dot{V}O_2$max,** in which $\dot{V}$ denotes volume per time. To determine $\dot{V}O_2$max, exercise scientists measure the maximal difference between the amount of O_2 in the inspired air and that in the expired air during carefully programmed and executed exercise tests to exhaustion. We express $\dot{V}O_2$max either in liters of oxygen per minute or, relative to body mass (which, naturally, affects it), in milliliters per kilogram per minute ($mL \cdot kg^{-1} \cdot min^{-1}$).

Studies have shown that $\dot{V}O_2$max depends on the total amount of hemoglobin in the body. We can calculate total hemoglobin if we multiply the blood hemoglobin concentration by the total blood volume (since concentration is amount per volume). Although it is easy to measure the hemoglobin concentration in a drop of blood by a spectrophotometric method, determining the total blood volume is more sophisticated, but nevertheless feasible. Athletes with high aerobic capacity have a combination of hemoglobin concentration and blood volume producing a high amount of hemoglobin, along with optimal state of other features that affect $\dot{V}O_2$max, such as cardiac function and body composition.

All the crucial features of hemoglobin would be absent had it the simpler structure of myoglobin. The ability to transport O_2 to every cell in the body and remove part of CO_2 and H^+ is due to hemoglobin's being composed of four subunits similar to myoglobin. Thus, the examination of these two proteins convinces us that complexity in biological systems is not a purposeless luxury; instead, it appears in the course of evolution under the pressure to serve new, more intricate needs.

3.13 Enzymes

Enzymes are biological catalysts, substances that accelerate biochemical reactions without being altered in the end. A reaction catalyzed by an enzyme is an *enzyme reaction.* The reactants of an enzyme reaction are called the **substrates.** Enzymes are proteins, with one intriguing exception that we will discuss in section 4.17.

Enzymes catalyze nearly all reactions taking place in living organisms. Thus, there is almost no biological substance that is not a substrate for an enzyme. Even enzymes themselves are formed and degraded by other enzymes (or by their very selves in some cases). Now, if this makes you wonder how the first enzyme came into existence, you will have to be patient until you get to section 4.17.

Enzymes are usually named after their substrate(s), or the kind of reaction they catalyze, or both. Their most frequent ending is -ase (see, for example, the two enzymes in figures 3.10 and 3.11), and biochemists like to write their names above the arrow(s) of the corresponding reaction. Thus, the conversion of substrate S to product P with the aid of enzyme E would be symbolized as

$$S \overset{E}{\rightleftharpoons} P$$ **(equation 3.3)**

Enzymes share three main features:

- *They increase reaction velocities* by 10^5 to 10^{17} times. Most of the reactions occurring in living organisms would be so slow without enzymes that they would be practically incapable of supporting life. Some enzymes make a reaction take place several hundred thousand times per second!

- *They have high specificity,* that is, selectivity. Each enzyme catalyzes a limited number of different reactions and recognizes as substrates only substances that are structurally similar. As a case in point, an enzyme usually distinguishes the isomers of a compound, recognizing only one as substrate.

- *Their catalytic power is regulated.* The velocity of an enzyme reaction is affected by more factors than those affecting the velocity of a common chemical reaction. This property renders the enzymes extremely versatile as catalysts and lets living organisms control how fast they synthesize or break down their components depending on their needs. (We will discuss several ways of regulating enzyme activity in the context of how exercise affects metabolism in part III.)

We encounter an example of isomer discrimination by enzymes in protein synthesis (see section 4.16): Enzymes involved in this process use only L-amino acids as substrates. This is how proteins come to contain amino acids of this form only.

3.14 The Active Site

The specificity of an enzyme is due to the complementary structure of part of its molecule in relation to the molecule(s) of the substrate(s). A relatively small area on the surface of the enzyme, called the **active site,** attracts the substrate(s) and facilitates product formation.

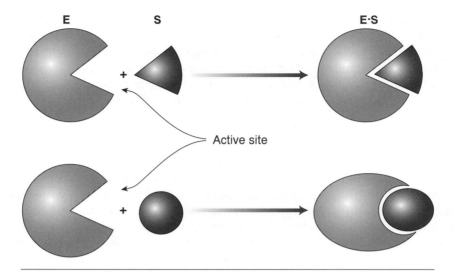

▶ **Figure 3.14** Mutual recognition of enzyme and substrate. Two ways of recognition between enzyme (E) and substrate (S) have been proposed. According to the lock-and-key model (top), S fits the active site of E as a key fits a lock. According to the induced-fit model (bottom), E and S interact and cause mutual conformational changes that result in S fitting the active site.

There are two hypotheses explaining how an enzyme and its substrate recognize each other (figure 3.14). The *lock-and-key model* maintains that the substrate fits exactly the active site just as a key fits the proper lock. In contrast, the *induced-fit model* maintains that the active site is not complementary to the substrate a priori. However, when the substrate touches the active site, their interaction modifies the shape of both in such a way that they fit each other accurately. This model seems to hold true for the majority of enzymes.

3.15   Enzymes Affect the Rate But Not the Direction of Reactions

There are two distinct sides in every chemical reaction. One is *which way and how far it goes;* the other is *how fast it gets there.* The former is examined by chemical thermodynamics, as presented in sections 2.1 and 2.2. The latter is examined by **chemical kinetics** and will be presented here.

A key term in chemical kinetics is *reaction velocity,* or *reaction rate.* Rate (to be exact, *average rate*) is *the change in the concentration of a reactant or product divided by the time within which this change is accomplished.* Thus, for reaction 3.3 (equation 3.3), rate V is

$$V = \frac{\Delta[P]}{\Delta t} = -\frac{\Delta[S]}{\Delta t}$$

(equation 3.4)

Note that the reason for the negative sign before the last fraction is that $\Delta[S]$ is negative (the substrate concentration decreases) whereas $\Delta[P]$ is positive (the product concentration increases). Rate is what an enzyme increases in a reaction. The increase in V attributable to an enzyme indicates its catalytic power and is referred to as **enzyme activity.**

The rate of a reaction is irrelevant to how favored the reaction is. For example, ATP hydrolysis is highly exergonic, but if the appropriate enzyme is missing, it is slow. The reason is that V is not connected to ΔG. Rather, V relates to the free energy of an intermediate and temporary state termed the *transition state*. This is an obligatory passage of high energy for the substrates—and the reactants of chemical reactions in general—on the way to the products. We may think of it as a mountain pass that one has to cross in order to get from one valley to another. The transition state is denoted by a double dagger (‡).

$$S \rightleftharpoons S^{\ddagger} \rightleftharpoons P \qquad\qquad\qquad \textbf{(equation 3.5)}$$

The free-energy change of the transition of S to S‡ is called the *free energy of activation* and is symbolized as $\Delta G^{\ddagger}$. $\Delta G^{\ddagger}$ is always positive. The higher its value, the harder the activation of the substrates is and the slower the reaction goes. This is exactly where enzymes—and catalysts in general—intervene. By binding the substrates at their active sites, they expedite the formation of the transition state, in effect lowering $\Delta G^{\ddagger}$ (figure 3.15).

In this way, enzymes increase reaction rates without, however, altering the outcome of the reactions. This is determined by the value of ΔG (without the double dagger), which, as discussed in section 2.2, depends on the nature and concentrations of substrates and products regardless of the route—catalyzed or uncatalyzed—that the reaction takes. It follows from this that if an enzyme catalyzes a certain reaction, it will also catalyze the reverse one, provided the concentrations of the participating substances are modified so as to change the sign of ΔG.

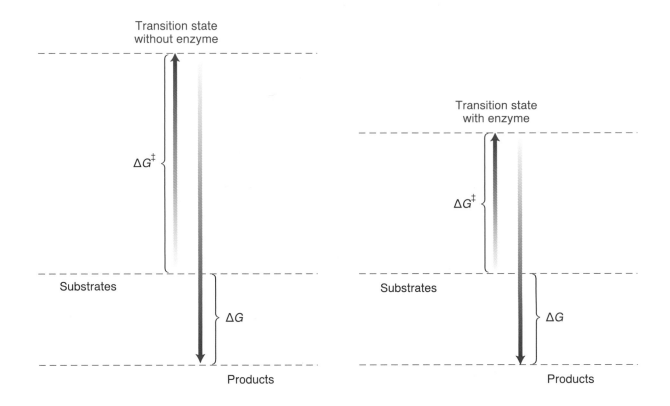

▶ **Figure 3.15** Decreasing the free energy of activation. Enzymes accelerate reactions by decreasing $\Delta G^{\ddagger}$. However, they do not determine the direction of reactions, since they do not affect ΔG.

3.16   Factors Affecting the Rate of Enzyme Reactions

The rate of an enzyme reaction depends on several factors including

> substrate concentration,
>
> enzyme concentration,
>
> temperature,
>
> pH, and
>
> ionic strength.

Let's examine them one by one.

The Michaelis constant is named after Leonor Michaelis, who, along with Maud Menten, made an important contribution to the study of enzyme kinetics nearly one century ago.

• *Substrate concentration.* The first factor affects the reaction rate in a positive manner: Increasing the substrate concentration (while keeping the enzyme concentration fixed) results in more substrate molecules reacting within a given time; therefore, the reaction rate increases. However, there is a limit to this increase: When the active sites of all available enzyme molecules are occupied by substrate molecules, any further increase in the substrate concentration has no effect on the reaction rate. If this happens, the enzyme is *saturated,* and *maximal rate* (V_{max}) is achieved. The dependence of reaction rate on substrate concentration is shown in figure 3.16. Two indices of the catalytic properties of an enzyme are the Michaelis constant and the turnover number. The **Michaelis constant**, symbolized as K_M, equals the substrate concentration corresponding to half-maximal rate (figure 3.16). Thus, an enzyme with low K_M requires a low substrate concentration to reach half its maximal activity. K_M ranges from 10^{-7} to 10^{-1} mol $\cdot$ L^{-1} for most enzymes. The **turnover number** is the number of substrate molecules converted to product by an enzyme molecule in a specified time (usually 1 s) when the enzyme is fully saturated with substrate. The turnover number shows how fast an enzyme dispatches a reaction, and ranges from 1 to $10^6 \cdot$ s^{-1} for most enzymes.

• *Enzyme concentration.* The rate of enzyme reactions is obviously influenced by enzyme concentration. Increasing it results in a proportional increase in reaction rate, since more active sites are available for substrate binding.

• *Temperature.* Raising the temperature increases the rate of chemical reactions because it increases the kinetic energy of molecules, thus raising the percentage of effective collisions (the collisions leading to reaction). The same holds true for enzyme reactions except that increasing the temperature above a certain limit—which is around 50° C for many enzymes—leads to denaturation (section 3.6). Then enzyme activity is decreased or even abolished (figure 3.17). The practice of sterilizing food and medical equipment by heating is based exactly on the thermal destruction of the enzymes that enable pathogens to survive and reproduce.

• *pH.* The catalytic power of enzymes depends on pH too. Changes in [H$^+$] cause the addition or removal of protons from an enzyme molecule and may thus alter the conformation of the active site. Enzymes are active in a more or less narrow pH range, usually exhibiting maximal activity at a value close to the physiological pH. Figure 3.18 presents an example of pH dependence of enzyme activity.

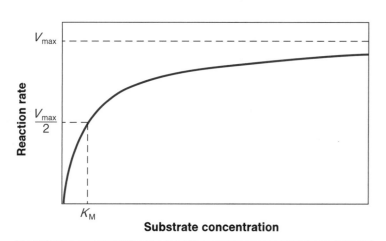

▶ **Figure 3.16**   Reaction rate–substrate concentration plot. The rate of an enzyme reaction increases with substrate concentration up to a maximal value corresponding to the saturation of the enzyme. K_M, the Michaelis constant, indicates the substrate concentration at which half-maximal rate is reached.

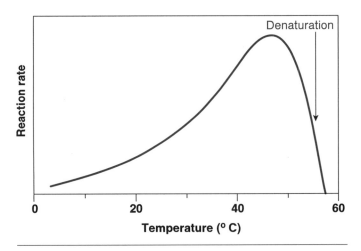

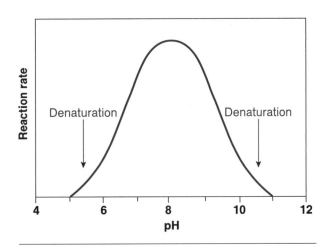

▶ **Figure 3.17** Enzyme activity–temperature plot. Enzyme activity increases with temperature up to the point of denaturation of the enzyme (and substrate, if it is susceptible to denaturation).

▶ **Figure 3.18** Enzyme activity–pH plot. The activity of an enzyme depends on the pH of the surrounding medium. The graph depicts glucose 6-phosphate phosphatase, an enzyme involved in glucose metabolism (see section 9.19).

- *Ionic strength.* Finally, enzyme activity is affected by the *ionic strength,* which is the sum of the concentrations of all ions in a solution. Ions interact with protein molecules through electrostatic bonds, thus modifying their tertiary and quaternary structures. Many enzymes are denatured when the ionic strength is shifted above or below the normal values.

You can find a further discussion of the factors affecting the rate of enzyme reactions, with a focus on the control of exercise metabolism, in the introduction to part III (section III.4).

Problems and Critical Thinking Questions

1. Consider a protein of 100 amino acids:
   a. How many peptide bonds are there in it?
   b. How many N-terminal amino acids does it have?
   c. How many C-terminal amino acids does it have?
   d. How many amino acids have bound amino groups?
   e. How many amino acids have bound carboxyl groups?

2. List ways in which muscle extracts O_2 from the blood.

3. Glucose 6-phosphate phosphatase (the enzyme of figure 3.18) catalyzes the reaction

   glucose 6-phosphate + H_2O → glucose + P_i

   What effect will the following have on the reaction rate?

   a. An increase in the concentration of glucose 6-phosphate
   b. An increase in pH from 7 to 8
   c. An increase in pH from 8 to 9
   d. An increase in temperature from 37° to 55° C
   e. An increase in the concentration of glucose 6-phosphate phosphatase

CHAPTER 4

Nucleic Acids and Gene Expression

The area of nucleic acids and gene expression is the "hottest" in modern biochemistry, the one with the most intense research activity. It is a relatively new area. Only half a century has passed since the discovery of the basic structure of DNA (the double helix). The genetic code was deciphered less than 40 years ago. Recognition of the roles played by the different kinds of RNA is even more recent, and a new kind is discovered every few years. The revolutionary laboratory techniques that signaled the birth of genetic engineering appeared about three decades ago. The sequencing of most of the human genome is almost yesterday's news. As for our knowledge on the control of gene expression, it resembles the knowledge of an infant starting to explore the world around it.

The present chapter introduces us to this magnificent world.

4.1   Introducing Nucleic Acids

Nucleic acids are the molecules of heredity. They are of two kinds:

> <u>D</u>eoxyribo<u>n</u>ucleic <u>a</u>cid, or **DNA**
>
> <u>R</u>ibo<u>n</u>ucleic <u>a</u>cid, or **RNA**

Nucleic acids are made of **nucleotides.** In particular, the building blocks of DNA are called **deoxyribonucleotides,** while those of RNA are called **ribonucleotides.** The terms nucleic and nucleotide are derived from the word nucleus. Indeed, DNA is the "trademark" of the (eukaryotic) cell nucleus, and this was where it was discovered. However, the nucleus is not the only home for nucleic acids. RNA—although it is synthesized in the nucleus—is also found in the cytosol and endoplasmic reticulum. In addition, nucleic acids are present inside the mitochondria. Thus, we are dealing with terms of historical rather than literal meaning.

As for the term acids, it reflects the acidic character of DNA and RNA owing to the presence of many phosphoryl groups in their molecules. Finally, the prefixes deoxyribo- and ribo- refer to the most striking difference in structure between DNA and RNA, one that we will discuss in due time.

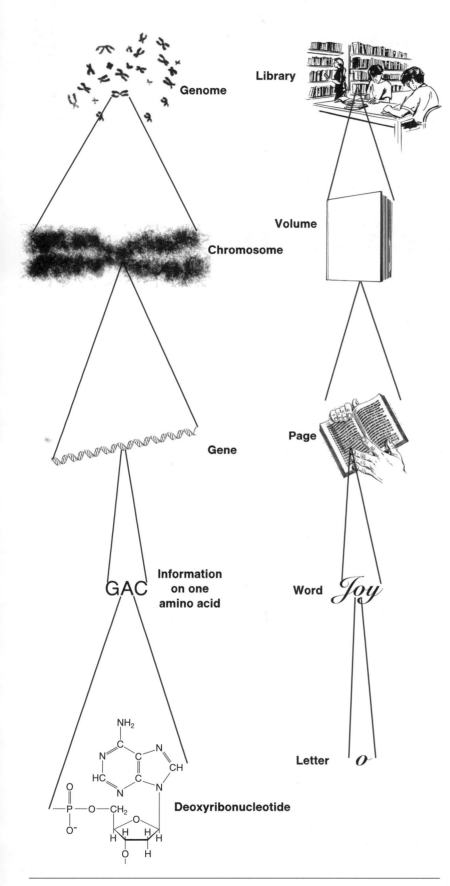

4.2   Flow of Genetic Information

We characterize nucleic acids as the molecules of heredity because they store and transmit genetic information (that is, information on how an organism is constructed and functions) from generation to generation. In fact, there is a division of labor between DNA and RNA: The former stores genetic information, and the latter transmits it. Much of the genetic information ends up in protein synthesis. Thus, the fundamental scheme of the flow of genetic information is

DNA → RNA → proteins

We call the sum of the genetic material of an organism its **genome.** The genome is like a library (figure 4.1): It consists of large DNA molecules, the **chromosomes**, just as a library consists of volumes. Every chromosome contains many **genes,** corresponding to the pages of a volume. *A gene is a piece of DNA containing an integral item of genetic information* such as the information on the amino acid sequence in a protein. The information on one amino acid is like a word on the page. Deoxyribonucleotides are the letters of that word.

4.3   Deoxyribonucleotides, the Building Blocks of DNA

Deoxyribonucleotides consist of three parts: a nitrogenous base; a deoxyribose unit; and one, two, or three phosphoryl groups (figure 4.2).

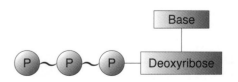

▶ **Figure 4.2**   Simplified structure of a deoxyribonucleotide. A deoxyribonucleotide is made up of a nitrogenous base, a deoxyribose unit, and one to three phosphoryl groups. The latter are each depicted as an encircled P. A ~ symbolizes a phosphoanhydride linkage.

DNA contains four different nitrogenous bases, which are presented in figure 4.3. Two bases consist of two rings each, one ring being six membered, the other being five membered. In all, the two rings contain five carbon and four nitrogen atoms. These bases are adenine and **guanine;** they belong to the category of **purines.** The other two bases belong to **pyrimidines;** their names are **thymine** and **cytosine.** They consist of one six-membered ring containing four C and two N. The rings of both purines and pyrimidines are planar. All four compounds in figure 4.3 are alkaline (that is why they are called bases). For the sake of brevity, we symbolize them by their initials (A, G, T, and C).

The second component of deoxyribonucleotides, β-D-**2-deoxyribose** (figure 4.4), is the one that gave DNA its name. β-D-2-Deoxyribose is derived from ribose (see figure 2.3) by the elimination of the oxygen connected to carbon 2 (thence the prefix 2-deoxy-). As for the prefix β-D-, I will explain its meaning in section 5.2. For the sake of simplicity, in most cases I will omit all prefixes of deoxyribose from now on.

Each base of DNA is connected to deoxyribose through a covalent bond formed either between nitrogen 9 of the purines and carbon 1' of deoxyribose or between N1 of the pyrimidines and, again, C1' of deoxyribose. Finally, the structure of a deoxyribonucleotide is completed by the addition of one, two, or three phosphoryl groups to C5'. The constitutional formula of a deoxyribonucleotide is shown in figure 4.5. It is called *deoxyadenosine monophosphate* and is symbolized as dAMP. Had it two phosphoryl groups, it would be *deoxyadenosine diphosphate* (dADP); had it three phosphoryl groups,

> A deoxyribonucleoside consists of a base tied to deoxyribose.

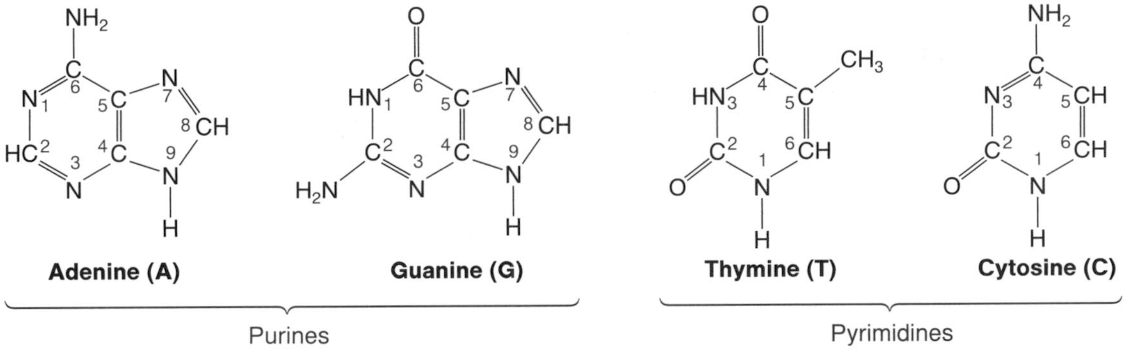

Adenine (A)   **Guanine (G)**   **Thymine (T)**   **Cytosine (C)**

Purines   Pyrimidines

▶ **Figure 4.3**   The bases of DNA. DNA contains four bases, two purines and two pyrimidines. Note the numbering of the atoms in the rings. Which one of the four bases have you met already in a previous chapter?

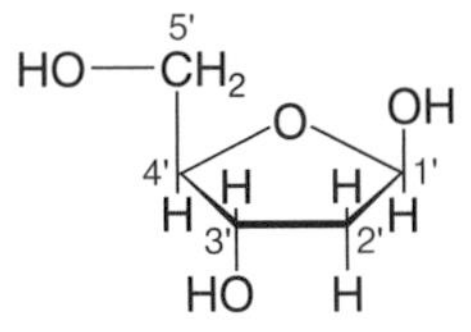

▶ **Figure 4.4**   Deoxyribose. β-D-2-Deoxyribose is a distinctive component of DNA. Four of the five atoms in its ring are carbon, but for the sake of simplicity they are not depicted (see also figure 2.3). The numbers of the carbons are followed by a prime (') to distinguish them from the numbers of the atoms in the rings of the bases (see figure 4.3), which are joined with deoxyribose in deoxyribonucleotides.

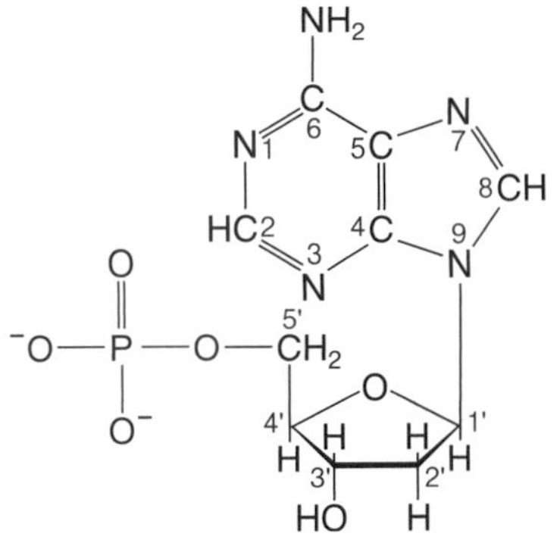

▶ **Figure 4.5**   dAMP. Deoxyribonucleotides like the dAMP shown here are the building blocks of DNA. The unit in black is deoxyadenosine (in general, a deoxyribonucleoside). What is the difference between this compound and AMP in figure 2.3?

The deoxyribonucleo-
sides of DNA are deoxy-
adenosine, deoxy-
guanosine, thymidine,
and deoxycytidine. The
prefix deoxy– is not
needed in thymidine,
as I will explain in sec-
tion 4.9.

it would be _deoxyadenosine triphosphate_ (dATP). In general, we distinguish deoxyribonucleotides according to the number of phosphoryl groups in them by calling them
deoxyribonucleoside monophosphates, diphosphates, or _triphosphates,_ and symbolizing
them as dNMP, dNDP, and dNTP.

4.4   Primary Structure of DNA

Deoxyribonucleoside monophosphates, joined in a row by covalent bonds, form DNA,
which is therefore referred to as a **polynucleotide chain.** The bond connecting two
deoxyribonucleoside monophosphates is formed between the phos-
phorus of one and the oxygen at position 3' of the other (figure 4.6). In
this way, every phosphoryl group in DNA (except the terminal one) is
linked to carbons 3' and 5' of two adjacent deoxyribose units. This link
is termed a **phosphodiester linkage.**

_The sequence of the deoxyribonucleotides composing a DNA molecule
is its primary structure._ The only variable groups in this sequence are
the bases. In contrast, the backbone, consisting of deoxyribose units
alternating with phosphoryl groups, does not vary among DNA molecules. Therefore, if one wants to describe the primary structure of
a DNA molecule, all one has to do is report its base sequence, say,
ACGTACCT. . . .

The bonding of each phosphoryl group of DNA with two different positions of deoxyribose (C5' and C3') results in the molecule's
having two distinct ends. On one end (top of figure 4.6), C5' of the
terminal deoxyribose unit does not participate in a phosphodiester
linkage, whereas C3' does. Conversely, on the other end, C3' of the
terminal deoxyribose unit does not participate in a phosphodiester
linkage, whereas C5' does. We therefore speak of the 5' and 3' _ends_ of
the polynucleotide chain, respectively. By convention, we write the
primary structure in the 5' → 3' direction unless the opposite direction
is preferable, in which case we indicate the two ends.

4.5   The Double Helix of DNA

Neighboring deoxyribonucleotides in a DNA molecule assume unique
conformations, which constitute its _secondary structure._ The most
common secondary structure of DNA is the _double helix._ Its discovery
by James Watson and Francis Crick in 1953 is a landmark in the history
of biological sciences.

We designate the double helix of Watson and Crick as _B DNA_ to
distinguish it from other DNA conformations that were discovered
afterward. The main features of B DNA (figure 4.7) are the following:

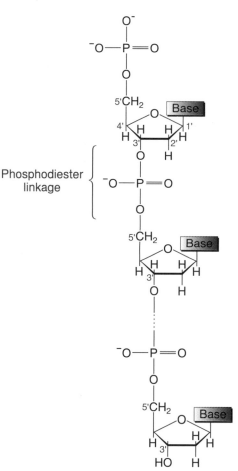

▶ **Figure 4.6**   DNA chain. A DNA molecule is a
chain of deoxyribonucleotides. Note the different
ends of the chain.

Why not distinguish the
ends of a DNA molecule
according to which one
bears a phosphoryl
group? A phosphoryl
group does not reside in
one end only (as figure
4.6 implies). Both ends
may have no phospho-
ryl groups, or one or
more.

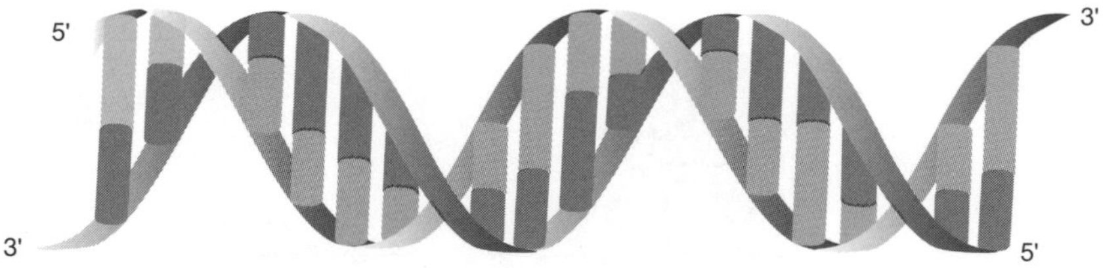

▶ **Figure 4.7**   The double helix. DNA usually has the shape of a right-handed helix containing
two polynucleotide chains of opposite directions (5' → 3', 3' → 5'), attracted through their bases.
In this simplified diagram, the chain backbones are represented by ribbons, and the base pairs
are represented by two-colored bars. In reality there is no empty space between base pairs.

1. DNA consists of two polynucleotide chains wrapped around each other in a right-handed fashion. The chains are also called *strands*. The diameter of the double helix is 20 Å.

2. The backbones of the two strands lie on the surface of the double helix, whereas the bases are sequestered inside. The planes of the bases are parallel to each other and perpendicular to the longitudinal axis of the helix. Each base in one strand faces a base in the other strand and is attracted to it by hydrogen bonds.

3. The two strands have opposite directions, that is, one runs 5' → 3' and the other 3' → 5 '. Thus, the 5' end of one strand and the 3' end of the other strand are present in each end of the double helix (if DNA is not circular; see next section).

4. The four bases pair in a strictly defined manner: adenine pairs with thymine, and guanine pairs with cytosine. Thus, each base pair (bp), be it A-T or G-C, contains one purine and one pyrimidine.

5. Each turn of the double helix encompasses 10.5 bp and occupies 36 Å on its longitudinal axis.

The matter of base-pairing in the double helix is worth discussing a bit further. The obligatory presence of adenine opposite thymine and of guanine opposite cytosine has a very important corollary: *The base sequence in one strand determines the base sequence in the other.* Because of this, we call the two strands *complementary.*

4.6   The Genome of Living Organisms

DNA molecules are the largest of all biological molecules. In fact, their size corresponds roughly to the complexity of an organism. The DNA of the virus *φX174* was the first to be sequenced in 1977, by Frederick Sanger and coworkers. It consists of 5,386 bp (although it exists in a single-stranded form as well). Its molecular mass is $3.3 \cdot 10^6$ Da, and its length is 1.8 μm. The DNA of φX174 is circular, as is the DNA of bacteria. One of them, *Escherichia coli,* is an inhabitant of our colon and is the most fully characterized cell biochemically. Almost all of *E. coli'*s genome is gathered in a single circular DNA molecule consisting of $4.6 \cdot 10^6$ bp.

The genome of eukaryotes is much larger and is organized in multiple chromosomes. Every chromosome is an open-chain (not circular) double-stranded DNA molecule. Human diploid cells contain 46 chromosomes organized in 23 pairs. In 22 of these pairs, the two chromosomes are duplicates called *autosomes.* The final pair consists of the *sex chromosomes:* an X and a Y chromosome in males and two X chromosomes in females. The DNA of our 46 chromosomes consists of approximately $6 \cdot 10^9$ bp (figure 4.8), and most of their sequence was reported in 2001.

> We use the term "circular" in DNA to signify that it has no ends, not that it has a circular shape.

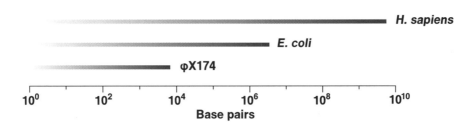

▶ **Figure 4.8**   Genome sizes. The comparison of the genetic material of a multicellular organism such as the human *(Homo sapiens)* to that of primitive life forms such as the *Escherichia coli* bacterium and the φX174 virus is overwhelming. Note that the scale is logarithmic (the number of base pairs increases by a factor of 10 from notch to notch). Were it proportional, the band corresponding to the human would have to be over 1 km long in order for the band corresponding to the virus to be barely visible (1 mm long)!

If we could stretch the DNA of a human diploid cell in a straight line, it would be 2 m long! Nevertheless, these 2 m are contained in a nucleus just 5 to 10 μm in diameter. How is this tremendous condensation achieved? Remember that the double helix is extremely thin (only 20 Å in diameter). DNA is wrapped in a very orderly manner with the aid of proteins to form the *chromatin* of the nucleus (see figure 1.5). This ordered wrapping of the genetic material leads to an astonishing compression. Suffice it to say that the final length of a chromosome is about 10,000 times less than the length of its DNA. In addition to saving space, the packaging of DNA is important for the transmission of genetic information.

4.7   DNA Replication

To develop and stay alive, an organism needs to produce new cells. Most of our cells are produced by **mitosis**, that is, the division of a cell into two "daughter" cells, each equipped with an exact copy of the genome of the "parent" cell. Therefore, cells need to duplicate their DNA. This is done in a process called **DNA replication.**

DNA replication takes place in eukaryotic cells before mitosis, during a stage of their life called the *S phase* (S stands for synthesis). Replication is a complex process requiring the combined action of a multitude of proteins. The leading part belongs to **DNA polymerase**, an enzyme catalyzing DNA polymerization, that is, the sequential addition of deoxyribonucleotides to a growing chain.

DNA is not duplicated in its double-stranded form; rather, the two strands have to be separated for replication to begin. This is performed by *helicase,* a protein that unwinds the double helix at internal points of the chromosome. Helicase hydrolyzes ATP to produce the energy necessary to separate the two strands. Then DNA polymerase uses each strand as a *template* to synthesize a new, complementary strand (figure 4.9). The substrates for this synthesis are deoxyribonucleoside triphosphates, which are synthesized in sufficient quantities to duplicate DNA.

Apart from deoxyribonucleoside triphosphates and a template strand, DNA polymerase requires a *primer,* as it is unable to start replicating DNA from scratch. The primer is a short RNA strand, complementary to the template strand and attached to it by base-pairing (as we will see shortly, RNA has bases that are almost identical to those of DNA). The primer is made by *primase,* an enzyme capable of synthesizing RNA by starting with one ribonucleotide. When replication is on its way, the primer is excised from the newly formed DNA.

Figure 4.10 shows how DNA polymerase initiates replication. First, the primer strand is attached to the template strand by hydrogen bonding between complementary bases. Note that the two strands have opposite directions, just as in the DNA double helix. The stage is set now for DNA polymerization to begin. A dNTP having a base complementary to the first unbound base of the template after the 3' end of the primer is attached to this base by hydrogen bonding. For example, if the base in the template is T, dATP binds to it. Next, the oxygen at the 3' end of the primer forms a covalent bond with the α phosphorus of the dNTP, and the two phosphoryl groups at the end of the molecule are detached. As a result, a dNMP unit is added to the primer and pyrophosphate is produced.

All this happens inside the active site of DNA polymerase (not shown in figure 4.10). When the addition of a deoxyribonucleotide is completed, the enzyme moves slightly toward the 5' end of the template strand and adds the next deoxyribonucleotide to the nascent strand by the mechanism just described. Thus, the polymerase "walks" along DNA, replicating it on the way.

It is evident from the preceding description that the parental strand is copied in a 3' → 5' direction, whereas the new strand is synthesized in a 5' → 3' direction. The

Why are dNTP used as substrates for DNA polymerization if only the dNMP part is incorporated in DNA? A reaction of the nascent strand with a dNMP would be endergonic, thus not favored. In contrast, the reaction with dNTP is exergonic because of the cleavage of a phosphoanhydride linkage. Moreover, as pointed out in section 2.3, the PP_i produced can be further hydrolyzed to two P_i, thus releasing additional energy.

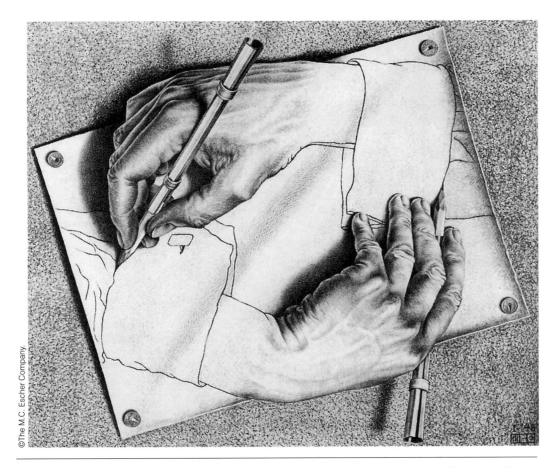

▶ **Figure 4.9** Complementary hands. During replication, each of the two complementary DNA strands dictates the base sequence of the other, just as one hand draws the sleeve of the other in this lithograph by the graphic artist of the bizarre, Maurits Cornelis Escher.

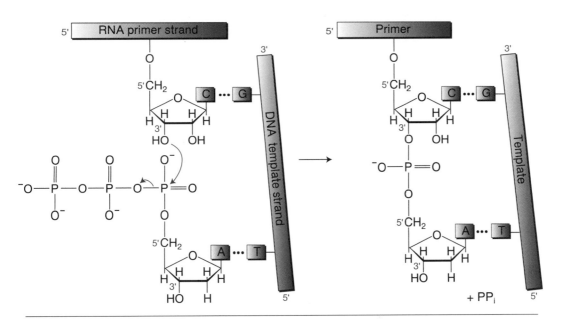

▶ **Figure 4.10** DNA replication. Each DNA strand serves as template to form a new strand during replication. DNA polymerase utilizes an RNA primer strand to add, one after another, deoxyribonucleotides that are complementary to the template strand. The colored arrows indicate the way in which a deoxyribonucleotide is added and PP_i is formed.

51

opposite directions and the complementarity of the two strands fit perfectly with the structure of the double helix. Thus, as the new strand grows, it wraps around the parental strand, forming a double-stranded DNA that is neither totally new nor totally old; instead, it is half new and half old. The same happens with the other parental strand that is being copied (figure 4.11).

In the end of replication, one parental DNA (chromosome) gives rise to two daughter DNA (chromosomes). Each one of the latter contains one strand from the parental

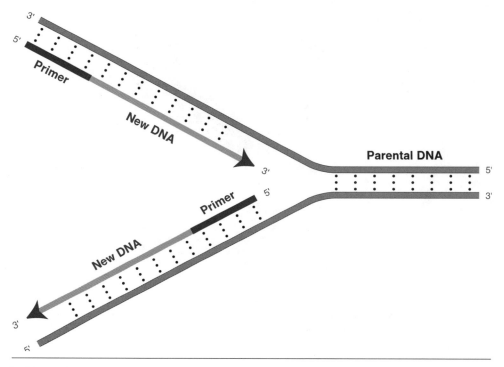

▶ **Figure 4.11** Replication fork. As helicase unwinds the parental DNA, DNA polymerase copies each strand in a 3' → 5' direction, forming a new strand in a 5' → 3' direction.

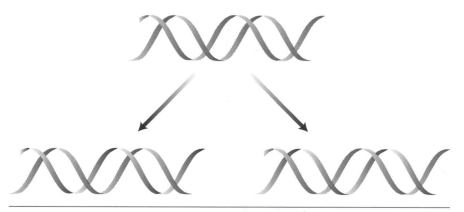

▶ **Figure 4.12** DNA replication is semiconservative. A parental chromosome (top) yields two daughter chromosomes, each of which contains one parental strand (gray) and a complementary new one (dark red).

DNA and one new strand (figure 4.12). When each of the two chromosomes—and each member of the other pairs of identical chromosomes formed by replication—has been enclosed in separate daughter cells (at the end of mitosis), the genome of the parent cell will have been equally divided between them. We express this by calling DNA replication *semiconservative.*

> In DNA replication, a double-stranded DNA yields two identical double-stranded DNA.

4.8　Mutations

DNA replication is exceptionally but not completely faithful. DNA polymerase, although extremely accurate, is not infallible. Thus, once in a while, it inserts a base that is not complementary to the one in the template. If this mistake escapes an inherent proofreading mechanism (which rarely happens), the newly formed strand will not be completely complementary to the parental strand. The frequency of error ranges from one in 10 million to one in 10 billion, depending on the organism. These changes in DNA base sequence are called **mutations.** Mutations can also be due to external causes such as ultraviolet irradiation and substances called **mutagens.**

There are basically three kinds of mutations:

Substitution of one base for another, this being the most frequent kind of mutation

Insertion of one or more bases within the normal sequence

Deletion of one or more bases from the normal sequence

We can understand these mutations more easily by relating them to similar "mutations" in children's talk:

Look at this aphlete running! *(substitution)*

I wouldn't like me as a friend becaurse I tell fibs. *(insertion)*

Shut your eyes, I've got a suprise for you. *(deletion)*

A mutation can modify the genetic information contained in a gene in such a way that it dictates a defective protein. This defect may cause disease or even death. Alternatively, a mutation can be "silent"; that is, it can take place at a point containing no genetic information or can modify the genetic information in a way that makes no difference to the organism. Finally, a mutation may render a protein more efficient in its biological role or even endow it with a new property. A mutation of this kind may confer a benefit to its bearer in relation to the rest of the population as far as survival is concerned. By improving the prospects of survival, this individual increases the chances of reproducing and passing the beneficial mutation to his or her or its offspring (if the mutation is present in the cells used for reproduction, such as an animal's germ line). Because the offspring too will have a higher probability of reproducing, the proportion of individuals bearing the mutation will increase generation after generation until they finally prevail and the novel genetic feature becomes a possession of the entire population.

The accumulation of beneficial mutations or more drastic changes in the genome (such as duplication and shifts of large DNA segments) may give rise to a new biological species. This is how species that are more complex and better adapted to their environment emerge on Earth continually. We can therefore say that *mutations are the vehicle of evolution.*

Genes for Fitness and Performance

Although the overwhelming majority of the genome is identical in humans of the same sex, approximately 0.1% differs because of mutations that have accumulated over thousands of generations since the appearance of *H. sapiens.* These differences in **genotype** are responsible for differences in **phenotype:** facial characteristics, colors, stature, blood group, and so on. Harmful mutations are responsible for hereditary diseases such as sickle cell anemia. Are there also mutations that are responsible for reduced physical performance or fitness in some individuals? Conversely, are there mutations that predispose for increased performance or fitness? A rapidly growing body of evidence is providing positive answers to both these questions.

The authors of a recent report scrutinized the relevant literature to identify 160 genes and other points in our genome that exhibit a *polymorphism* (variation in DNA base sequence) associated with performance and health-related fitness (see sidebar figure). In simple terms, this means that a certain variant of a gene may dictate an RNA or protein that causes higher performance or fitness than another variant does. For example, the gene dictating a protein called *angiotensin I-converting enzyme* displays two variants, I and D; the former appears more frequently in elite endurance athletes compared to nonathletes. The same variant appeared more frequently in exercising than in sedentary patients with hypertension.

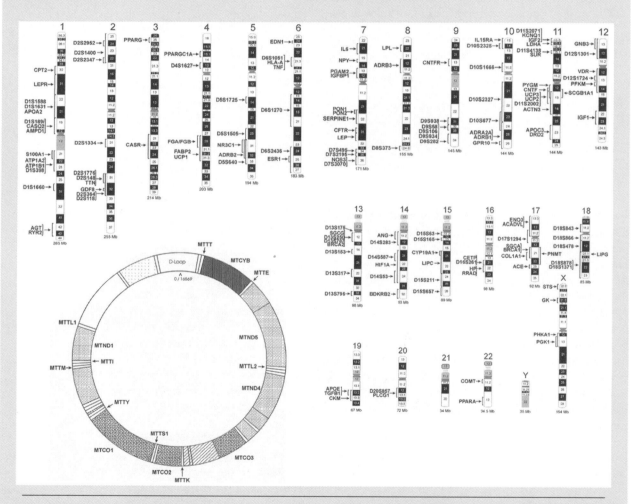

Gene map for human performance and health-related fitness phenotypes. Much like a map showing the leisure and sporting facilities in a city, this one displays the positions of 160 genes that have been related to performance and health-related fitness on the 22 autosomes, the X chromosome, and the much smaller circular mitochondrial DNA drawn in larger scale at the bottom left (more on the mitochondrial DNA in section 12.6). The genes are given in abbreviated form. For example, CPT2, the first entry, stands for carnitine palmitoyltransferase II, an enzyme that we will consider in section 10.4. The map offers a comprehensive view of the genetic basis for athletic achievement and fitness, although it may overwhelm the uninitiated reader with its complexity.

Reprinted, by permission, from B. Wolfarth et al., 2005, "The human gene map for performance and health-related fitness phenotypes: the 2004 update," *Medicine and Science in Sports Exercise* 37(6): 881-903.

The 160 genes and genetic loci fall under 10 classes of phenotypes, some of which we will consider in parts III and IV: endurance; strength and anaerobic performance; hemodynamics; familial cardiac arrhythmias; anthropometry and body composition; insulin and glucose metabolism; lipids, inflammation, and hemostatics; chronic disease; exercise intolerance; and physical activity. Containing only 29 genes in 2000, the list is expected to grow year after year and will certainly have become considerably longer by the time you read these lines.

4.9   RNA

Let's turn now to the other nucleic acid, RNA. Its building blocks, the ribonucleotides, resemble the deoxyribonucleotides of DNA (as the similarity in the names suggests). They differ in two aspects (figure 4.13). First, RNA contains no thymine (with one exception that I will present in section 4.15) but a structurally related compound, **uracil.** (Because of this, we do not need the prefix deoxy- in the deoxyribonucleoside thymidine, as already mentioned in section 4.3.) Second, instead of β-D-2-deoxyribose, RNA contains **β-D-ribose,** hence the name ribonucleic acid. In contrast to deoxyribose, ribose possesses oxygen at position 2'.

A ribonucleoside consists of a base tied to ribose. The ribonucleosides of RNA are *adenosine, guanosine, uridine,* and *cytidine.*

By analogy to DNA, each RNA base is linked to ribose, and ribose is linked to one, two, or three phosphoryl groups to form ribonucleotides specified as *ribonucleoside monophosphates, diphosphates,* and *triphosphates,* or NMP, NDP, and NTP. You have already met such compounds: In chapter 2 we considered adenosine monophosphate, adenosine diphosphate, and adenosine triphosphate (AMP, ADP, and ATP) as the main participants in energy exchange in biological systems. At the end of the present chapter we will examine the significance of the fact that our energy currency is a ribonucleotide.

An RNA molecule (figure 4.14) is made of ribonucleoside monophosphate units joined in a row by phosphodiester linkages

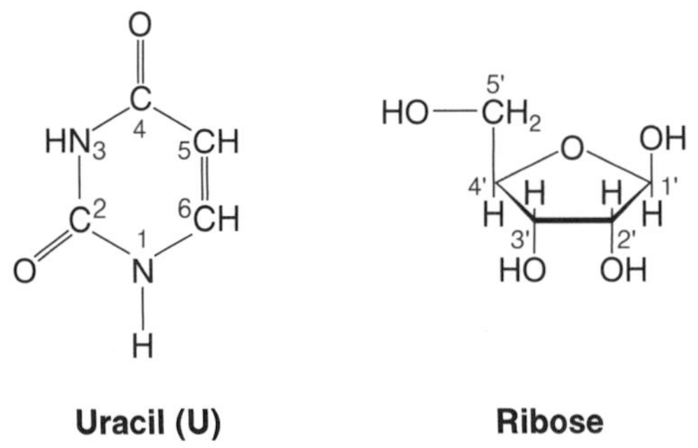

▶ Figure 4.13 Unique features of RNA. RNA shares three bases (adenine, guanine, and cytosine) with DNA but contains uracil instead of thymine. (How do these two differ?) Additionally, RNA has ribose in place of DNA's deoxyribose.

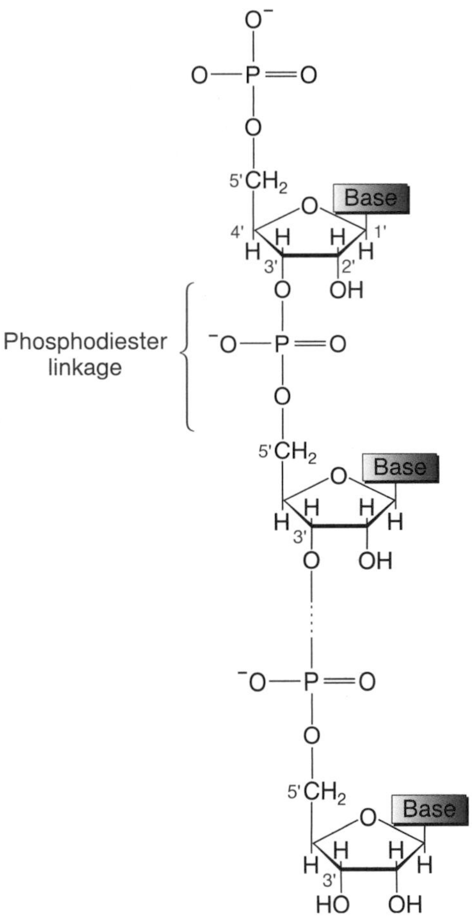

Uracil (U)   Ribose

▶ Figure 4.14 RNA chain. An RNA molecule is a chain of ribonucleotides.

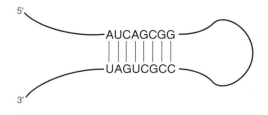

5'

AUCAGCGG
||||||||
UAGUCGCC

3'

▶ **Figure 4.15** Base-pairing in an RNA strand. Short complementary base sequences in an RNA molecule may be attracted by hydrogen bonds to form a double helix. Such helices may confer a particular shape, which is important for RNA function.

identical to those in DNA. The conventional direction of an RNA chain is 5' → 3' too, and its ribonucleotide sequence is signified by the initials of the bases, A, G, U, and C.

RNA molecules vary in size and secondary structure. They are mostly single stranded but may form short double helices intramolecularly (figure 4.15), if complementary base sequences of opposite direction happen to exist at different sites of a molecule. The complementary base pairs in RNA are A-U and G-C. The size of RNA molecules ranges from a few tens of ribonucleotides to 100,000. This heterogeneity in structure is the basis for diversity in function. In the following pages, we will have the opportunity to explore several kinds of RNA and their biological roles.

4.10 Transcription

The information recorded in DNA is transmitted to RNA by **transcription.** Transcription differs from replication in that it takes place ceaselessly during a cell's lifetime. In contrast (as we saw), replication takes place only once, when the cell is about to divide. Another difference between the two processes is that replication encompasses the entire DNA, whereas transcription is performed on parts of the DNA.

RNA polymerase stars in transcription, much as DNA polymerase stars in replication. RNA polymerase catalyzes the synthesis of RNA, DNA being the template. Ribonucleoside triphosphates serve as substrates in this reaction. As with replication, the two strands of DNA need to unfold locally, at the site where transcription is to begin. RNA polymerase possesses this ability in itself (it does not require helicase). Moreover, RNA polymerase does not require a primer strand.

The way RNA polymerase acts (figure 4.16) is similar to the way DNA polymerase acts: While one of the separated DNA strands is in the active site, two NTP having bases complementary to two adjacent bases in the template are attached to them by

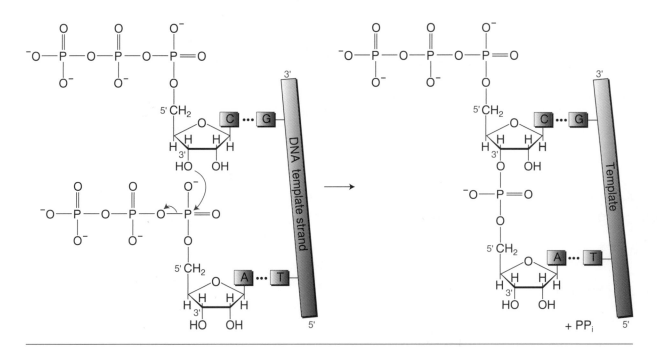

▶ **Figure 4.16** Transcription. A single strand of DNA is transcribed to RNA by the sequential joining of ribonucleotides having bases complementary to the ones in the template. Look for differences between this figure and figure 4.10.

hydrogen bonding. A covalent bond is then formed between the oxygen at 3' of one NTP and the α phosphorus of the other. As a result, the two ribonucleotides are linked and PP_i is formed.

Repetition of the previous reaction, with ribonucleotides being added one after another opposite the next bases of DNA, leads to the formation of an RNA molecule (often called a *transcript*) having a base sequence complementary to that of the template strand. Moreover, as is evident in figure 4.16, whereas DNA is transcribed in a 3' → 5' direction, RNA grows in a 5' → 3' direction. Because of this complementarity and direction, *the RNA base sequence is that of the DNA strand that is not transcribed,* since both the RNA strand and the DNA strand that is not transcribed are complementary to the DNA strand that is transcribed and both have a direction that is opposite to the direction of the DNA strand that is transcribed (figure 4.17). The only difference in base sequence is the presence of uracil in place of thymine.

The question now arises: If transcription takes place in pieces, how are the pieces of DNA to be transcribed selected? There are DNA base sequences named *promoters* and *enhancers,* which are recognized by proteins termed **transcription factors.** Transcription factors bind to these sequences and serve as docking sites for RNA polymerase. Likewise, there are DNA base sequences and proteins that force the polymerase to stop transcribing.

In transcription, RNA is formed that is complementary to a DNA strand.

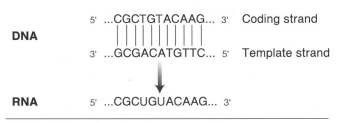

▶ **Figure 4.17** Complementarity of RNA to DNA. RNA formed by transcription of DNA has a base sequence identical to the strand that is not transcribed (with U in place of T), which is termed the coding strand. In the example shown, the lower DNA strand is transcribed; therefore, RNA has the base sequence of the upper strand.

The term *factor* next to a biological process is used to describe a protein controlling the process (see also growth factors in section 12.2).

4.11   Genes and Gene Expression

In section 4.2, we defined the gene roughly as a piece of DNA containing an integral item of genetic information. After our acquaintance with RNA and transcription, we are in a position to define the gene more accurately as *every region of DNA that is transcribed to functional RNA.*

The number of genes in an organism relates to its complexity. *E. coli* has "only" 3,000 genes, whereas the human genes are estimated to number between 20,000 and 25,000. The size of each gene ranges from hundreds to tens of thousands of base pairs. It is remarkable that genes constitute a minor part of the genome of eukaryotes. The rest consists of sequences that control transcription, play a structural role, or serve no apparent purpose. Whether living organisms accumulated useless genetic material in the course of evolution or whether the utility of these sequences is yet to be discovered is something that only future research can tell.

The presence in a cell, organ, or tissue of RNA or protein synthesized according to the information contained in a gene is called **gene expression.** When this happens, we say that *the gene is expressed* and we refer to RNA and protein as **gene products.** Only part of the gene repertoire of a multicellular organism is expressed in a single cell. As a result, there are many kinds of cells—in terms of both form and function—in an organism. For example, a muscle cell differs from a liver cell, which differs from a cell in adipose tissue, because each expresses some genes that the other does not. Gene expression is controlled primarily, though not exclusively, at the level of transcription. In chapter 12, we will discuss how exercise modifies gene expression, thus causing many of the known adaptations of the body to training.

Differences in gene expression can be likened to differences in recipe selection from a cookbook. Although many cooks may have the same book, each one selects and prepares those recipes that suit his or her idiosyncrasies, budget, the preferences of family or clients, and the circumstances (season of the year, holiday, etc.). Likewise, although all cells in an organism have the same genome, each one expresses genes according to its type, energy status, signals from other cells, and the phase that the cell itself or the organism is in.

4.12   Messenger RNA

Some of the RNA molecules synthesized through transcription carry the information to build proteins. However, RNA synthesized by RNA polymerase in eukaryotes undergoes extensive processing inside the nucleus before it becomes ready to transmit this information. Processing involves mainly the excision of sequences from the interior of the primary transcript and the splicing of the segments that are left (figure 4.18). We call the segments that are excised *introns* and those that are left, *exons.* Exons form **messenger RNA (mRNA)**, which contains the precise information on the amino acid sequence in a protein. A primary transcript may have many introns; in fact, the total length of the introns may be greater than that of mRNA.

After their formation, the mRNA molecules cross the pores of the double nuclear membrane, called *nuclear envelope,* and transfer the information for protein synthesis to the cytosol and endoplasmic reticulum.

Readers who are careful with numbers may have traced the following contradiction: How can the number of human proteins range from 100,000 to 200,000 (section 3.3) when our genes number only 20,000 to 25,000? The answer lies mainly in the *alternative splicing* of exons, that is, the fact that different introns may be excised from a primary transcript at a time (figure 4.19). This results in the production of more than one mRNA species from a single gene. Another reason for the existence of more proteins than genes is that the protein dictated by an mRNA may be cut after translation, producing two new proteins. Additionally, both DNA strands are transcribed (in

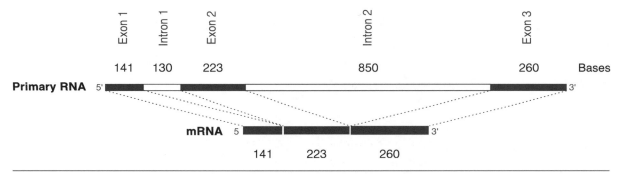

▶ **Figure 4.18** RNA splicing. The genetic information carried by a transcript is discontinuous, since the ribonucleotide sequence contains introns that have to be removed accurately in order for the information-containing pieces, the exons, to be united. Shown are the primary RNA and mRNA of the gene directing the synthesis of human β-globin (see section 3.11).

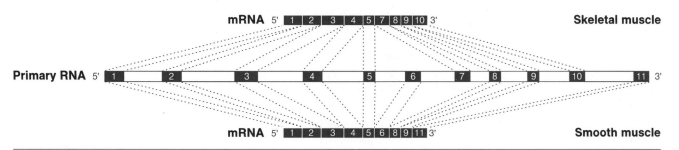

▶ **Figure 4.19** Alternative splicing. A primary RNA may produce more than one mRNA if its exons are spliced in different ways. The example shown is a gene directing the synthesis of tropomyosin (a muscle protein that we will consider in section 7.9) in chicken. The primary transcript contains 11 exons that are spliced differently in skeletal and smooth muscle. In skeletal muscle, exons 6 and 11 are omitted, whereas in smooth muscle exons 7 and 10 are omitted. Note that the size of the exons has been exaggerated for diagrammatic purposes. In reality, their total size is only 1/11 the size of the introns.

opposite directions) in certain genes, thus doubling the available genetic information. Finally, *RNA editing*, the changing of the RNA base sequence after transcription by processes other than splicing, is an effective means of modulating the genetic information contained in a gene.

Because proteins are so much more numerous and diverse than genes, *proteomics* (the study of protein structure and function) is much more complex than *genomics* (the study of genes).

4.13 Translation

The synthesis of proteins according to the base sequence of mRNA is called **translation** and takes place in the cytosol and endoplasmic reticulum. Translation is performed by **ribosomes**, huge—in molecular terms—complexes of RNA and protein having a mass of 4,200 kDa! Four RNA molecules occupy the bulk of each eukaryotic ribosome (that is why it is called a ribosome). This **ribosomal RNA (rRNA)** is mainly responsible for ribosome coherence and function. In addition, each ribosome contains over 50 proteins.

Ribosomes consist of two **subunits**, a large one and a small one (figure 4.20), which join to translate an mRNA molecule in a 5' → 3' direction, synthesizing a protein in

> In translation, proteins are synthesized on the basis of instructions contained in mRNA.

> Do not confuse ribosomal subunits with protein subunits. The former are much larger, and each contains many proteins (along with RNA).

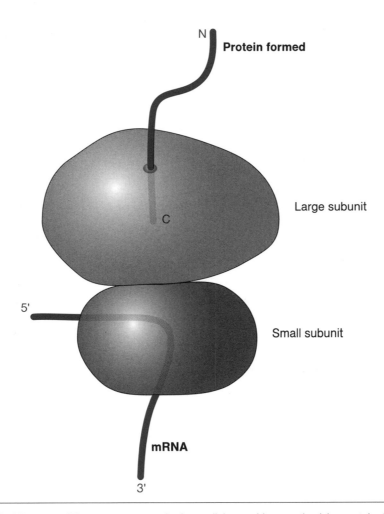

N
Protein formed

Large subunit

C

5'

Small subunit

mRNA

3'

▶ **Figure 4.20**   Ribosome. Ribosomes are complex intracellular machines synthesizing proteins based on information kept in mRNA. A eukaryotic ribosome consists of two subunits and is about 35 nm tall (1 nm = 10 Å).

60   Exercise Biochemistry

an N → C direction. The subunits dissociate when translation is complete. Now, to follow the entire process in detail, we need to first consider the genetic code and the molecules serving as translators.

4.14   The Genetic Code

How does the four-letter alphabet of the RNA bases dictate the 20 amino acids of proteins? Evidently, there cannot be a one-to-one correspondence between bases and amino acids. How about a correspondence of two bases to one amino acid? The number of possible pairs of four items is 16 (4^2), which is not sufficient either. Therefore, we need three bases to correspond to one amino acid, even though the resulting number of combinations ($4^3 = 64$) is much larger than necessary. The quest for the correspondence between all the possible triplets of RNA bases and the amino acids was one of the most exciting adventures in the history of research in the biological sciences. The fruit of this quest was the discovery of the **genetic code.**

To discuss the genetic code we need to introduce some terminology. The triplets of bases are called **codons.** To convey that a specific amino acid corresponds to a specific codon, we say that *the codon encodes the amino acid.* By extension, we say that a gene or an mRNA encodes a protein. Other terms will follow later in this section.

Orthogonal in its classical form and circular in a modern, handier version (figure 4.21), the genetic code lets us find which amino acid is encoded by every codon. To do this, we have to locate the bases of the codon one after another. Suppose we are interested in (5')AGC(3'). We seek the first base (A) in the inner circle and find it in the lower left quadrant. Next we seek the second base (G) within the zone surrounding this quadrant; it is located at the left end of the zone. Finally, we seek the third base (C) within the segment of the outer zone delimited by the preceding G and find that it corresponds to serine.

Because the number of codons is more than three times the number of amino acids, most amino acids are encoded by more than one codon. Thus, there is no one-to-one correspondence between codons and amino acids. We express this by saying that the genetic code is *degenerate.* An additional feature of the genetic code is its *universality:* It is the same for all living organisms, with very few exceptions. *The universality of the genetic code is strong evidence for the evolution of all forms of life on Earth from a common ancestor.*

Of the 64 codons, one signals the initiation of translation for all proteins. This *initiation codon* is AUG, and it encodes methionine. The second amino acid in a polypeptide chain is encoded by the next triplet of bases in mRNA (there is no "punctuation" between codons) and so on until one of three *termination codons* appears. These are UAA, UAG, and UGA; they encode no amino acid and signal the termination of translation.

An example of a short peptide encoded by an imaginary sequence of mRNA bases follows. (Note that this is purely for practice; in reality mRNA encodes polypeptide chains.)

5'...AGGAGGUAUGCCAGUUAGCCUACCCAAGUGAAUC...3'
Met–Pro–Val–Ser–Leu–Pro–Lys

What you have to do—and what a ribosome does—to translate an mRNA is, first, locate the initiation codon by scanning the base sequence from the 5' to the 3' end. Once you find it, write down Met. Then continue translating based on the genetic code until you reach a termination codon. Write down nothing; translation is over.

In everyday non-scientific, and even scientific, speech and writing, the genetic code is often confused with the genetic material. For example, the recent sequencing of the human genome has been heralded as the "discovery of the human genetic code." However, the genetic code is the same for humans, plants, and bacteria alike, and its discovery dates back to 1966.

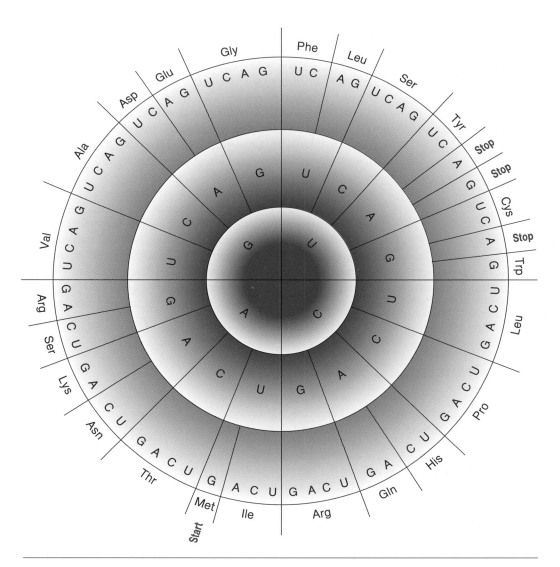

▶ **Figure 4.21** The genetic code. The genetic code relates all possible triplets of RNA bases to an amino acid or signal to stop translation. Look for the first base (5' end) of a triplet in the inner circle, the second base in the middle zone, and the third base in the outer zone.

4.15 Transfer RNA

Amino acids and codons do not recognize each other directly but by the intervention of yet another kind of RNA, **transfer RNA,** or **tRNA,** so called because it transfers amino acids. There is at least one tRNA for each amino acid. To specify the various tRNA, we write the amino acid (in its three-letter abbreviation) as superscript. For example, tRNA[Ala] is the tRNA specialized in transferring alanine.

Transfer RNA molecules contain roughly 80 ribonucleotides including several unusual ones, which are formed by modification of standard ribonucleotides after transcription. For example, thymine, which is otherwise absent from RNA, is present in some tRNA. Transfer RNA molecules have the characteristic shape shown in figure 4.22.

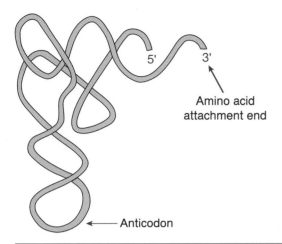

▶ **Figure 4.22** Transfer RNA. Transfer RNA translates the language of nucleic acids into the language of proteins. Its three-dimensional structure is usually depicted as an inverted L. Shown in this simplified diagram are the ends of amino acid attachment and codon recognition (anticodon).

Their chain folds to bring together complementary bases, which form short stretches of double helix. The resulting L-shaped molecule has two distinct ends. One (in the upper right corner of the figure) contains both the 3' and 5' ends of the polynucleotide chain. The amino acid carried by tRNA is attached to the 3' end by a covalent bond between the carboxyl group of the amino acid and the hydroxyl group at position 2' or 3' of the terminal ribose of tRNA. Formation of these *aminoacyl-tRNA,* or simply *charged tRNA,* is catalyzed by highly specific *aminoacyl-tRNA synthetases.* There is at least one of these enzymes for each amino acid.

The other end of the L corresponds approximately to the middle of the polynucleotide chain. Here a triplet of bases is complementary to the mRNA codon, which, according to the genetic code, corresponds to the amino acid carried by the particular tRNA. This triplet of tRNA is called **anticodon.** Thus, the tRNA molecule, having at one end the amino acid and at the other the ability to recognize the cognate codon, bridges the world of ribonucleotides with the world of amino acids.

4.16   Translation Continued

When mRNA, ribosomes, and charged tRNA are available, translation may begin. The steps of this process (in simplified form) are as follows.

1. A tRNA charged with methionine binds to the small subunit of a ribosome. This tRNA is specialized for the initiation of translation and is abbreviated as Met-tRNA$_i$ (i stands for initiation). The ribosomal subunit then binds to the 5' end of mRNA and proceeds along it toward the 3' end. When it reaches the initiation codon AUG, the anticodon of Met-tRNA$_i$ recognizes it (since they have complementary sequences) and binds to it by hydrogen bonding. As in other cases of complementary binding between strands of nucleic acids, the chains of mRNA and tRNA have opposite directions. Finally, the large ribosomal subunit joins the small subunit and everything is ready for translation to continue (figure 4.23*a*).

2. Ribosomes have three binding sites for tRNA, spanning both subunits and distinguished by the initials P (for peptide), A (for amino acid), and E (for exit). The P site is occupied by Met-tRNA$_i$ when translation begins. It is also the site where the polypeptide chain grows, amino acid by amino acid. The A site is occupied by a tRNA charged with the amino acid encoded by the next codon in mRNA (figure 4.23*b*). It is also the site where charged tRNA enter to provide the amino acids for protein synthesis. We will talk about the E site in a moment.

3. The two amino acids are now ready to be linked: The amino group of the second amino acid attacks the carboxyl group of methionine and removes it from Met-tRNA$_i$, forming a peptide bond (figure 4.23*c*). Thus, the second tRNA acquires a dipeptide (becoming a *peptidyl-tRNA*) and occupies the P site of the large subunit, although the area around its anticodon is still in the A site of the small subunit.

4. The ribosome moves by one codon toward the 3' end of mRNA (figure 4.23*d*). As a result, tRNA$_i$ moves to the E site and leaves the ribosome, while the peptidyl-tRNA occupies the P site entirely.

A repetition of steps 2 to 4 follows. That is, a new tRNA, charged with the amino acid encoded by the third codon, enters the vacant A site. This amino acid is then added to

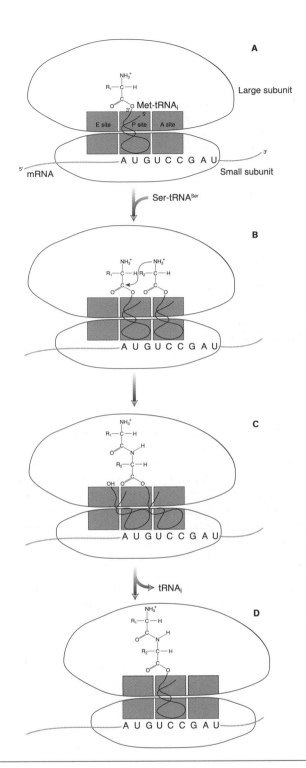

▶ **Figure 4.23** Translation. Follow the synthesis of a protein by a ribosome based on instructions from mRNA in four stills. Synthesis begins with the entrance of the special initiation tRNA charged with methionine (Met-tRNA$_i$) to the P site in the ribosome and with the location of the initiation codon (AUG) in the mRNA base sequence *(a)*. The special tRNA charged with the amino acid encoded by the second codon enters the A site *(b)*. This amino acid is serine in the hypothetical sequence of the figure. Next, serine forms a peptide bond with methionine *(c)*. The ribosome moves by one codon in a 5' → 3' direction *(d)*, resulting in tRNA$_i$ being dislodged to the E site (through which it leaves the ribosome) and Met-Ser-tRNA^{Ser} occupying the P site. Notice, by comparing stills *a* and *d* vertically, that while mRNA did not move, the ribosome moved to the right. Also note the similarity of stills *a* and *d*: Both have charged tRNA in the P site, whereas the A site is vacant. R$_1$ and R$_2$ are the side chains of Met and Ser, respectively. For diagrammatic purposes, in this representation we have not retained the proportions of the ribosome's parts.

the dipeptide of the tRNA in the P site. The ribosome moves by yet another codon and the uncharged tRNA in the P site moves to the E site (leaving the ribosome), while the tRNA carrying the tripeptide occupies the P site. With each cycle of this process, the polypeptide chain is elongated by one amino acid.

If any of the three termination codons appears in the A site, the addition of amino acids stops, as there is no tRNA having an anticodon complementary to these codons. The polypeptide chain is liberated from tRNA by hydrolysis of the bond between the carboxyl group of the last amino acid and ribose. Transfer RNA and mRNA detach from the ribosome, which then dissociates to its subunits and is ready to synthesize another protein molecule.

It is evident from the preceding description that the synthesis of a polypeptide chain proceeds from the free amino group to the free carboxyl group, that is, in an N → C direction. Finally, I have to stress that protein synthesis in the ribosomes requires a substantial input of energy. This energy is supplied by the hydrolysis of ATP and GTP (guanosine triphosphate) to AMP and GDP (guanosine diphosphate), respectively (you may practice writing the equation of GTP hydrolysis). Approximately four ∼P are broken down per amino acid added to a polypeptide chain, although this rather brief presentation of translation did not let me pinpoint which endergonic processes the energy released from ∼P hydrolysis propels.

4.17 In the Beginning, RNA?

To put life in perspective, Earth is 4.6 billion years old and the universe has recently been estimated to be 13.7 billion years old.

The origin of life has puzzled humans for thousands of years. Throughout the centuries, the human mind has tried to provide answers based on the available knowledge. However, scientifically sound views on the descent of living organisms began to appear only during the second half of the 19th century. The theory of *evolution by natural selection,* founded by Charles Darwin at that time, is nowadays supported by a huge body of evidence and is universally accepted by scientists. Our current view on the origin of life is that modern complex multicellular organisms are descendants of simpler ones, which in turn originated from unicellular organisms that appeared in the primordial lakes and seas of our planet at least 3.5 billion years ago.

The deeper we travel into the past, the fuzzier the picture becomes. How did the first cell arise? Did it have the features of modern cells? Going even further, we leave the realm of *biological evolution* and enter *chemical evolution,* which must have been taking place on Earth during the first billion years after its formation. Which was the first molecule of life? "Protein!" one might answer, since almost no biological process can happen without it. But how was the first protein formed? The instructions for protein synthesis reside in DNA. Even if the first proteins were put together in a way different from the one we know today, in some way that did not necessitate DNA, how did they reproduce? Polypeptide chains lack the ability to serve as templates for the formation of like chains. On the other hand, DNA (which possesses this ability) is unable to duplicate without the aid of enzymes. Thus, we are in a vicious cycle with no obvious way out (figure 4.24a).

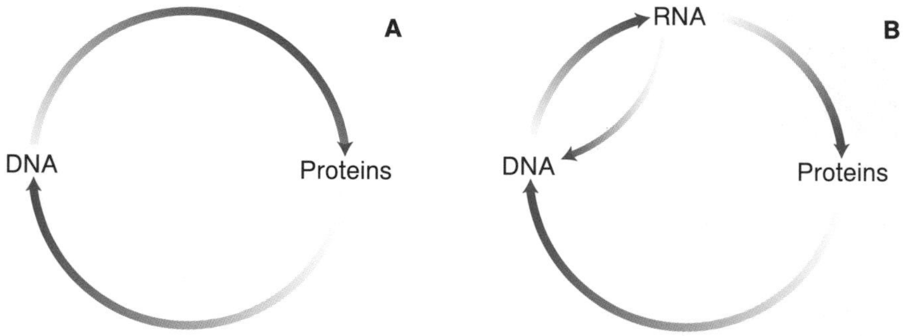

▶ **Figure 4.24** A chicken-and-egg problem. Which was the first biological molecule? *(a)* DNA cannot be synthesized without proteins, but proteins cannot be synthesized without instructions from DNA. *(b)* A primitive RNA capable of both reproducing itself and catalyzing reactions may have preceded DNA and proteins. RNA may then have driven the formation of these compounds and relinquished the functions of genetic information storage and catalysis, reserving the role of go-between.

A probable solution to the conundrum appeared in the 1980s, when Thomas Cech and Sidney Altman independently discovered biological catalysts that were not proteins but RNA. Today we know of over 100 catalytic RNA, which have been aptly named *ribozymes* to reflect the fact that they are ribonucleic acids that can act as enzymes. Ribozymes are endowed with the ability to cut, splice, and transpose RNA segments. There is also evidence that they can recognize and handle amino acids directly.

The coexistence of the catalytic activity of enzymes and the reproductive ability of nucleic acids in one molecule overcomes the problem of which one—DNA or proteins—preexisted: The earth at the dawn of life could have been a world of ribonucleic acids where every bit of genetic information was contained in RNA and every reaction in cells was catalyzed by RNA also.

A primary role of RNA is boosted by its intense presence and that of its components (the ribonucleotides) in biological processes, as summarized in the following:

- RNA is omnipresent in the transmission of genetic information. It carries the instructions from the genes (mRNA), forms the bulk of ribosomes (rRNA), and matches ribonucleotides with amino acids (tRNA). Moreover, the very formation of peptide bonds during translation is catalyzed by an rRNA in the large ribosomal subunit!

- A ribonucleotide (ATP) is the universal energy currency of cells.

- Another ribonucleotide, GTP, participates in metabolic processes as a high-energy compound (sections 4.16, 9.9, and 9.19) and in signal transduction pathways (section 9.3).

- Of the remaining ribonucleoside triphosphates, UTP is utilized in the activation of molecules participating in carbohydrate synthesis (section 9.1), while CTP plays a similar role in lipid biosynthesis.

- An unusual ribonucleotide, cyclic AMP, mediates the transduction of the messages of many hormones (section 9.3). Cyclic GMP serves a similar function.

- Some compounds holding central positions in metabolism contain ribonucleotide units in their structure. NAD and FAD (section 2.5), as well as coenzyme A (section 9.7) are such compounds.

It is thus possible that the first system of cell function and reproduction was based on RNA (figure 4.24*b*). Later, with small modifications (of ribose to deoxyribose, and uracil to thymine), DNA appeared as a more appropriate depository of genetic information because it is more stable. At the same time, catalytic RNA molecules acquired protein segments, which improved their function, until, in the end, proteins displaced RNA in this area as better and more versatile catalysts. The ability of RNA to recognize amino acids made it the link between DNA and proteins and granted it more than one role in the transmission of genetic information. Finally, the participation of ribonucleotides in a multitude of other cellular functions shows that some primitive solutions to the problems of life were so good that they withstood the trial of evolution.

How was the first RNA molecule created? Laboratory experiments have been aimed at discovering what kinds of substances could have formed on Earth under the conditions that, we assume, prevailed before the emergence of life, that is, 4.6 to 3.5 billion years ago. According to available evidence, these conditions were quite different from contemporary conditions and completely inhospitable to modern forms of life. The atmosphere contained hydrogen, methane, ammonia, and water, but hardly any oxygen. It was stunned by lightning, thunder, and volcanic explosions and was also stormed by ultraviolet radiation from space. Therefore, there must have been ample energy for the ingredients of the primordial atmosphere to react. Experiments simulating those conditions have yielded many of today's cellular components, including amino acids, nucleic acid bases, simple carbohydrates, fatty acids, peptides, and RNA-like compounds.

Thus, it appears that circumstances on *prebiotic* (before the emergence of life) Earth favored a chemical evolution from simple to complex organic compounds capable of supporting two fundamental functions of contemporary biomolecules (catalysis and reproduction). This gave life on Earth the needed spark. Finally, it is also possible that the chemical precursors of life arrived on Earth from space, on meteorites bombarding the planet at the formation of our solar system.

Problems and Critical Thinking Questions

1. Twenty-three percent of the bases of a double-helical DNA segment are cytosine. What are the percentages of the other bases?

2. Biochemists often study metabolism in a living system (for example, cells in culture) by adding a metabolite that contains one or more radioactive isotopes in its structure and examining which other compounds become radioactive through metabolic reactions. A researcher wants to study DNA replication in multiplying cells by labeling DNA but not RNA. Which radioactive compound should he or she add to the culture medium?

3. Does the following base sequence belong to DNA or RNA? Write its complementary sequence.

   5' ACUAGCGCUA 3'

4. Consider the following piece of DNA:

   5' . . . ACGACATGTT . . . 3'

   3' . . . TGCTGTACAA . . . 5'

   Write the RNA base sequence resulting from the transcription of the lower strand.

5. How many mRNA, rRNA, and tRNA molecules are involved in the synthesis of a protein molecule?

6. Consider the following piece of synthetic RNA:

   5' . . . GAGCGAUGCGCAGUUACCCUACCCAAUGAAUGAA . . . 3'

   Write the peptide resulting from its translation.

7. A mutation involving a single base in the previous sequence results in the production of the following peptide.

   Met-Arg-Cys-Ala-Val-Thr-Leu-Pro-Asn-Glu

   Which mutation is it?

CHAPTER 5

Carbohydrates and Lipids

Our acquaintance with the remaining classes of biological substances will be briefer than that with proteins and nucleic acids, not because they are less important but because we will deal with them extensively in part III. One can distinguish proteins and nucleic acids from carbohydrates and lipids in that the former are not considerable energy sources whereas the latter are. Since exercise is interwoven with increased energy demand, in part III we will discuss how energy is extracted from carbohydrates and lipids. In the present chapter, we will examine only their structure and some elements related to their utility.

5.1   Carbohydrates

Carbohydrates are compounds of the molecular formula $C_nH_{2n}O_n$ or are derived from such compounds. If one takes n as a common factor, the formula becomes $(CH_2O)_n$ and gives the impression that carbon atoms and water molecules are present in equal numbers. This coincidence is the source of the name carbo-hydrate. In reality, however, there are no water molecules inside a carbohydrate molecule.

Carbohydrates serve a multitude of needs in living organisms:

- They are, as we already said, sources of energy. Such is the role of sugar (probably the most familiar and popular carbohydrate) and glucose (the most common energy source of cells).
- They offer cells external protection. Cellulose is the support (literally) of the entire plant kingdom.
- Attached to proteins and lipids, they help cells to recognize molecules or other cells in their surroundings.
- They are part of every building block of nucleic acids (as ribose or deoxyribose).

Depending on their size, carbohydrates are divided into monosaccharides, oligosaccharides, and polysaccharides.

A protein having a covalently attached carbohydrate unit is a *glycoprotein,* while a lipid having a covalently attached carbohydrate unit is a *glycolipid.* As we will see, the prefixes glyco- and gluco- dominate in carbohydrate nomenclature. They are derived from the Greek word *glykýs,* meaning "sweet," which is what many carbohydrates are.

Organic compounds bearing a carbonyl group at a terminal position are called *aldehydes,* whereas those bearing a carbonyl group at an intermediate position are called *ketones.*

5.2 Monosaccharides

Monosaccharides are the simplest of carbohydrates. A monosaccharide usually has three to seven carbon atoms and is called specifically a *triose* (3 C), *tetrose* (4 C), *pentose* (5 C), *hexose* (6 C), or *heptose* (7 C). Let's meet some of them.

Trioses

D-Glyceraldehyde

Dihydroxyacetone

▶ Figure 5.1 Trioses. Glyceraldehyde and dihydroxyacetone are the simplest carbohydrates. Note the numbering of carbon atoms. The carbonyl group is marked in color.

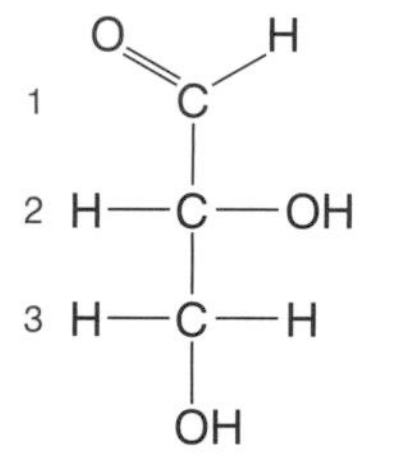

The trioses *glyceraldehyde* and *dihydroxyacetone* (figure 5.1) are the smallest monosaccharides. The two compounds are isomeric, since they share the molecular formula $C_3H_6O_3$. The way they differ is in the position of the *carbonyl group.*

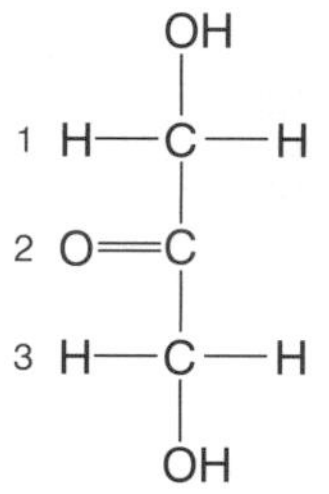

Like amino acids, monosaccharides with the exception of dihydroxyacetone exhibit those particular isomeric forms (enantiomers) characterized by the prefixes D- and L- (cf. section 3.1). In contrast to amino acids, however, the carbohydrates that prevail in biological systems are of the D form.

Pentoses

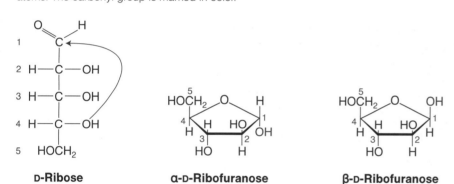

D-Ribose

α-D-Ribofuranose

β-D-Ribofuranose

▶ Figure 5.2 Ribose. D-Ribose exists in open-chain and cyclic form. The arrow shows how the former is converted to the latter. We distinguish the cyclic form from the open-chain form by inserting the term furano. There are two cyclic isomers, α and β, differing in the position of the hydroxyl group at C1.

We move now from trioses to a familiar pentose, ribose (the component of ribonucleotides). Ribose has the molecular formula $C_5H_{10}O_5$ and can be either open chain or cyclic (figure 5.2). In cyclic ribose—the predominant form in biological fluids—the oxygen of C4 forms a covalent bond with C1, while the aldehyde oxygen of the open-chain form is converted to hydroxyl. The resulting five-membered ring, bearing one oxygen and four carbons, is called a *furanose.* C5 stays outside the ring. If we want to distinguish the cyclic from the open-chain form, we call the former *ribofuranose.*

The cyclization of ribose gives rise to two possible positions of the newly formed hydroxyl group at C1: one above and the other below the ring. Thus, there are two possible isomers. As we will see shortly, discerning them is important; that is why we introduce two additional prefixes, α- and β-. When looking at the formula from above (so that the numbering of the carbons follows a clockwise direction), we have α-ribose if the hydroxyl group is below the plane of the ring and β-ribose if the hydroxyl group is above the plane.

Hexoses

We conclude our acquaintance with monosaccharides by examining two hexoses, glucose and **fructose,** which occur in fruits and account for their sweetness. Both have the molecular formula $C_6H_{12}O_6$ and, like ribose, appear in open-chain and cyclic forms, the latter predominating in biological systems. Open-chain glucose cyclizes when the O of C5 attacks C1 (figure 5.3). The aldehyde oxygen of C1 is then converted to hydroxyl. The resulting six-membered ring with one O and five C is called a *pyranose,* while the cyclic glucose is called *glucopyranose* (if we want to distinguish it from the open-chain form). C6 stays outside the ring. Again, depending on the position of the hydroxyl group at C1, we have two isomers, α and β.

Fructose bears a carbonyl group at position 2 in its open-chain form. Because of this, a five-membered furanose ring forms when the O at C5 attacks C2 (figure 5.4). Two carbon atoms (C1 and C6) are left outside the ring. Alternatively, the O at C6 may attack C2, yielding a six-membered pyranose ring.

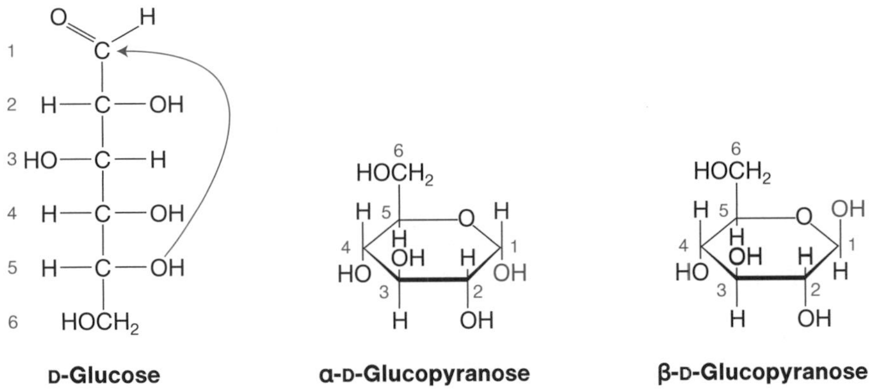

D-Glucose   **α-D-Glucopyranose**   **β-D-Glucopyranose**

▶ **Figure 5.3** Glucose. D-Glucose, a fundamental energy source for cells, exists in open-chain and cyclic forms. We distinguish the latter from the former by inserting the term pyrano. There are two cyclic isomers, α and β, differing in the position of the hydroxyl group at C1.

D-Fructose   **α-D-Fructofuranose**   **β-D-Fructofuranose**

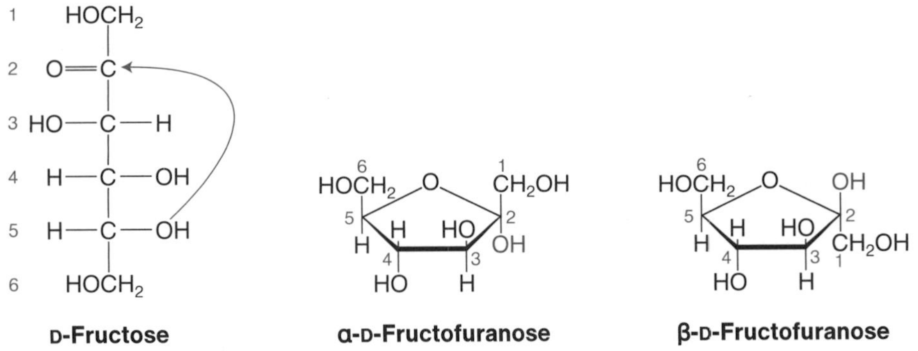

▶ **Figure 5.4** Fructose. D-Fructose differs from D-glucose in having a carbonyl group at position 2, rather than 1. As a result, the closure of the O at C5 on C2 produces a furanose ring. The assignment of the α or β configuration to fructofuranose is based on the position of the hydroxyl group at C2. Try to draw fructopyranose, produced when the O at C6 closes on C2.

5.3   Oligosaccharides

When 2 to 10 monosaccharides are linked covalently, **oligosaccharides** result. The linkages between monosaccharide units in an oligosaccharide are called glycosidic. A **glycosidic linkage** consists of an oxygen atom joined to two carbon atoms and is formed from two hydroxyl groups upon the removal of a water molecule (figure 5.5).

Let's meet two *disaccharides,* that is, oligosaccharides with two monosaccharide units. *Maltose* (figure 5.5) is a breakdown product of starch, a polysaccharide to be presented shortly, and consists of two α-D-glucosyl units linked at C1 and C4. That is why the glycosidic linkage in maltose is characterized as α1 → 4. *Sucrose,* the other disaccharide

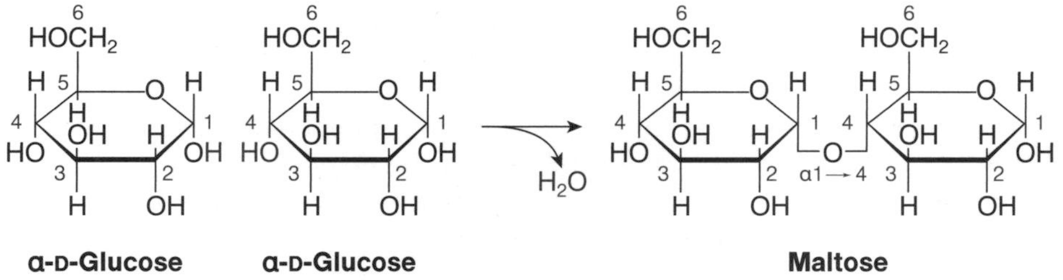

▶ **Figure 5.5** Glycosidic linkage. Two monosaccharides can be linked through their hydroxyl groups when they shed one molecule of water. In this example, two molecules of α-D-glucose are joined by an α1 → 4 linkage to form the disaccharide maltose. The bonds between O and C in the glycosidic linkage have been bent for diagrammatic purposes (to allow drawing the two monosaccharide units side by side); in reality, they are straight, just like every other covalent bond.

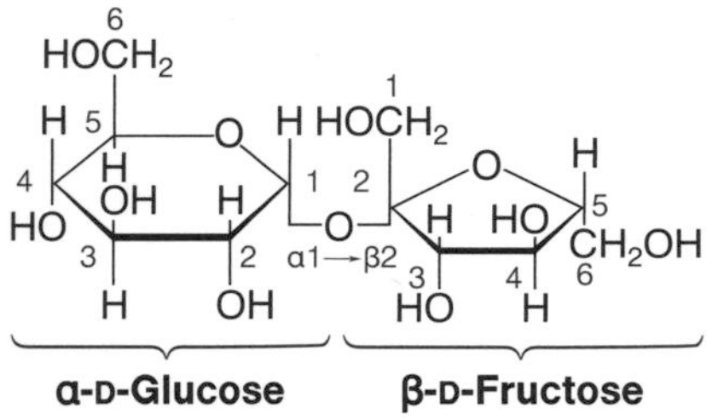

▶ **Figure 5.6** Sucrose. Our most familiar sweetener is a disaccharide of glucose and fructose. Note that the β-D-fructosyl unit has been turned upside down relative to figure 5.4.

(figure 5.6), is the common sugar. It consists of one α-D-glucosyl unit and one β-D-fructosyl unit linked at C1 and C2, respectively. The glycosidic linkage is therefore α1 → β2.

5.4   Polysaccharides

Polysaccharides are the most abundant category of carbohydrates. They contain more than 10 monosaccharide units and, depending on their composition, are divided into *homopolysaccharides* and *heteropolysaccharides*. The former are polymers of a single monosaccharide, whereas the latter are composed of various monosaccharides. We will deal only with three homopolysaccharides consisting of thousands to tens of thousands of D-glucosyl units. These polysaccharides are cellulose, starch, and glycogen.

Cellulose

Cellulose is a polymer of β-D-glucosyl units linked in a row by β1 → 4 linkages (figure 5.7). Thus, one end of a cellulose molecule has a free hydroxyl group at C1 and a bound hydroxyl group at C4, whereas at the other end things are the opposite. The former end is called *reducing,* because it has the ability to reduce other substances; the latter end is called *nonreducing.* Remember that proteins and nucleic acids also have distinct ends.

Cellulose is found in plants and is the main solid constituent of their leaves and wood. It helps to support and protect plant cells because its glycosidic linkages give rise to straight chains forming bundles through hydrogen bonding. Cellulose is one of the most abundant organic compounds on Earth. Each year, approximately one trillion tons of it is synthesized in the biosphere from glucose, the product of photosynthesis (reaction 3.1 [equation 3.1]). In terms of human nutrition, cellulose is a constituent of *dietary fiber.* Large quantities of it are found in fruits and vegetables. Cellulose is the main component of the book you are reading and, no doubt, some of your clothing, since paper and cotton are almost pure cellulose.

In contrast to the structural role of cellulose, the role of the other two polysaccharides is to store energy—starch does so in plants and glycogen does so in animals. Starch and glycogen differ from cellulose in that they consist of α-D-glucosyl instead of β-D-glucosyl units. Because of this the chains of starch and glycogen assume curved, compact shapes, which render them suitable for energy storage.

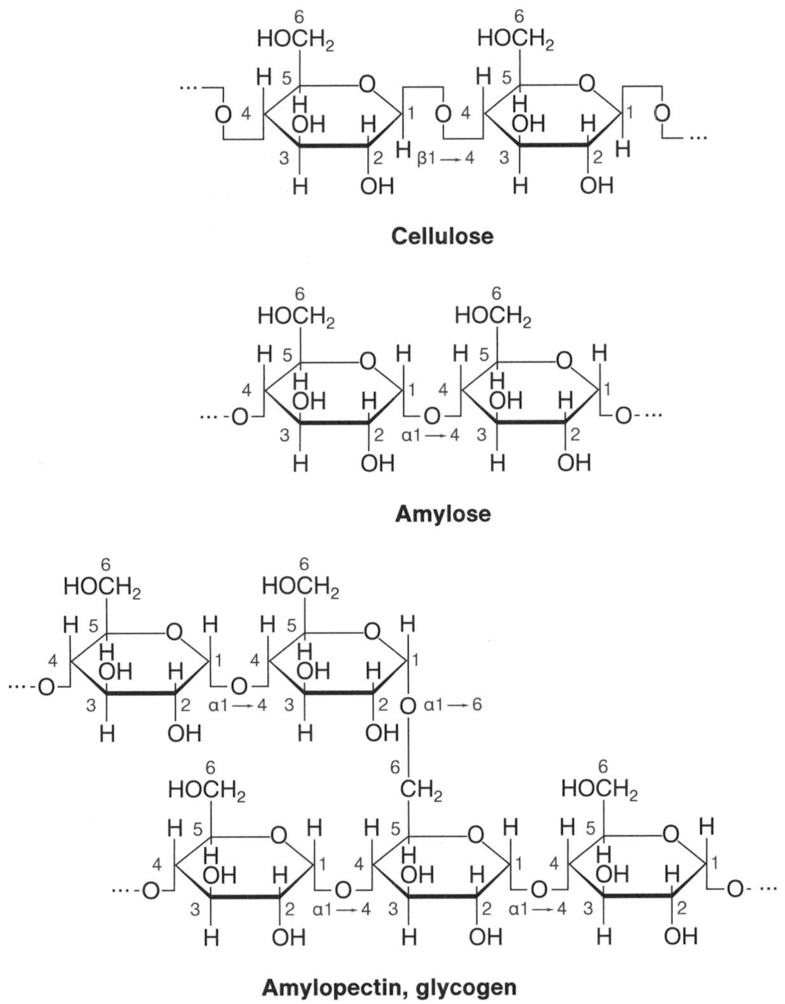

▶ **Figure 5.7** Polysaccharides. Cellulose, starch, and glycogen, three polysaccharides of glucose, differ in the way their monomers are connected. Only a few glucosyl units from each polymer are presented, since the linkages are repeated throughout their molecules. Cellulose is unbranched with $\beta 1 \rightarrow 4$ linkages. Starch is a mixture of amylose, which is unbranched with $\alpha 1 \rightarrow 4$ linkages (note the similarity with maltose in figure 5.5), and amylopectin, which in addition has branches with $\alpha 1 \rightarrow 6$ linkages. Glycogen resembles amylopectin but is bushier. Reducing ends are on the right-hand side of the structures.

Starch

Starch is found in the seeds of plants. Among these seeds, cereals, legumes, and nuts are the most important for human nutrition. Flour consists of starch mainly, and potatoes are great sources of it as well. Starch is a mixture of two substances, *amylose* and *amylopectin*. Amylose is unbranched, its glucosyl units being linked in an $\alpha 1 \rightarrow 4$ configuration (figure 5.7). In amylopectin, on the other hand, a branch appears every 20 to 30 glucosyl units. The glycosidic linkage is $\alpha 1 \rightarrow 6$ at the branch point but remains $\alpha 1 \rightarrow 4$ inside the branch. It is possible to have a branch point in the branch, another branch point in the new branch, and so on, so that the molecule acquires a bushy appearance. As a result, amylopectin has only one reducing end but many nonreducing ends.

Glycogen

Glycogen is found mainly in the liver and muscle. Its structure (figures 5.7 and 5.8) is analogous to that of amylopectin: It is branched, having the same glycosidic linkages, but its branches appear more frequently and are shorter. Therefore, a glycogen molecule has more nonreducing ends than an amylopectin molecule of equal mass (although glycogen molecules are usually larger than amylopectin molecules). The existence of many branches enables a faster breakdown for energy production, since, as we will see in section 9.2, glycogen is degraded by the sequential removal of glucosyl units from the nonreducing ends of its branches. The potential for a fast breakdown of glycogen contributes to its prominent place in exercise metabolism, as we will see in part III.

The presence of so many hydroxyl groups in the molecules of carbohydrates renders them hydrophilic, although because of their size polysaccharides are not water soluble. Proof of the hydrophilicity of carbohydrates is the great solubility of sugar in water and the great water absorbency of cotton and paper.

5.5 Lipids

Lipids present a great variety of structures and functions. This class of biological compounds encompasses the main constituent of a large tissue (adipose tissue), membrane components, hormones, vitamins, and a plethora of other molecules possessing important biological properties. The common feature of all these compounds is the low solubility in water. Considered more or less inert chemical entities until recently, they seem to get the reputation they deserve—that of active participants in a multitude of biochemical processes—in our day.

We will consider four lipid categories: fatty acids, triacylglycerols, phospholipids, and steroids.

▶ **Figure 5.8** Glycogen structure. Glycogen has relatively dense branches, which make it more susceptible to degradation, so that it can produce energy more easily. In this diagram of a small part of its structure, the hexagons depict glucosyl units. All ends are nonreducing, since the sole reducing end is buried in the core of the molecule. Can you spot the α1 → 6 linkages?

5.6   Fatty Acids

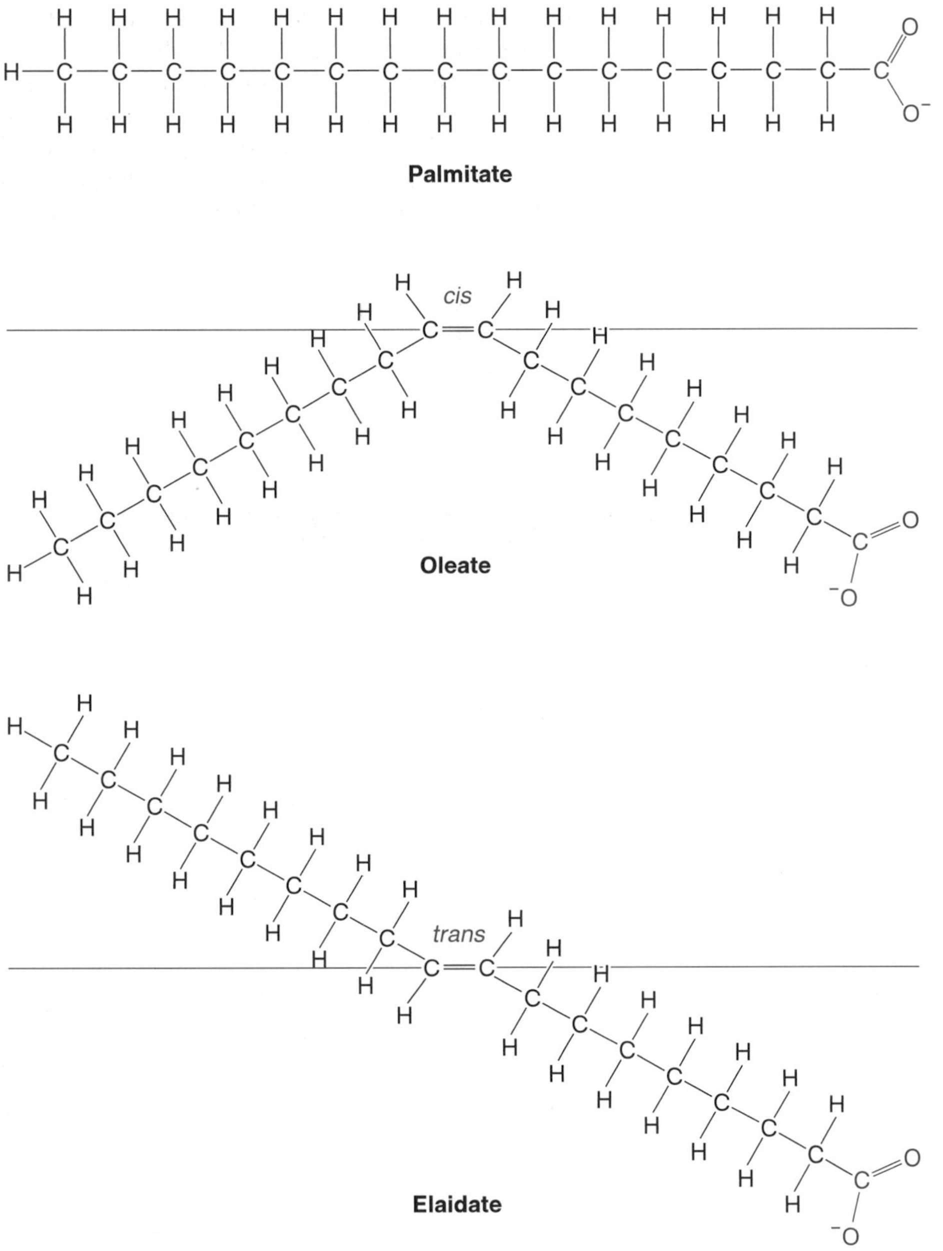

Fatty acids are organic acids containing relatively large numbers of carbon atoms (usually 12-26). They consist of a long carbon chain with hydrogens all around (organic chemists call such a chain *aliphatic*) and a carboxyl group at one end (figure 5.9). As with amino acids, the carboxyl group lacks a proton at physiological pH; thus, it carries one negative charge. The carboxyl group is hydrophilic, whereas the aliphatic chain is hydrophobic. The two characteristics are combined in a fatty acid to produce an **amphipathic** (meaning "passionate for both" in Greek), or **amphiphilic** ("loving both"), molecule. Amphipathic compounds are poorly soluble in water.

Most fatty acids have even numbers of carbon atoms because they are synthesized by the joining of acetyl groups, which have two carbons (see section 10.6). Apart from length, they differ in number of double bonds. If the carbons of a fatty acid are linked by single bonds only, the fatty acid is called **saturated** because there is no room to add any more hydrogens. Conversely, if there are double bonds between some of the carbons, the fatty acid is called **unsaturated** because a double bond may receive two H to become single. If we wish to distinguish the fatty acids with one double bond from those with more double bonds, we use the terms **monounsaturated** and **polyunsaturated**, respectively. The double bonds of a fatty acid do not exceed six and are almost invariably spaced three carbons apart. There are no triple bonds in fatty acids.

Figure 5.9 presents a saturated fatty acid with 16 carbons, named *palmitate*, and two monounsaturated fatty acids with 18 carbons, named *oleate* and *elaidate*. A double bond divides the carbon chain into two parts, which may lie either on the same side or on

Palmitate

Oleate

Elaidate

▶ **Figure 5.9**   Fatty acids. A fatty acid contains an aliphatic chain (in black) and a carboxyl group at the end. Fatty acids may or may not have double bonds between their carbons. Depending on the location of the two parts into which a double bond separates the carbon chain relative to the line passing through the carbons of the double bond, double bonds are assigned a *cis* or *trans* configuration. Palmitate and oleate are among the most abundant fatty acids in animal tissues, whereas elaidate is a minor fatty acid.

opposite sides of the imaginary line passing through the two carbons participating in the double bond. The former configuration (appearing in oleate) is called *cis;* the latter (appearing in elaidate) is called *trans.* Most unsaturated fatty acids have double bonds of the *cis* configuration. Thus, when referring to unsaturated fatty acids, I will mean unsaturated fatty acids with *cis* double bonds.

Figure 5.9 illustrates the remarkable change introduced by a *cis* double bond in the structure of a fatty acid: Whereas the molecule of a saturated fatty acid is straight, the molecule of an unsaturated fatty acid is bent. This difference affects the physical properties of fatty acids and those lipid categories that contain fatty acid units, since the straight molecules of saturated fatty acids can lie closer to each other than the crooked molecules of unsaturated fatty acids. Thus, more hydrophobic interactions (section 3.5) develop among the molecules of saturated than of unsaturated fatty acids. Because of this, saturated fatty acids have higher melting points than unsaturated fatty acids with the same number of carbon atoms, being solid at room temperature as opposed to most unsaturated fatty acids, which are liquid.

Like many biological compounds, most fatty acids are better known by their empiric than by their systematic nomenclature. Some of the most common ones are presented in table 5.1.

The main utility of fatty acids is energy production both at rest and during exercise, as we will discuss in chapter 10. Their concentrations in biological fluids are generally low. In contrast, much higher quantities of fatty acids are enclosed in other lipid categories such as triacylglycerols and phospholipids.

▶ **Table 5.1**   Usual Fatty Acids in Humans and Animals

Name	Carbon atoms	Double bonds	Position of double bonds[a]
Laurate	12	0	
Myristate	14	0	
Palmitate	16	0	
Palmitoleate	16	1	9
Stearate	18	0	
Oleate	18	1	9
Linoleate	18	2	9, 12
α-Linolenate	18	3	9, 12, 15
Arachidonate	20	4	5, 8, 11, 14
Eicosapentaenoate	20	5	5, 8, 11, 14, 17
Docosahexaenoate	22	6	4, 7, 10, 13, 16, 19

[a]Numbers indicate the carbon after which there is a double bond. Carbon numbering starts at the carboxyl end.

5.7 Triacylglycerols

Triacylglycerols, or **triglycerides,** are the most abundant lipid category. They are the main component of animal and human fat, as well as vegetable oils. Triacylglycerols serve mainly as energy depots. They consist of a glycerol unit and three fatty acid units. Glycerol (figure 5.10), also known as *glycerin,* is a compound of three carbons and three hydroxyl groups. Each hydroxyl group is linked to the carboxyl group of a fatty acid through an **ester linkage.** Thus, every triacylglycerol (figure 5.10) contains three ester linkages, which makes it a *triester.*

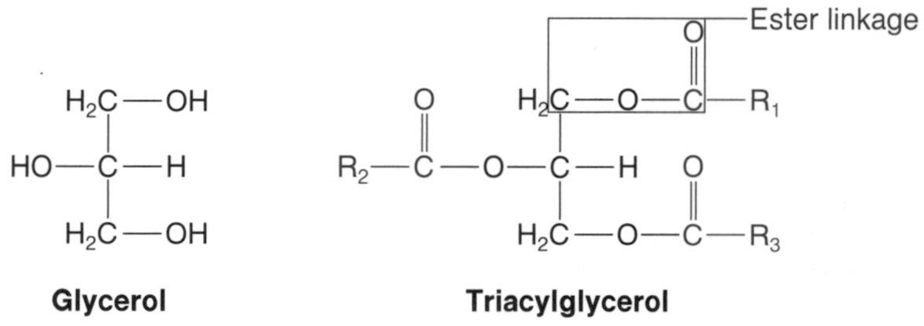

Because there are many different fatty acids and because each hydroxyl group of glycerol can be linked to any fatty acid, we get a great variety of triacylglycerols. All

Glycerol   **Triacylglycerol**

▶ **Figure 5.10**   Glycerol and triacylglycerol. Triacylglycerols, the largest energy depot in living organisms, are triesters of glycerol and fatty acids. R_1, R_2, and R_3 represent the aliphatic chains of the fatty acids, which usually differ. Acyl groups are shown in color.

are hydrophobic, which is evident by the immiscibility of fats or oils with water. Moreover, they have a low thermal conductivity, rendering the subcutaneous fat of animals an efficient insulator of their internal organs against cold (figure 5.11).

The part of the fatty acid connected to a glycerol oxygen in a triacylglycerol is called an **acyl group.** This is where the term tri-acyl-glycerol comes from, making this term more accurate than triglyceride. The acyl groups are derived from the ion of a fatty acid by removal of O⁻, and they bear the name of the fatty acid with the ending -oyl instead of -ate. Thus, the acyl group of palmitate is called the *palmitoyl group.*

The difference in melting point between saturated and unsaturated fatty acids described in the previous section is reflected in triacylglycerols: The more saturated acyl groups they contain, the higher their melting point is. Triacylglycerols of animal origin have a high content of saturated acyl groups, which is why animal fat is solid at room temperature. Conversely, plant triacylglycerols have a high content of unsaturated acyl groups, which is why vegetable oils are liquid.

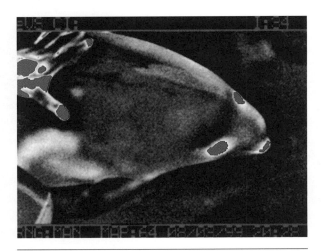

▶ **Figure 5.11** A great insulator. Infrared photograph of a harbor porpoise, the smallest cetacean of the North Atlantic, reveals what an effective insulator fat is. Body heat is emitted only through the blowhole, snout, and eye (colored dots). Heat is also lost from the nails and a large part of the palm of the person holding the harbor porpoise.

D. Ann Pabst. University of North Carolina.

5.8 Phospholipids

Phospholipids are a category of lipids presenting a remarkable structural variety. The most common phospholipids have a glycerol backbone (like triacylglycerols) and are called **glycerophospholipids** or **phosphoglycerides.** The simplest glycerophospholipid has a phosphoryl group attached to the terminal hydroxyl group of glycerol and an acyl group attached to each of the remaining groups (figure 5.12). It is called *phosphatidate* and is the parent compound of glycerophospholipids. Note that phosphatidate—as well as all glycerophospholipids—is not a single compound but a group of compounds, since its two acyl groups may be derived from a variety of fatty acids.

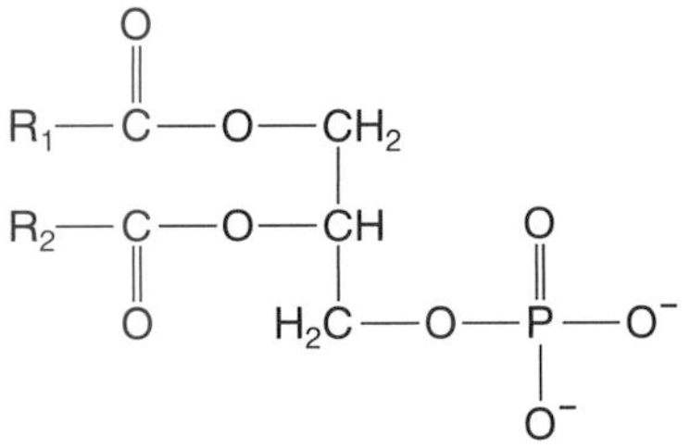

Phosphatidate is a minor glycerophospholipid of cells. The major glycerophospholipids are derived from phosphatidate by the attachment of an alcohol to its phosphoryl group. The alcohols most frequently encountered in glycerophospholipids are *choline, ethanolamine,* the amino acid serine (considered an alcohol because of the hydroxyl group at its side chain), and *inositol.* These alcohols are presented in the left column of figure 5.13. The corresponding glycerophospholipids are called *phosphatidyl choline, phosphatidyl ethanolamine, phosphatidyl serine,* and *phosphatidyl inositol* (right column of figure 5.13).

Animal tissues contain small amounts of a different phospholipid, *sphingomyelin* (figure 5.14). Sphingomyelin contains *sphingosine,* an amino alcohol with a long aliphatic chain, rather than glycerol. One acyl group and a phosphoryl choline unit are attached to sphingosine to produce sphingomyelin.

Like fatty acids, phospholipids are amphipathic. Their aliphatic chains are hydrophobic, whereas the remainder of their molecules are hydrophilic. The hydrophilic part of a phospholipid is frequently referred to as the *polar head group.* In fact, the size of this group (much larger than that of the carboxyl group in fatty acids) makes the amphipathic character of phospholipids more evident. This renders phospholipids ideal constituents of cell membranes, as we will see in section 5.10. To form membranes is thus the main function of phospholipids.

▶ **Figure 5.12** Phosphatidate. Glycerophospholipids, the main group of phospholipids, are derivatives of phosphatidate. How does this compound differ from a triacylglycerol?

Alcohols are organic compounds with hydroxyl groups.

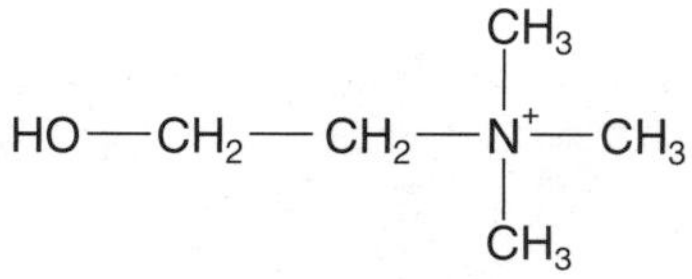

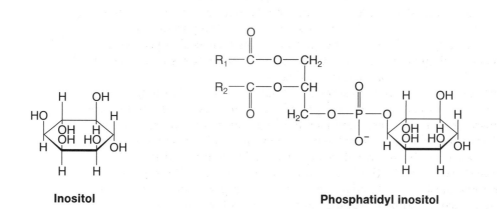

Choline

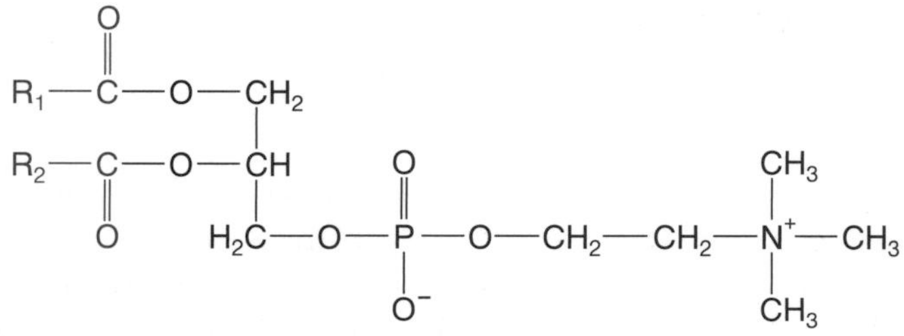

Phosphatidyl choline

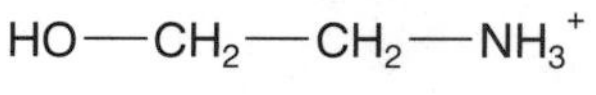

Ethanolamine

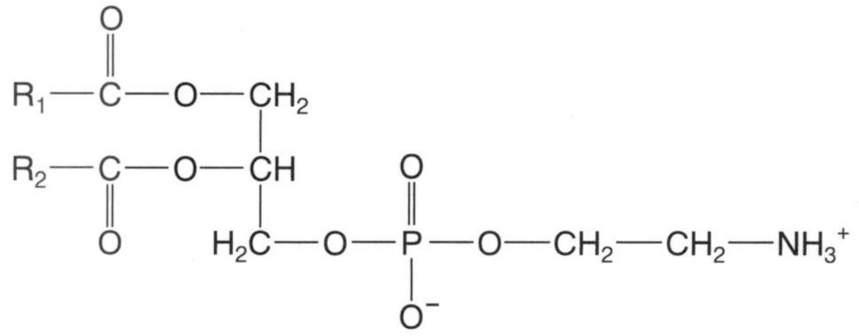

Phosphatidyl ethanolamine

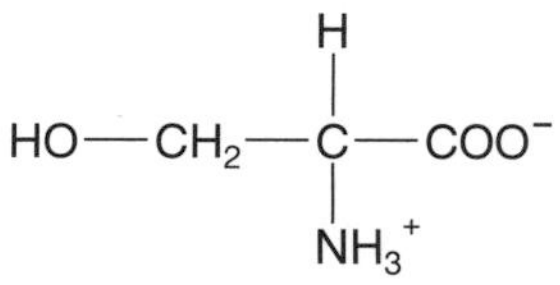

Serine

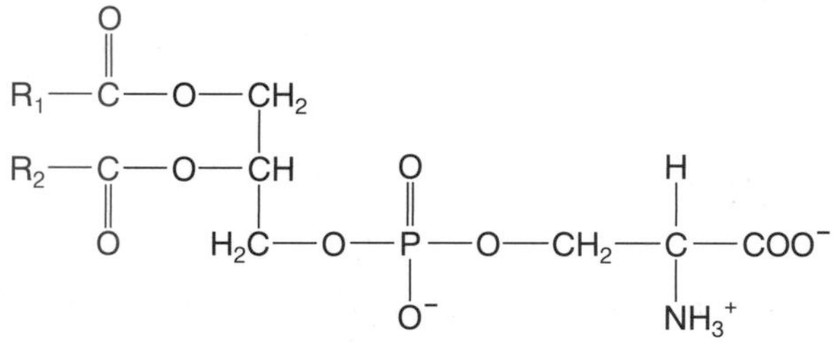

Phosphatidyl serine

Inositol

Phosphatidyl inositol

▶ **Figure 5.13** Glycerophospholipids. The linking of phosphatidate to choline, ethanolamine, serine, and inositol yields the most common glycerophospholipids of animals.

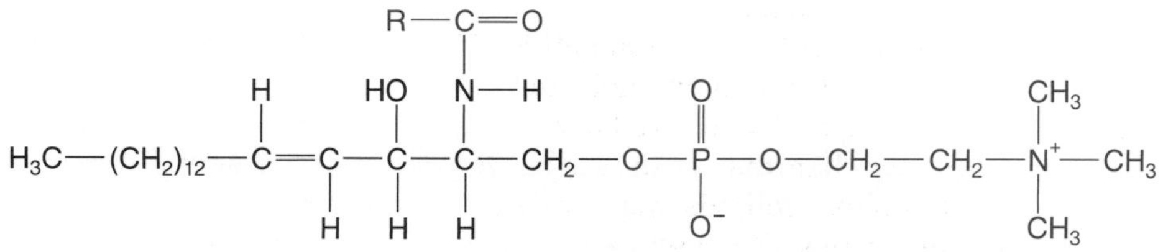

▶ **Figure 5.14** Sphingomyelin. A minor phospholipid of animals is composed of a sphingosine unit (in black), an acyl group, a phosphoryl group, and a choline unit.

5.9 Steroids

Steroids are another complex lipid category, encompassing hormones (see section 16.4) and other specialized compounds. All are derivatives of **cholesterol,** a compound containing 27 carbon atoms and four rings (figure 5.15). The arrangement of these rings is the "trademark" of steroids. Cholesterol too is amphipathic: Most of its structure is hydrophobic, whereas the hydroxyl group at its end is hydrophilic; this is also called a polar head group. Thanks to its amphipathic character and size, cholesterol participates in membrane formation next to phospholipids.

If cholesterol is linked to a fatty acid by an ester linkage between the hydroxyl group of the former and the carboxyl group of the latter, a hydrophobic **cholesterol ester** is produced. Cholesterol esters serve as a cholesterol reservoir. Cholesterol and its esters are found in animal tissues but are absent from plants, although plants contain other *sterols,* that is, steroids with hydroxyl groups.

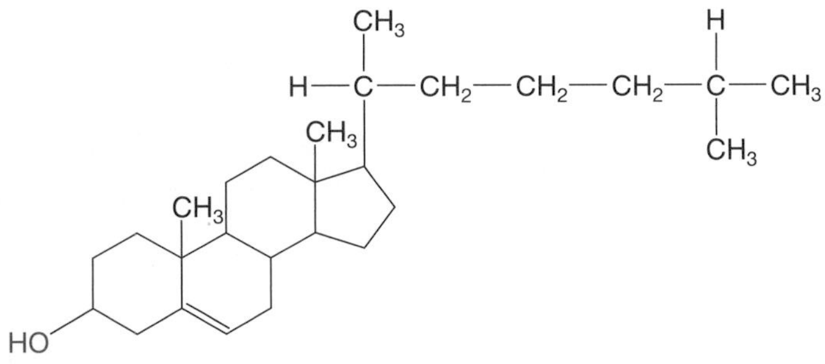

▶ **Figure 5.15** Cholesterol. Cholesterol is the parent compound of all steroids and a component of the membranes of animal cells.

5.10 Cell Membranes

Cell membranes are wonders of molecular architecture. Remember that they surround cells and intracellular organelles like the nucleus and mitochondria (section 1.11). Membranes separate aqueous solutions (for example, the cytosol from the extracellular fluid or the cytosol from the nuclear content). Because they are in touch with water, they need to have some affinity for it. At the same time, however, they need to be insoluble in water or they will vanish. Living organisms overcame this contradiction with membranes consisting of two phospholipid layers placed in such a way that their hydrophobic parts make contact, being attracted by hydrophobic interactions and excluding water. Their hydrophilic parts, on the other hand, face outward, interacting with water (figure 5.16). This arrangement creates a *bilayer* that is 6 to 10 nm thick.

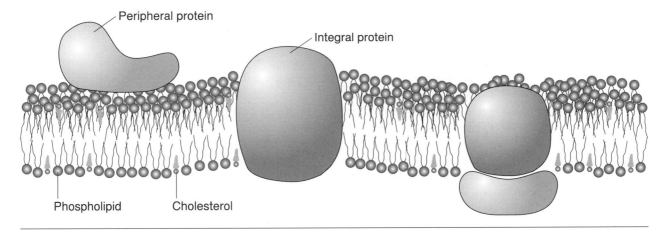

▶ **Figure 5.16** Cell membrane. Cell membranes are bilayers of phospholipids and cholesterol. The polar head groups of these amphipathic lipids (depicted as colored spheres) face the aqueous environment, whereas their hydrophobic tails (depicted as two lines in phospholipids and a wedge in cholesterol) are hidden inside the bilayer. The coherence of the membrane is primarily due to hydrophobic interactions among the tails. Membranes also contain peripheral and integral proteins.

Apart from phospholipids, membranes (especially the plasma membrane of animal cells) contain cholesterol. This is placed parallel to the phospholipids so that its small polar head group is aligned to the polar head groups of the phospholipids, and its hydrophobic body is aligned to the acyl groups of the phospholipids. Finally, membranes are studded with proteins, which control the communication of the two sides and permit the recognition of cells by substances or other cells. Many of these proteins are tightly bound to the membrane, usually spanning the bilayer and appearing on both faces of it. These are called *integral proteins.* Other membrane proteins adhere less tightly to one face of the membrane and are called *peripheral proteins.*

The Versatile Roles of Vitamins and Minerals

Carbohydrates, lipids, proteins, and nucleic acids cannot fulfill their biological roles without a helping hand from vitamins and minerals, two classes of nutrients that we need to obtain in small amounts daily. We have already considered the roles of two vitamins, niacin and riboflavin, as precursors of NAD and FAD (section 2.5). Then, in sections 3.10 through 3.12, we saw that atmospheric oxygen is harnessed, transported, and stored in the body bound directly to iron, a mineral so important for health and sport that we will devote an entire chapter to it in part IV.

Other vitamins and minerals serve important functions too. To begin with vitamins, some of the so-called *water-soluble vitamins* are used by our cells to synthesize **coenzymes,** organic compounds present in the active sites of many enzymes to help them catalyze reactions. *Thiamine (vitamin B₁)* is converted in the body to *thiamine pyrophosphate,* a coenzyme of enzymes participating in carbohydrate metabolism and the citric acid cycle. Thiamine deficiency results in *beriberi,* manifested by damage to the nervous system and the heart. *Pyridoxine (vitamin B₆)* yields *pyridoxal phosphate,* present in the active sites of enzymes metabolizing glycogen and amino acids. *Pantothenate* contributes to the structure of coenzyme A (see figure 9.18), a key molecule in the metabolism of carbohydrates, lipids, and proteins (figure 2.10). *Biotin* is a part of enzymes involved in glucose and fatty acid synthesis. *Folate* is precursor to *tetrahydrofolate,* needed by enzymes synthesizing the bases adenine, guanine, and thymine. Finally, *ascorbate (vitamin C)* serves as a reducing agent (or *antioxidant*) in the synthesis of collagen, the main protein of our connective tissue. The importance of vitamin C is evident in *scurvy,* a malady caused by vitamin C deficiency and manifested by hemorrhages under the skin, falling gums, weakness, and heart failure.

Of the *fat-soluble vitamins, retinol (vitamin A)* is precursor to retinal, a compound that responds to light in the retina of our eyes and initiates a nervous response that becomes vision (see section 6.6). Another derivative of retinol, *retinoic acid,* turns on the transcription of genes related to growth and development. *Vitamin D* is synthesized in the skin as *cholecalciferol* from a cholesterol derivative with the energy of the sun's ultraviolet radiation. Cholecalciferol is then converted to *calcitriol,* which activates the transcription of genes controlling the metabolism of calcium and phosphorus. Thus, adequacy of vitamin D is crucial for bone health. *α-Tocopherol (vitamin E)* serves as an antioxidant protecting the unsaturated fatty acids in the phospholipids of cell membranes from oxidation and damage by radicals such as the hydroxyl radical (section 1.4). Finally, *phylloquinone (vitamin K)* is required in the synthesis of proteins mediating blood clotting; its absence or inhibition of action causes hemorrhage.

What about minerals? Three of the most abundant ones in the body, *sodium, potassium,* and *chloride,* are the main *electrolytes* (ions conducting electricity) in body fluids, regulating the water content of the different compartments, blood pressure, the electrical transmission of nerve signals, and muscle contraction (see part II for the latter two processes). *Calcium* and *phosphorus* are major components of the bones and teeth. In addition, calcium controls a multitude of cell functions, from muscle contraction to gene expression, many of which we will consider in the rest of the book. Phosphorus, on the other hand, is omnipresent in organic biological compounds such as high-energy compounds, nucleic acids, and phospholipids (as we have already seen), as well as carbohydrates and proteins (as we will see in part III). *Magnesium* is in the active site of hundreds of enzymes, notably those utilizing ATP (see section 7.5). Other less abundant minerals, such as *copper, manganese, molybdenum, selenium,* and *zinc,* are also parts of enzymes (see figure 3.11). *Fluorine* is incorporated in the bones and teeth, protecting the latter against erosion by acids, and *iodine* participates in the structure of the *thyroid hormones,* which increase the expression of genes encoding key enzymes in catabolism.

Problems and Critical Thinking Questions

1. What are the differences among the three natural fibers, cotton, wool, and silk, used in clothing?

2. How does corn become popcorn?

3. Shortenings are made from vegetable oils through a chemical process that turns them from liquid to solid at room temperature. What might that process be?

4. Rank triacylglycerol, phosphatidyl inositol, sphingomyelin, and cholesterol ester in order of increasing number of acyl groups.

5. Oil and water do not mix, yet they get along perfectly in mayonnaise, which is made with oil, vinegar or lemon juice or both, and eggs or egg yolks or both. How does the egg reconcile oil and water?

Part I Summary

Metabolism

Life is characterized by complex interactions and reactions of molecules that take place within cells and in the extracellular fluids of multicellular organisms. The sum of the chemical reactions taking place in living organisms is metabolism. Reactions are accompanied by energy changes. A reaction is favored if its free-energy change (ΔG) is negative. Such a reaction is called exergonic, and its ΔG can be used to produce useful work.

Metabolism is divided into catabolism (which includes degradation processes) and anabolism (which includes synthetic processes). Catabolism serves to produce raw materials for anabolism and energy required for biological processes such as anabolism. Anabolism, in turn, is necessary for the development and maintenance of an organism, as well as for the formation of energy depots.

Whereas catabolic processes produce energy, anabolic processes require energy. Energy produced by the former is channeled to the latter through ATP, a compound of high energy content. By its breakdown to ADP or AMP, ATP yields energy capable of driving biosynthetic reactions forward.

Catabolic processes are usually accompanied by oxidation of metabolites, whereas anabolic processes are accompanied by reduction of metabolites. NAD^+, $NADP^+$, and FAD serve as oxidizing agents, whereas NADH, NADPH, and $FADH_2$ serve as reducing agents. The oxidizing agents have the capacity to acquire hydrogens, whereas the reducing agents have the capacity to donate hydrogens. Thus, they oxidize and reduce, respectively, their partners in oxidation–reduction reactions. Oxygen is the ultimate oxidizing agent in animals.

There are four large classes of biological compounds: proteins, nucleic acids, carbohydrates, and lipids.

Proteins

Proteins are macromolecules playing pivotal roles in all biological processes. They consist of amino acid residues, which are joined, by the hundreds or thousands, through peptide bonds to form polypeptide chains. Cells use only 20 different amino acids to synthesize their proteins. These amino acids differ in the side chain, which can be hydrophobic, hydrophilic, uncharged, negatively charged, or positively charged at physiological pH. The side chains affect the three-dimensional structure of a protein and, thus, its function.

Short stretches of a polypeptide chain can form regular shapes such as the α helix and β pleated sheet. The hydrogen bond, the electrostatic bond, the van der Waals interactions, the disulfide bond, and the hydrophobic interactions contribute to the final conformation of a protein. All these forces develop among atoms of both the side chains and the backbone of a polypeptide chain, resulting in a defined three-dimensional figure. Certain proteins consist of more than one polypeptide chain; these chains also have a defined arrangement in space.

The tens of thousands of existing proteins serve a multitude of purposes in living organisms, such as the acceleration of chemical reactions, the transport and storage of substances, movement, the mechanical support of cells, defense against intruders, signal transduction, and control of development. Of this entire medley, we examined two oxygen carriers and the enzymes.

Hemoglobin and myoglobin have been assigned the crucial task of carrying atmospheric oxygen in the blood and muscles, respectively. Hemoglobin is more complex than myoglobin, consisting of four polypeptide chains rather than one. Each chain carries a prosthetic group of heme with an iron at its center. Iron is where O_2 binds. The presence of four chains in hemoglobin accounts for its allostery, that is, the dependence of its affinity for a substance on whether other molecules of the same or another substance are already bound to it. In particular, the binding of O_2 to hemoglobin is favored by the presence of already bound O_2, whereas the binding of O_2 is not favored by the presence of bound CO_2 or H^+ and vice versa. These interactions facilitate the saturation of hemoglobin with O_2 in the lungs, the delivery of O_2 to the tissues, the removal of CO_2 and H^+ from the tissues, and the exhalation of CO_2. The amount of hemoglobin in the human body has been correlated with aerobic capacity.

Enzymes catalyze biochemical reactions by increasing their rate spectacularly. Each enzyme catalyzes only a few reactions thanks to the specialized structure of its active site. Enzymes do not affect the outcome of reactions; they only help them to go faster in the direction dictated by their ΔG. The rate of enzyme reactions depends on substrate (reactant) and enzyme concentrations, temperature, pH, and ionic strength.

Nucleic Acids

Nucleic acids are biological macromolecules consisting of nucleotides. Each nucleotide of deoxyribonucleic acid (DNA) contains one of the four bases adenine, guanine, thymine, and cytosine; the carbohydrate deoxyribose; and a phosphoryl group. Each nucleotide of ribonucleic acid (RNA) contains one of the four bases adenine, guanine, uracil, and cytosine; the carbohydrate ribose; and a phosphoryl group.

Nucleotides are linked by merely tens to as many as hundreds of millions through phosphodiester linkages between positions 5' and 3' of their ribose or deoxyribose units. DNA is the genetic material of cells and usually has the shape of a double right-handed helix, in which two strands

of opposite directions (5' → 3', 3' → 5') are held together by hydrogen bonding. The bonds develop between the base pairs adenine-thymine and guanine-cytosine; these bases face each other in the two strands. Thus, the DNA strands have complementary base sequences.

There are several analogies between nucleic acids and proteins. These are summarized in table I.1 below.

The complementarity of DNA strands is essential for its semiconservative replication. During replication (which precedes cell division) the enzyme DNA polymerase, using each strand as a template, constructs a new strand with complementary base sequence. Each daughter chromosome contains a parental and a newly formed strand. In this way daughter cells ensure an exact copy of the parental DNA. Errors in replication happen less frequently than once in a million. They are called mutations, and they are the cause of disease or evolution.

Gene Expression

The genetic information contained in DNA is transmitted to RNA through transcription. RNA polymerase, using a DNA strand as a template, constructs an RNA molecule with complementary base sequence. DNA is transcribed in pieces called genes. Genes are a minor part of the total DNA in higher organisms. Humans have 20,000 to 25,000 genes, some of which have been associated with physical performance and health-related fitness. Some of the synthesized RNA molecules, called messenger RNA or mRNA, carry the instructions for protein synthesis. The instructions are "read" by ribosomes, which are complexes of proteins and another kind of RNA, ribosomal RNA or rRNA.

Protein synthesis based on the mRNA base sequence is called translation and obeys the

▶ **Table I.1**   Analogies Between Nucleic Acids and Proteins

Feature	Protein	Nucleic acid
Building block (monomer)	Amino acid	Nucleotide
Number of different monomers	20	4
What links monomers	Peptide bond	Phosphodiester linkage
Conventional direction of chain	N → C	5' → 3'
Periodic secondary structure	α helix, β pleated sheet	Double helix
What stabilizes the secondary structure	Hydrogen bond	Hydrogen bond

genetic code, which matches every possible triplet of RNA bases with one amino acid or a termination signal. Of the 64 possible triplets, or codons, one signals the initiation of translation and encodes the amino acid methionine. Three codons signal the termination of translation. The remaining 60 codons encode the other 19 amino acids. The matching of codons and amino acids is mediated by transfer RNA (tRNA) molecules, which bind an amino acid at one end and recognize the codon at the other. Messenger RNA is translated in a $5' \rightarrow 3'$ direction, and a protein is synthesized in an N $\rightarrow$ C direction.

The primary role of RNA in the flow of genetic information, the participation of nucleotides in a multitude of metabolic reactions, and the existence of RNA molecules with catalytic properties are evidence for its appearance in the history of life on Earth before DNA and proteins.

Carbohydrates

Carbohydrates are diverse biomolecules divided into monosaccharides, oligosaccharides, and polysaccharides. Monosaccharides such as glucose are joined by glycosidic linkages to form oligosaccharides and polysaccharides. The disaccharide sucrose (sugar), consisting of a glucosyl and a fructosyl unit, is the best-known oligosaccharide. Cellulose, starch, and glycogen are polysaccharides consisting of thousands of glucosyl units. Cellulose is the main structural component of plants, whereas starch and glycogen are energy depots in plants and animals, respectively. Glucosyl units are linked by $\beta 1 \rightarrow 4$ linkages in cellulose, and $\alpha 1 \rightarrow 4$ as well as $\alpha 1 \rightarrow 6$ linkages in starch and glycogen.

Lipids

Lipids are a large class of biological compounds whose molecules are dominated by hydrophobic aliphatic groups. Fatty acids are the simplest lipids and components of other lipid categories. Fatty acids can be saturated (with no double bonds between their carbon atoms) or unsaturated (containing as many as six double bonds). Triacylglycerols are the most abundant lipid category; they are the main constituent of adipose tissue and the largest energy depot in the body. Another lipid category, phospholipids, has an amphipathic (hydrophilic and hydrophobic) character, which makes them suitable for cell membranes. Membranes also contain sterols (such as cholesterol) and proteins. In addition to being a membrane component, cholesterol is the parent compound of steroids.

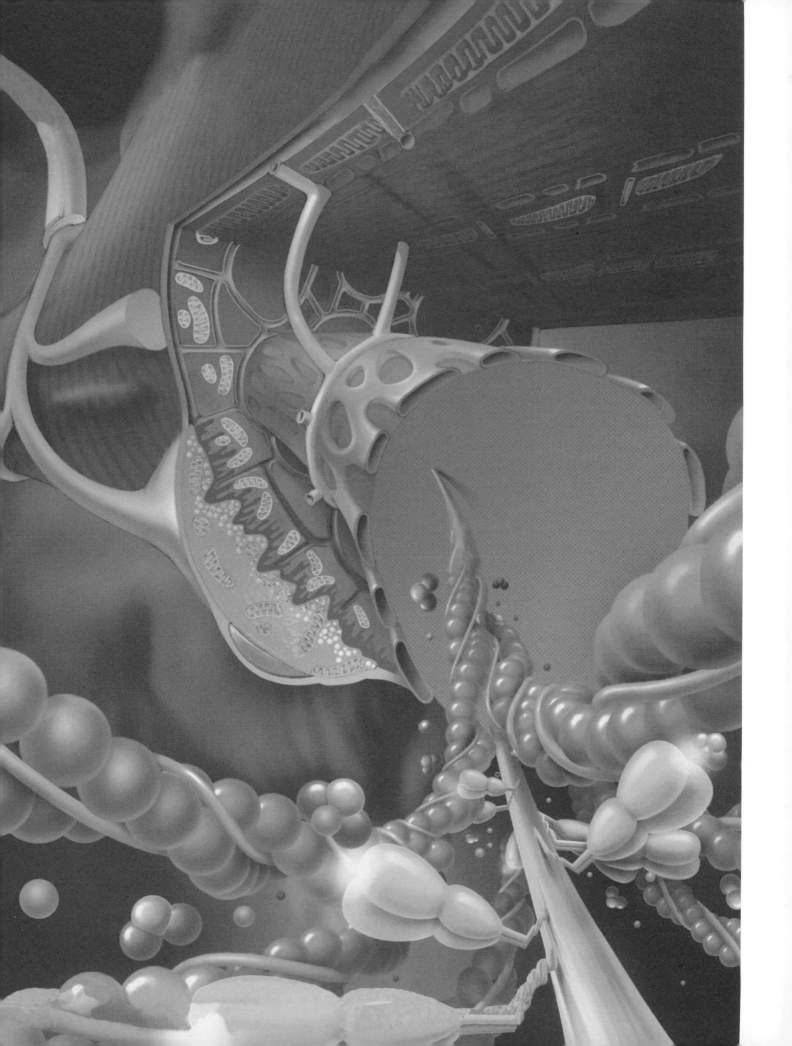

Biochemistry of the Neural and Muscular Processes of Movement

The inertia of objects is deceptive. The inanimate world appears static, "dead," to humans only because of our neuromuscular chauvinism. We are so enamored of our own activity range that we blind ourselves to the fact that most of the action in the universe is unfolding outside our range, occurring at speeds so much slower or faster than our own that it is hidden from us as if by a . . . a veil.

—Tom Robbins, *Skinny Legs and All*

Movement is an integral trait of matter, from its largest to its smallest scale. Galaxies spin in the universe but also electrons move around the nuclei of atoms. Movement is particularly evident in living matter, plants and animals. Plants move as they grow, open, and close the petals of their flowers, or—the carnivorous ones—entrap insects. However, the movement of plants doesn't measure up to that in the animal kingdom. Here movement dominates as a way of life, endowing animals with a vital advantage: the ability to go from one place to another, where the living conditions (such as the abundance of food, the absence of enemies or competitors, the environment, and the climate) are more favorable and thus the chances for survival are higher.

Muscle was the tool that granted animals fast movement. Its "invention" through evolution over 600 million years ago must have brought about an unimaginable revolution in life on Earth. By the power of their muscles, animals—in an incredible variety of forms, sizes, and capabilities—conquered

▶ The introductory figure of part II (facing page) takes you on an imaginary trip to the inner space of a motor unit. At the very top left is the motor neuron, which transmits the signals for contraction. Its axis is wrapped in myelin up to a point, after which it splits into terminal branches that stick to the muscle fiber. The cutaway view of a terminal branch and the muscle fiber reveals an undulating interface, a wealth of synaptic vesicles and mitochondria on the side of the neuron, and a wealth of mitochondria and transverse tubules on the side of the muscle fiber. The transverse tubules embrace the myofibrils, one of which projects. It is wrapped in sarcoplasmic reticulum and is full of thick and thin filaments, which dominate the lower part of the picture. One thick filament is surrounded by six thin filaments. Double myosin heads project from the thick filament and split ATP, releasing the energy (flashes) required for motion. The thin filaments consist mainly of actin (the double winding necklace); tropomyosin (the double lace) also appears. The elongated bulge between two terminal branches of the neuron is a satellite cell (introduced in part III, section 12.5). Come back to the figure after you have read all of part II and identify its components without reading the caption.

the oceans and lands of our planet. The climax of this course was the appearance of *Homo sapiens.* Based mainly on the abilities of their hands and guided by another astonishing tool, the brain, human beings managed to become the sovereigns of the earth in a blink of a (geological) eye.

The nervous and muscular systems star in part II. We will examine, in as many details as are fitting in a textbook, the chain of biochemical processes that convert the signal for muscle contraction to a visible effect.

CHAPTER 6

Neural Control of Movement

Movement as a result of muscle contraction is impossible without the transmission of signals from the nervous system to the muscles. Most of our movements are involuntary and reflexive: They constitute immediate responses to information provided by the sensory organs or the muscles through appropriate nerves to the spinal cord, which, in turn, transmits signals for muscle contraction without "consulting" the brain. Even signals coming from the brain down the spinal cord act largely by affecting reflex routes.

The area of the brain mainly involved in controlling voluntary movement is the *motor cortex,* a strip across the middle of the cerebral cortex, roughly where the frame of a headset rests when we listen to music in private (figure 6.1). The motor cortex sends signals to the spinal cord through the *brain stem,* which also sends "reports" to the *cerebellum.* This structure integrates these reports with sensory information to coordinate the contraction of the different muscles involved in a movement. As a result, the movement is smooth and the body maintains balance while moving.

The biochemical events that permit the transmission of nerve signals to the muscles are the same regardless of their origin and the route they follow. These events are the topic of the present chapter.

Instantly retracting our hand from a hot surface is an example of a reflex.

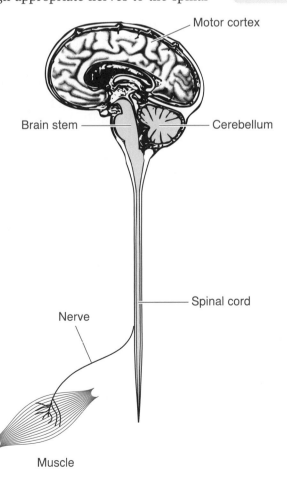

▶ **Figure 6.1** Nervous motorways. Skeletal muscles receive signals to contract from nerves originating in the spinal cord. The spinal cord generates such signals in response to sensory input from all over the body. It also receives signals for motor activity from the brain, notably the motor cortex and the brain stem. The cerebellum helps to coordinate these signals and produce smooth motion.

6.1 Nerve Signals Are Transmitted in Two Ways

To travel from its origin to its destination, a nerve signal like the signal directing a muscle to contract has to go through **neurons,** that is, cells of the nerve tissue. A neuron is highly asymmetric (figure 6.2). It consists of the *cell body,* where the nucleus is, and two kinds of processes: the *dendrites,* which bring nerve signals to the cell body, and a very long *axon,* which takes the signals away from the cell body. The axon, in turn, ends in numerous thin branches that can make contact with other neurons or even other parts of the same neuron.

A relatively inert substance called **myelin** surrounds most of the axon's length in many neurons of vertebrates. Myelin is formed by two types of specialized cells: *Schwann cells* in the peripheral nervous system (the nerves) and *oligodendroglial cells,* or *oligodendrocytes,* in the central nervous system (the brain and spinal cord). These cells roll their plasma membrane repeatedly around the axon (figure 6.3), forming something like a "pig in a blanket" (a sausage wrapped in dough).

Many Schwann or oligodendroglial cells lay their rolls of plasma membrane one next to the other along an axon in order to cover it. The coverage is not complete, as small areas of the axon's plasma membrane between adjacent rolls remain naked. These areas are called *nodes of Ranvier* (figure 6.2). As we will see in section 6.4, the myelin sheath and the nodes of Ranvier are crucial for the rapid propagation of nerve signals along the axon.

Nerve signals are transmitted differently inside neurons and from one neuron to another:

- Inside each neuron, signals are transmitted electrically (as electric current).

- From one neuron to another, signals are usually transmitted chemically. Signals are also transmitted chemically from a **motor neuron** to muscle.

Let's examine each of these ways.

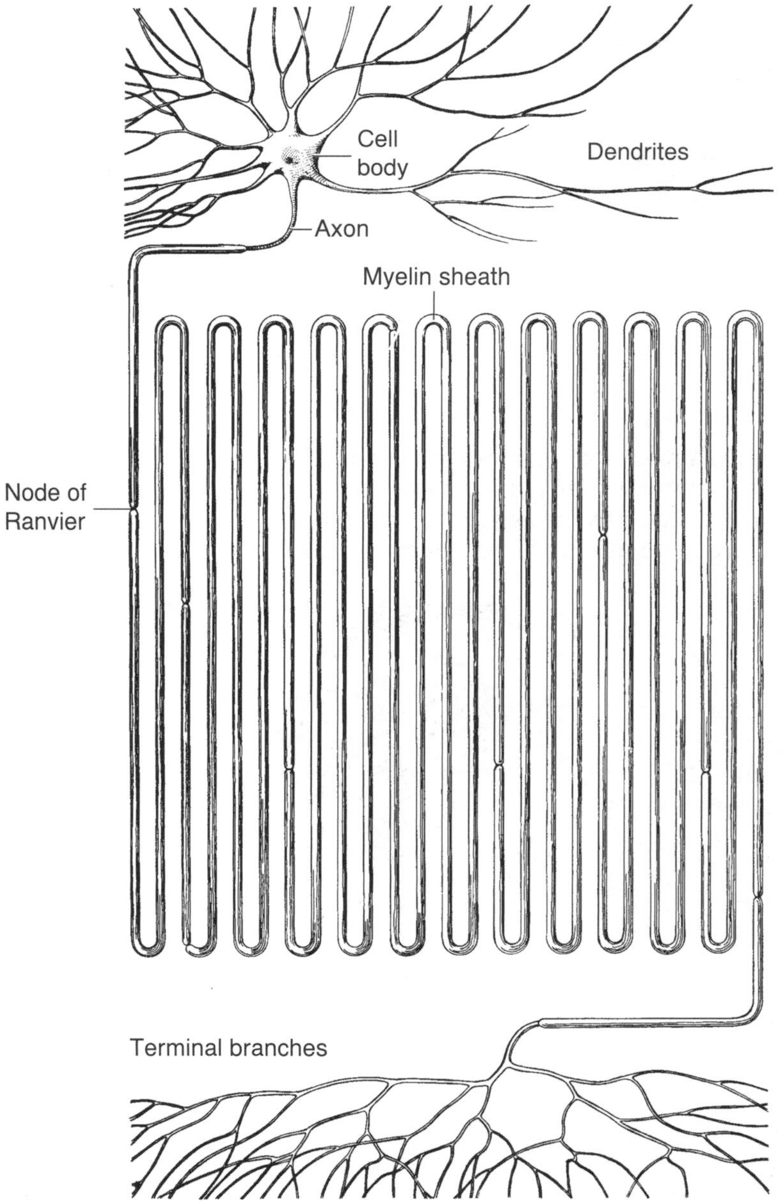

▶ **Figure 6.2** Neuron. Nerve signals are transmitted through neurons like the one depicted here enlarged 200 times, with its parts drawn to scale. The actual length of the axon, which is folded in order to fit onto the page, is 1 cm. However, there are axons over 1 m long. The dendrites and the axon's terminal branches may number in the hundreds or thousands. The axon is covered by myelin, which serves as an electrical insulator, accelerates signal transmission, and saves energy. The direction of the nerve signals is from the dendrites to the cell body to the terminal branches.

Courtesy of Carol Donner.

Outer loop

Axon

Myelin sheath

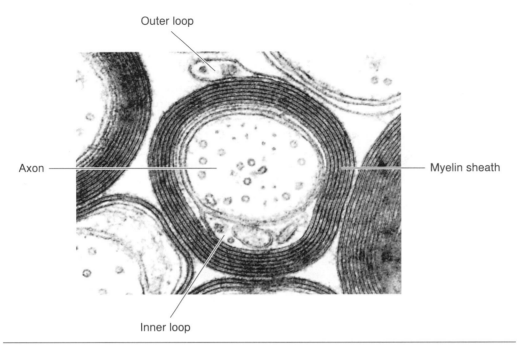

Inner loop

▶ **Figure 6.3** Myelin. An oligodendrocyte, the maker of myelin in the central nervous system, wraps its plasma membrane tightly around the axon of a neuron in this cross-sectional view produced by an electron microscope. The beginning (outer loop) and end (inner loop) of the myelin sheath are marked.

Photograph courtesy of Dr. Cedric S. Raine.

▶ Use the idea of a pig in a blanket to picture the way myelin surrounds an axon.

6.2   The Resting Potential

The electrical transmission of nerve signals is due to the movements of sodium and potassium ions across the plasma membrane of neurons (by contrast, the current in an electrical cable is due to the flow of electrons through the metal lattice of the wire). These movements are made possible by existing *concentration gradients* (that is, unequal distributions) between the cytosol and the extracellular fluid. Concentration gradients characterize not only neurons but also other cells and are vital to their operation. Table 6.1 presents the concentrations of Na⁺ and K⁺ in some body fluids, along with the concentrations of some other important ions, to furnish a more complete picture of ion concentration gradients in the body.

▶ **Table 6.1** Ion Concentrations in Human
Body Fluids (mmol · L⁻¹)

Ion	Cytosol	Extracellular fluid	Plasma[a]	Sweat
Na^+	3-10	133-144	136-145	40-80
K^+	148-156	3.8-4.4	3.5-5.1	4-8
Ca^{2+}	0-1	≤1.7	2.2-2.5	1.5-2
Mg^{2+}	15-20	0.5-1.7	0.6-1.1	0.5-2
Cl^-	2-3	112-115	98-107	30-70
HCO_3^-	7-10	21-28	21-29	0-36

mmol · L⁻¹: thousandth of a mole per liter.
[a]Plasma is the fluid of the blood outside its cells. See more on plasma in
the introduction to part IV (section IV.1).

To remember where the
[Na^+] is higher, you may
compare the extracel-
lular fluid to seawater,
which was probably
the environment of the
first cells that appeared
on Earth and which has
a high salt (Na^+Cl^-) concentration.

Na^+ has a higher concentration in the extracellular fluid
(about 140 mmol · L⁻¹) than in the cytosol (10 mmol · L⁻¹).
In contrast, K^+ is less in the extracellular fluid (4 mmol ·
L⁻¹) than in the cytosol (150 mmol · L⁻¹). The two fluids are
separated by the plasma membrane, which prevents Na^+
and K^+—and many other substances—from moving spontaneously down their concentration gradients, that is, from
the compartment of high concentration to the compartment
of low concentration. This is so because the ions are polar,
whereas the core of the lipid bilayer is nonpolar.

How are the Na^+ and K^+ concentration gradients formed
in the first place? They are formed by a "pump," which
throws Na^+ out of the neuron and brings K^+ into the
neuron. The pump transports Na^+ and K^+ "uphill," that
is, from a compartment of low concentration to a compartment of high concentration. This transport requires an
input of energy and is therefore called **active transport.**
Energy comes from the hydrolysis of ATP according to
equation 2.5.

What is the pump that moves Na^+ and K^+ in opposite directions? It is an integral
protein of the plasma membrane (figure 6.4) named the **sodium-potassium pump
(Na^+-K^+ pump).** The same protein is endowed with the ability to hydrolyze ATP;
thus, it exhibits enzyme activity. That is why it is also referred to as sodium-potassium

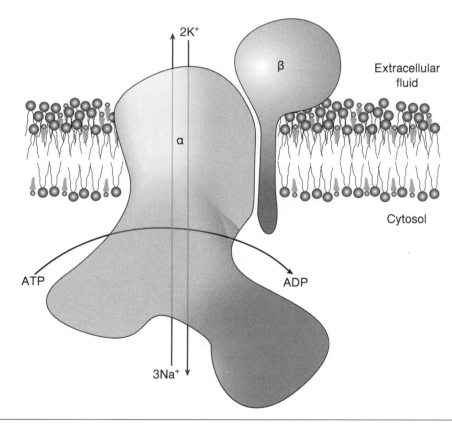

▶ **Figure 6.4** Na^+-K^+ pump. The Na^+-K^+ pump spans the plasma membrane and maintains an unequal distribution of Na^+ and K^+ between the inside and outside of cells at the expense of energy from the hydrolysis
of ATP. The protein is an αβ dimer. The protein and membrane are drawn to scale.

adenosine triphosphatase, or **Na⁺-K⁺ ATPase.** For every ATP hydrolyzed, three Na⁺ exit and two K⁺ enter the cytosol within 10 ms (thousandths of a second).

The pump is composed of two subunits designated α and β (110 and 55 kDa, respectively). Both subunits span the membrane bilayer, but the bulk of each is on opposite sides (α in the cytosol and β in the extracellular space). Although the α subunit carries out both of the protein's tasks (ion transport and ATP hydrolysis), the β subunit is required for proper function.

The plasma membrane is studded with molecules of the pump. The cells expend high amounts of energy for the maintenance of the Na⁺ and K⁺ concentration gradients. This is particularly true for nerve and muscle cells, since, as we will see, their excitation results in the disturbance of these gradients.

Because the concentration gradients across the plasma membrane involve charged solutes, they result in the generation of an electrical potential difference, or voltage, symbolized as ΔV. This is called the **resting potential**, since it characterizes the resting, not the excited, state. The resting potential is 60 to 70 mV negative inside relative to outside and forms the basis for the electrical transmission of a nerve signal, which I describe in the next section (a millivolt [mV] is a thousandth of a volt). We say that the plasma membrane

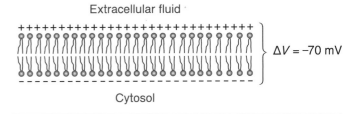

> ▶ **Figure 6.5**   Polarized membrane. The unequal distribution of ions between the cytosol and the extracellular fluid polarizes the plasma membrane and gives rise to a resting potential of about –70 mV.

is *polarized* because we can depict it as an electrical dipole having its negative pole in the cytosol and its positive pole in the extracellular space (figure 6.5).

6.3   The Action Potential

An electrical nerve signal is an instantaneous *depolarization* (reversal of the polarity) of the plasma membrane (figure 6.6). The depolarization is due to the influx of Na⁺ from the extracellular space to the cytosol through pores in the plasma membrane. The pores are formed by molecules of another transmembrane protein, the **sodium channel (Na⁺ channel).**

The Na⁺ channel is closed most of the time, hindering the entry of Na⁺ to the cytosol and contributing to the maintenance of the resting potential. However, for reasons that we will discuss in sections 6.4 and 6.6, the membrane voltage around the channel can increase. If the increase is large enough, exceeding what is termed the *excitation threshold,* the shape of the Na⁺ channel changes and a narrow pore forms in its interior. Because it opens when the voltage changes, the Na⁺ channel is called a **voltage-gated channel.** The excitation threshold is at least 15 mV (meaning that a resting potential of –65 mV will have to increase to at least –50 mV for the Na⁺ channel to respond), and its precise value differs from one neuron to another. The larger the threshold, the more difficult it is for a neuron to be excited.

When the Na⁺ channel opens, Na⁺ ions pass in a direction dictated by their concentration gradient: They move from where the concentration is high (the extracellular fluid) to where the concentration is low (the cytosol). This form of "downhill" movement is called **passive transport.**

At this point it is worth remembering that we have encountered three ways by which molecules or ions cross membranes. These are (in order of increasing complexity):

- *Simple diffusion,* as in the case of O₂ crossing cell membranes down its concentration gradient (sections 3.11 and 3.12). O₂ needs no transporter because it is nonpolar like the core of the lipid bilayer.

- *Passive transport* (also called *facilitated diffusion*), as in the case of Na⁺ entering a cell down its concentration gradient with the help of the Na⁺ channel.

- *Active transport,* as in the case of Na⁺ and K⁺ crossing the plasma membrane against their concentration gradients through the action of the Na⁺-K⁺ pump, powered by ATP hydrolysis.

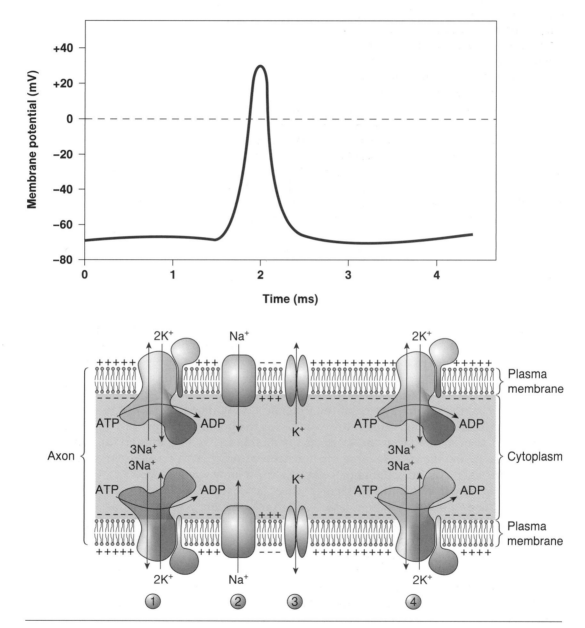

▶ **Figure 6.6** Electrical transmission of a nerve signal. The nerve signal is transmitted electrically in a neuron thanks to a momentary leap of the plasma membrane potential from a negative to a positive value (upper panel). The negative value (resting potential) is maintained by the sodium-potassium pump (lower panel, 1), whereas the positive value (action potential) appears when the sodium channel opens (2) and Na⁺ ions rush into the cytosol. The potassium channel (3) lets K⁺ ions out and reinstates the resting potential. The Na⁺-K⁺ pump restores the disturbance in the [Na⁺] and [K⁺] gradients (4). The position of the proteins in the lower panel corresponds to the different phases of the membrane potential in the upper panel. In reality, however, there is no orderly arrangement of the three proteins. Rather, they are scattered all over the membrane.

Coming back to the Na$^+$ channel, its opening and the subsequent influx of Na$^+$ result in a reversal of the membrane potential to values that are positive inside relative to outside (up to 30 mV). The new voltage is called the **action potential.** The depolarization is extremely short; in less than 1 ms the Na$^+$ channel closes spontaneously. At the same time, the change in membrane potential triggers the opening of another integral protein of the plasma membrane, also a voltage-gated channel. This one is a **potassium channel (K$^+$ channel)** that lets K$^+$ ions exit the cytosol toward the extracellular space. In effect, the membrane potential returns to the resting value. The K$^+$ channel also closes spontaneously in a few milliseconds.

The generation of an action potential changes the Na$^+$ and K$^+$ concentrations in a neuron minimally (by approximately one-millionth). However, even these insignificant changes have to be reversed because their accumulation after the passage of successive action potentials (as long as the signal lasts) may make the membrane nonexcitable, that is, unable to be depolarized. The Na$^+$-K$^+$ pump takes over the task of ridding the cytosol of the Na$^+$ ions that intruded and recalling the K$^+$ ions that left, at the expense of ATP.

6.4   Propagation of an Action Potential

An action potential is not designed to remain stagnant but to travel along the neuron. The instantaneous depolarization at a point in the plasma membrane causes neighboring molecules of the Na$^+$ channel to open, resulting in the depolarization of the adjacent area. This, in turn, causes the membrane to depolarize a bit farther, and so on. Thus, an action potential is propagated along the neuron like a wave caused by a pebble falling into a pond. That is why a nerve signal is often referred to as a *nerve impulse.*

The propagation of an action potential along axons encased in myelin is somewhat different. Because myelin bars the axon from the extracellular fluid, an action potential cannot run along the axon's myelinated segments. However, it can jump from one node of Ranvier to another the way a stone skips across the water. Remember that the plasma membrane is exposed to the extracellular fluid at the nodes. Both the Na$^+$ and K$^+$ channels are concentrated there—in fact, much more densely than in the nonmyelinated axons. The propagation of an action potential from one node of Ranvier to another is termed *saltatory conduction* (figure 6.7), as opposed to the *continuous conduction* that is characteristic of nonmyelinated axons.

We are now ready to discuss the role of myelin—the role of electrical insulator (similar to the insulation of electrical wires). By confining the electrical activity that develops along an axon to certain points (the nodes of Ranvier), myelin minimizes the effect on neighboring axons. Another utility of myelin is that it accelerates the propagation of nerve impulses. Indeed, saltatory conduction is much faster than continuous conduction. Finally, myelin saves energy. Because the movement of Na$^+$ and K$^+$ ions through the respective channels is limited to the nodes of Ranvier, the Na$^+$-K$^+$ pump has much less work to do in order to bring the concentrations of these ions back to resting levels. Thus, the neuron expends less ATP.

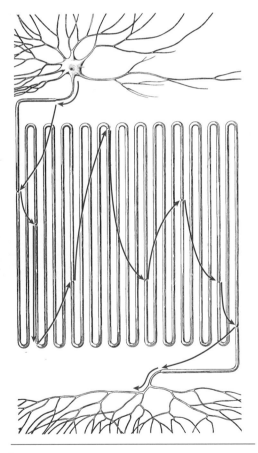

▶ **Figure 6.7**   Saltatory conduction. A nerve impulse jumps from one node of Ranvier to another along a myelinated axon, thus saving time and energy.

Courtesy of Carol Donner.

In summary, myelin

electrically insulates the axons of neurons,

accelerates the propagation of nerve signals, and

saves ATP.

The great importance of myelin for the proper function of the nervous system becomes evident when myelin is destroyed from pathological causes. This leads to the development of degenerative diseases such as *multiple sclerosis,* which is characterized by the gradual degradation of myelin in certain regions of the central nervous system. This slows down the conduction of nerve impulses and causes loss of motor coordination and partial paralysis.

6.5 Transmission of a Nerve Impulse From One Neuron to Another

When, after having traveled down the axon, an action potential reaches the terminal branches, it needs to jump a small but important hurdle in order to convey the signal to another neuron: It must cross the space separating the two neurons. This is where the second means of transmission of nerve signals, the chemical one, comes into play.

The point of contact between neurons is called a **synapse** (figure 6.8). At the synapse, the plasma membrane of one neuron comes very close to the plasma membrane of

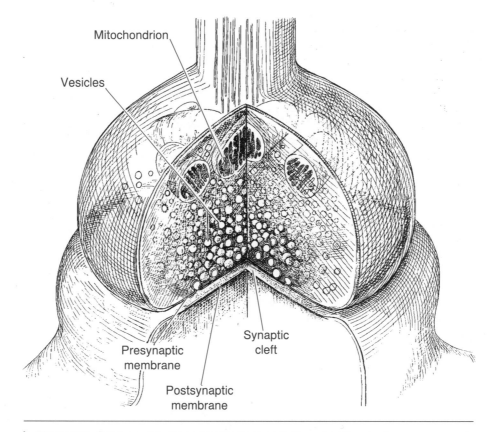

▶ **Figure 6.8** Synapse. A synapse permits the propagation of an impulse from one neuron to another. The bulged ending of the neuron transmitting the signal (top) and the surface of the neuron receiving the signal (bottom) come close to each other. The former is filled with vesicles containing neurotransmitter molecules.

Courtesy of Carol Donner

another. The membrane of the neuron transmitting the impulse is termed *presynaptic,* whereas the membrane of the neuron receiving the impulse is termed *postsynaptic.* The space separating them is the **synaptic cleft;** it varies from 20 to 200 nm in width and is filled with extracellular fluid.

The transmission of a nerve impulse across the synaptic cleft is accomplished by compounds collectively called **neurotransmitters.** Several of these are known today. Each has a specific distribution in the nerve tissue and specific ways of action, thus contributing to the inconceivable variety of messages handled by our nervous system.

Despite their variety, neurotransmitters share several features. For one, following their synthesis in the neurons they are stored in **synaptic vesicles** inside the terminal branches, very close to or even in contact with the presynaptic membrane. Another common feature of neurotransmitters is the way in which they are liberated: The arrival of an action potential at the presynaptic membrane causes the instantaneous (within hundredths of a millisecond) fusion of the vesicles with the membrane and the discharge of their contents into the synaptic cleft in a process termed **exocytosis.**

How does the action potential accomplish this feat? The key to neurotransmitter release is the influx of calcium ions to the cytosol. Ca^{2+} is another ion with unequal distribution across the plasma membrane. Its concentration is higher in the extracellular fluid than in the cytosol (see table 6.1). The plasma membrane contains many molecules of a voltage-gated Ca^{2+} **channel,** which open when the membrane is depolarized. Through the channel, Ca^{2+} rushes to the cytosol and causes the membrane of the synaptic vesicles to fuse with the plasma membrane (figure 6.9). Researchers believe that *synaptotagmin 1,* an integral protein of the vesicle membrane, senses the rise in

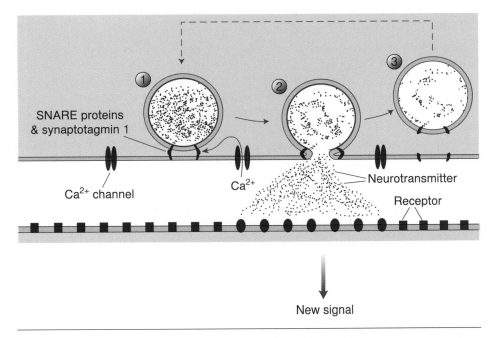

▶ **Figure 6.9** Chemical transmission of a nerve impulse. (1) A population of synaptic vesicles, full of neurotransmitter molecules and ready to release their contents, is tethered to the presynaptic membrane when a neuron is not excited. (1 → 2) When an action potential arrives at the synapse, a voltage-gated calcium channel opens. The incoming Ca^{2+} binds to synaptotagmin 1, which is complexed with SNARE proteins and causes the vesicle membrane to fuse with the plasma membrane. (2) Neurotransmitter molecules are emptied into the synaptic cleft and bind to receptor molecules in the postsynaptic membrane, causing a change in their shape (schematically, from square to oval) and the generation of a new signal. (2 → 3) When action potentials stop arriving at the synapse, the Ca^{2+} channel closes. (3) The vesicle is reconstituted with a small amount of neurotransmitter that was not discharged. (3 → 1). The vesicle is refilled by the biosynthetic machinery of the neuron and prepared to release its contents again.

intracellular $[Ca^{2+}]$ and undergoes a crucial change in its interaction with membrane phospholipids and a protein complex called *SNARE proteins*, which triggers the fusion of the two membranes.

The neurotransmitter molecules diffuse to the postsynaptic membrane and bind to receptors. Remember that receptors are proteins receiving molecular messengers (section 3.8). A neurotransmitter receptor is highly specialized in recognizing and binding a neurotransmitter. Binding modifies the conformation of the receptor, causing a change in its biological activity.

This is where the common features of neurotransmitters end, as their receptors play a multitude of different roles. Some, for example, serve as ion channels, whereas others regulate the rate of reactions taking place in the cytosol. Thus, one cannot give a general scheme describing the outcome of the binding of a neurotransmitter to a receptor other than the coarse distinction between *excitatory* and *inhibitory* neurotransmitters. The former facilitate the propagation of a signal from one neuron to another, whereas the latter inhibit it.

How is the action of a neurotransmitter reversed when a nervous stimulus ceases? When action potentials stop arriving at the presynaptic membrane (or arrive at a slower rate), synaptic vesicles stop fusing with it (or fuse at a slower rate). Those neurotransmitter molecules present in the synaptic cleft are degraded by enzymes or are taken back up by the presynaptic neuron. Thus, the neurotransmitter concentration in the synaptic cleft drops. This encourages the dissociation of the neurotransmitter from the receptor and the subsequent degradation or reuptake of the neurotransmitter. Following this, the receptor returns to its initial (neurotransmitter free) state.

> The binding of a neurotransmitter N to a receptor R is reversible and concentration dependent.
>
> $N + R \rightleftharpoons NR$
>
> When the [N] decreases, the reaction is shifted to the left.

6.6   Birth of a Nerve Impulse

In sections 6.3 and 6.4, we saw how an action potential is formed and propagated along an axon as a result of the opening of Na^+ channel molecules in its membrane, which, in turn, is triggered by a depolarization of the membrane in their vicinity. The obvious question arises: What triggers the very beginning of electrical activity in a neuron? The answer could be the chemical signal from a preceding neuron along the path of transmission of a command (in our context, the command to contract). But this just shifts the question one neuron back, and another neuron back, and so on until we reach the neuron(s) where the signal directing a muscle to contract originated. Then we are cornered. . . .

The cause of a signal to contract can be a decision born in our mind without a particular external stimulus (for example, while standing, we decide to start walking). Conversely, it could be a stimulus from one or more of our sensory organs (for example, we watch a tennis ball coming our way and decide to hit it with our racket). Thanks to research findings of recent years, the answer to what triggers the beginning of electrical activity leading to an effect like muscle contraction has begun to take shape in increasingly satisfactory molecular detail. Because the complete coverage of the matter is outside the scope of this book, I will describe an example only, that of the birth of a visual stimulus.

A visual stimulus begins when a photon (the elementary quantity of light) causes a chemical change in *retinal*, the prosthetic group of *rhodopsin* and other *photoreceptor* (light receiving) proteins located in the retina. The retina of vertebrates features two kinds of photoreceptor cells, *rods* and *cones*. Rods contain rhodopsin and convey the perception of light, whereas cones contain three other photoreceptor proteins and convey the perception of red, green, and blue. The biochemistry of vision has been especially well studied in rods, which greatly outnumber cones in the retina.

A rod (figure 6.10) is a highly specialized cell with an *outer segment* pointing toward the brain and an *inner segment* ending in a *synaptic body,* full of neurotransmitter-laden vesicles. The synaptic body is in touch with nerve cells in the retina, which in turn communicate with the optic nerve. The outer segment contains a stack of flattened membrane-bound *discs;* rhodopsin is a transmembrane protein of these discs.

When a photon illuminates rhodopsin, it causes a conformational change, which activates a series of enzyme reactions leading to the closure of a cation channel in the surrounding plasma membrane. The channel is open in the dark, permitting the influx of Na$^+$ from the extracellular space. This influx, combined with the operation of the

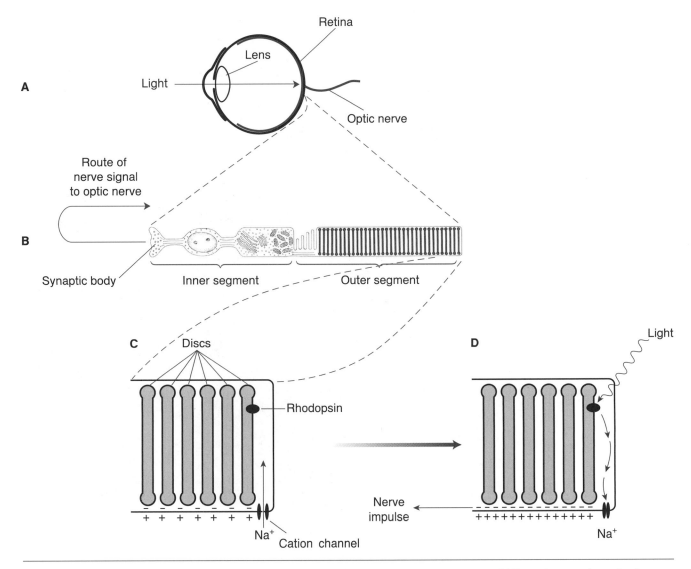

▶ **Figure 6.10** A signal is born. *(a)* A visual signal is generated when light hits the retina in the eye. *(b)* The retina contains rod cells separated into an outer and an inner segment. The outer segment, which is the part that senses light, contains a pile of membrane-bound discs. *(c)* The disc membrane is crowded with molecules of rhodopsin, an integral protein. In the dark, a cation channel in the plasma membrane remains open, permitting an influx of Na$^+$. *(d)* When light excites rhodopsin, it triggers a series of reactions that cause the cation channel to close. As a result, the membrane is hyperpolarized (depicted by more plus and minus signs), and this generates a nerve signal.

Figure 6.10b reprinted, by permission, from L. Stryer, 1998, *Biochemistry,* 3rd ed. (New York: W.H. Freeman and Company/Worth Publishers), 1028.

Na$^+$-K$^+$ pump located in the plasma membrane of the inner segment, maintains the membrane voltage at about –45 mV (inside relative to outside). However, the closure of the channel inhibits Na$^+$ entry, causing the membrane voltage to become even more negative. This *hyperpolarization* of the plasma membrane is transmitted to the synaptic body and reduces the rate of neurotransmitter release to the contiguous nerve cells. As the neurotransmitter is inhibitory, the lowering of its concentration in the synaptic cleft excites the postsynaptic membrane. This excitation is in turn transmitted to the optic nerve and finally to the brain for processing as a visual stimulus.

6.7　The Neuromuscular Junction

As a motor neuron approaches a muscle, it splits into hundreds or thousands of branches, each of which makes contact with a muscle cell. The motor neuron and the muscle cells that it innervates form a **motor unit.** A nerve signal for muscle contraction ends in the **neuromuscular junction**, the interface between a branch of a motor neuron and a muscle cell. Each neuromuscular junction contains many synapses like the one in figure 6.11. These synapses are similar to the synapses between neurons. In fact, the synapse at the neuromuscular junction has been studied in more detail and is the source of most of our knowledge about synapses in general.

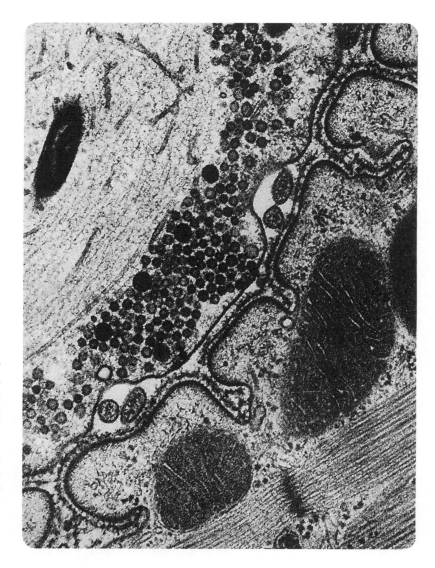

▶ **Figure 6.11**　Synapse@neuromuscular.junction. A narrow synaptic cleft, running diagonally across this electron micrograph, separates a neuron (up) from a muscle cell (down). Tens of synaptic vesicles, filled with acetylcholine, crowd the ending of the neuron, very close to or in contact with the presynaptic membrane. On the opposite bank, the membrane of the muscle cell forms invaginations called transverse tubules (see section 7.10). Further inside, you can make out two mitochondria (the dark, nearly elliptic structures) and a myofibril with a Z line (the bundle with the dark ribbon), which we will examine in section 7.1.

Courtesy of Dr. John Heuser.

A particular neurotransmitter, **acetylcholine**, operates at the neuromuscular junction. Acetylcholine is a relatively small compound consisting of an acetyl group and a choline unit (figure 6.12). You have already met the acetyl group as the product of stage 2 of catabolism (section 2.6), and choline as a component of phospholipids (section 5.8). Choline contains a nitrogen atom connected to four carbons; this is called a *quaternary amino group*. Quaternary amino groups carry a positive charge; thus, acetylcholine is a cation.

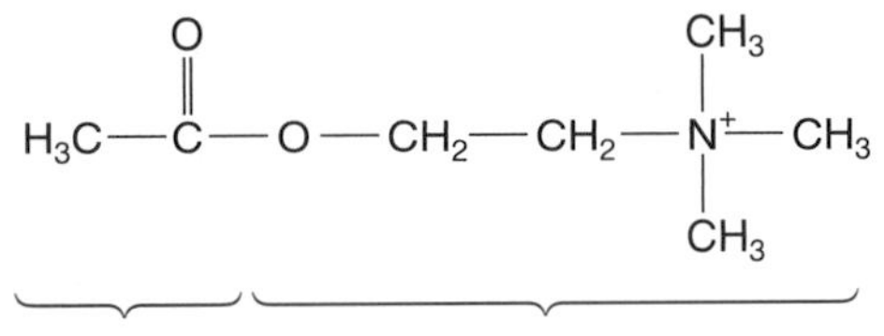

Acetylcholine is formed in neurons from acetyl coenzyme A (also introduced in section 2.6) and choline by the catalytic action of *choline acetyltransferase*.

acetyl coenzyme A + choline → acetylcholine + coenzyme A   **(equation 6.1)**

After being synthesized, acetylcholine is gathered in the synaptic vesicles (each about 18 nm in diameter and containing about 10,000 molecules) and is discharged upon the arrival of action potentials at the presynaptic membrane. Approximately 300 vesicles per synapse empty their contents into the synaptic cleft, raising the acetylcholine concentration by 50,000 times (from 0.01 to 500 $\mu mol \cdot L^{-1}$) in less than 1 ms ($\mu mol \cdot L^{-1}$ is a millionth of a mole per liter).

Acetylcholine carries the signal for contraction from a nerve to a muscle thanks to the *acetylcholine receptor,* which is located in the plasma membrane of muscle cells. We call this type of acetylcholine receptor *nicotinic* to distinguish it from the *muscarinic receptor,* which dominates in the autonomic nervous system. Acetylcholine serves as a neurotransmitter there too, but the muscarinic receptor functions differently from the nicotinic receptor. The names stem from *nicotine* (the known component of tobacco smoke) and *muscarine* (a substance found in some mushrooms). Investigators have used these compounds as ligands for distinguishing the two receptors (a ligand is a molecule that binds specifically to a larger molecule, for example a neurotransmitter that binds to a receptor).

Let's focus our attention on the nicotinic acetylcholine receptor. This is an integral protein of the plasma membrane of muscle cells (figure 6.13). It has a molecular mass of 280 kDa and consists of five similar subunits; in fact, two are identical. Thus, its structure is $\alpha_2\beta\gamma\delta$. The five subunits surround a pore at the center.

The receptor has two binding sites for acetylcholine located on its extracellular surface, one in each α subunit at or near the α-γ and α-δ subunit interfaces. When two molecules of acetylcholine bind to these sites, the conformation of the receptor changes, resulting in a widening of the pore within about 20 μs (figure 6.14). Na^+ and K^+ ions then move through the pore, down their concentration gradients: Na^+, from the extracellular space to the cytosol, and K^+, vice versa.

The pore is almost equally permeable to Na^+ and K^+. However, many more Na^+ ions enter than K^+ ions exit because the existing membrane potential (negative inside) favors the influx and inhibits the efflux of cations. Thus, the membrane of the muscle cell is depolarized and a new action potential, analogous to the ones dashing through neurons, appears. This is often termed *postsynaptic potential*. Then, as happens in the neurons, a voltage-gated K^+ channel opens, lets K^+ out of the cytosol, and brings the membrane potential back to its resting value.

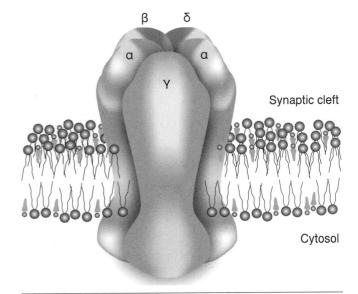

▶ **Figure 6.12**   Acetylcholine. The neurotransmitter of the neuromuscular junction is derived from the attachment of an acetyl group to choline.

The nervous system can be divided into *somatic* (serving the skeletal muscles) and *autonomic* (serving the internal organs).

▶ **Figure 6.13**   Nicotinic acetylcholine receptor. The receptor spans the plasma membrane of muscle cells. It is formed by five cylindrically arranged subunits, which protrude markedly toward the extracellular space and leave a narrow pore in the center. The protein and membrane are drawn to scale.

In contrast to the voltage-gated channels that we encountered in sections 6.3 and 6.5, the acetylcholine receptor is a **ligand-gated channel.** Its shape, position in the membrane, and function are reminiscent of magnificent gates that led in and out of cities at a time when cities were surrounded by walls (figure 6.15).

The change in the receptor's structure caused by acetylcholine is reversible. After about 1 ms, the pore shrinks. Free (unbound) acetylcholine in the synaptic cleft is hydrolyzed by the enzyme *acetylcholinesterase,* which is covalently linked to the post-synaptic membrane. The hydrolysis reaction is

acetylcholine + H_2O → acetate + choline + H^+ **(equation 6.2)**

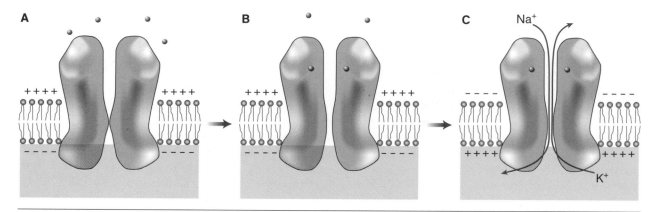

▶ **Figure 6.14**   Signal transmission across the neuromuscular junction. Acetylcholine causes the generation of an action potential at the plasma membrane of a muscle cell by interacting with the acetylcholine receptor, of which we have represented only two subunits in order to make the pore visible. *(a)* Acetylcholine molecules (colored balls) are released into the synaptic cleft. *(b)* Two acetylcholine molecules bind to the receptor and dilate the pore. *(c)* Many Na^+ ions flow to the cytosol and fewer K^+ ions flow to the extracellular space through the pore, resulting in the depolarization of the membrane (note the reversal of charges).

▶ **Figure 6.15**   The Ishtar gate. Named in honor of the great Babylonian and Assyrian goddess of fertility, this gate adorned one of the entrances/exits of ancient Babylon. Built in the sixth century BCE, it is kept in excellent condition and exhibited in the Pergamon Museum, Berlin. The gate resembles the acetylcholine receptor not only in shape but also in function, since the receptor operates as a gate of controlled entry to and exit from cells that are enclosed in membranes much as ancient cities were enclosed in walls.

Reprinted, by permission, from A. Janssen, 2001, "Treasure Island," *Lufthansa Magazin* 4: 18-19.

The elimination of free acetylcholine from the synaptic cleft leads to the dissociation of bound acetylcholine from the receptor according to what we said in section 6.5. Thus, the receptor returns to its initial state, ready to receive a new swarm of acetylcholine molecules when the synapse is excited again. In the meantime, the Na^+-K^+ pump (which is located in the plasma membrane of muscle cells too) restores the resting Na^+ and K^+ concentrations as well as the resting potential, so that a new postsynaptic potential can be generated.

The malfunctioning of the acetylcholine receptor is related to *myasthenia gravis* (the name is a blend of a Greek and a Latin term meaning "severe muscle disease"). In most cases of this disease, the body produces antibodies against the acetylcholine receptor. Less often, mutations prevent the receptor from operating properly. Problems like these result in inability of the muscles to receive nerve signals, muscle weakness, and muscle atrophy.

We have reached the end of the nervous transmission of the signal to contract. Postsynaptic potentials in the plasma membrane of muscle cells trigger a series of delicate biochemical processes culminating in the contraction of a muscle. These are the subject of the next chapter.

6.8 A Lethal Arsenal at the Service of Research

An unusual danger lurks in the swamps and streams of the vast Amazon basin. In places where other fish barely survive because of the low oxygen content of water, an eel answering to the telltale name *Electrophorus electricus* thrives. *E. electricus* is a living battery, with the positive pole at its head and the negative one at its tail. When immobile, it produces no electricity, but when it starts moving it produces discharges of as much as 600 V! Small animals in its vicinity (or, worse, in contact with it) are killed, while larger mammals get dizzy and may drown. Humans can withstand a single shock.

© Getty Images

▶ The skin and liver of the puffer fish, or fugu, is poisonous.

A similar unpleasant surprise awaits the fish that swim in temperate and tropical seas. This one is an electric ray of the *Torpedo* genus, which falls upon its prey, wraps the prey in its pectoral winglike fins, and stuns it with a discharge reaching 200 V. In antiquity, Roman physicians used the electric ray as a therapeutic agent.

The fugu, or puffer fish, is a culinary delicacy in Japan. However, if the person preparing the food does not meticulously remove the skin and liver, the fish is going to be someone's last meal. Fugu is armed with *tetrodotoxin,* an extremely potent poison, of which a mere 1 to 2 mg is estimated to be lethal to an adult human.

Equally horrible is *saxitoxin,* a poison produced by marine microorganisms called dinoflagellates. Humans are not usually affected directly but can be affected after consuming filter-feeding mollusks (shellfish) such as mussels and clams, which concentrate the toxin in their flesh. Less than 1 mg of saxitoxin is believed to be fatal to a human! Dinoflagellates can grow at amazingly rapid rates under certain environmental conditions, giving rise to the spectacular phenomenon known as red tides.

What do all of these examples have to do with the subject of this chapter? The biological warfare that I just described relies on the electrical transmission of nerve impulses or its inhibition. The electric eel and electric ray, to begin with, are endowed with electric organs containing *electroplaxes,* that is, stacks of flattened cells called *electrocytes,* which have evolved from muscle cells. Electrocytes have lost their contractile apparatus but have maintained the electrically excitable plasma membrane.

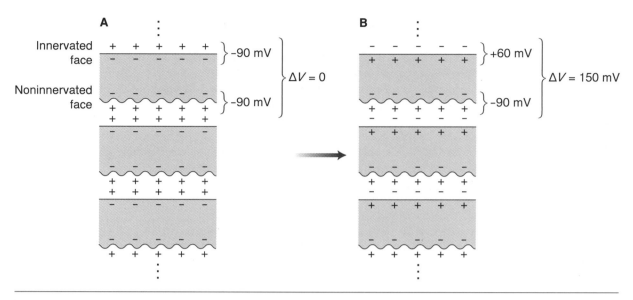

▶ **Figure 6.16** Living battery. The electric organs of certain fish contain piles of electrocytes with two different faces. One is relatively flat, innervated, and thus excitable; the other is intensely folded, noninnervated, and thus nonexcitable. When the cells are not excited *(a)*, there is no potential difference between the two faces. However, when a nerve impulse elicits an action potential on the innervated face *(b)*, a large Δ*V* develops. The stacking of thousands of electrocytes in an electric organ results in the generation of stunning electric shocks.

An electrocyte has two distinct faces (figure 6.16). One receives nerve endings and can be excited electrically from a resting potential of –90 mV to an action potential of +60 mV. The other is nonexcitable and remains at –90 mV. Thus, when the brain sends a nerve impulse to the innervated face, a voltage of +60 – (–90) = 150 mV develops between the two faces. Electrocytes are placed one on top of the other, so that the voltages are additive. As a result, an electroplax of 4,000 electrocytes can cause a 600 V discharge.

Thanks to their high content of excitable membranes, the electric organs are rich sources of the proteins that are responsible for the generation of action potentials. Thus, researchers have used them as starting materials for the isolation of the Na⁺ channel and the acetylcholine receptor. In fact, a large portion of our knowledge about these proteins comes from the study of purified Na⁺ channel from the electric eel and purified acetylcholine receptor from the electric ray.

Tetrodotoxin and saxitoxin, on the other hand, owe their toxicity to their blocking of the electrical transmission of impulses along the axons of neurons, which leads to inability of the muscles to contract. Death is caused by suffocation, when the diaphragm and the other muscles that control respiration become paralyzed. Both neurotoxins bind very strongly and selectively to the Na⁺ channel, thus preventing Na⁺ ions from flowing into the cytosol and triggering action potentials. Both contain positively charged *guanidinium* groups in their structure. Guanidinium forms an electrostatic bond with a (negatively charged) carboxyl group at the extracellular opening of the Na⁺ channel. In this way the toxin molecule blocks the opening and does not allow Na⁺ ions to pass. Thanks to their high affinity and selectivity for the Na⁺ channel, tetrodotoxin and saxitoxin have proven to be valuable tags for measuring the content of membranes in the Na⁺ channel and monitoring the purification of the Na⁺ channel in the laboratory.

Other poisons have similarly high affinity and selectivity for the acetylcholine receptor. The cobra and bungarus (a South Asian snake) kill their prey thanks to *cobratoxin* and *α-bungarotoxin,* respectively. These neurotoxins are small proteins (around 7 kDa) bearing a high percentage of amino acid residues with positively charged side chains.

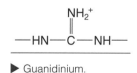

▶ Guanidinium.

The toxins bind to the acetylcholine receptor in the plasma membrane of muscle cells, thus inhibiting the binding of acetylcholine and hence neuromuscular communication. Both have been utilized to locate the acetylcholine receptor in membranes and track its isolation.

Finally, a plant-derived poison has also proven useful in biochemical research. Curare is a complex plant extract that the Amazon natives used to dip their arrow tips into before hunting or fighting. Curare's most active ingredient, *d-tubocurarine,* is produced by the *Chondodendron tomentosum* plant and was isolated from a form of curare that used to be transported in bamboo tubes (hence the name). d-Tubocurarine, like the snake toxins mentioned earlier, competes with acetylcholine for binding to the acetylcholine receptor. The d-tubocurarine molecule contains two quaternary amino groups similar to the ones in acetylcholine, which explains why the former competes successfully with the latter. This is another proof of the fundamental principle of modern biochemistry

▶ The poison with which a cobra kills its prey is a neurotoxin.

that *structure determines function,* or, in the words of one of the most eminent figures in the history of art, technology, and science, "The shape is the plastic representation of function" (Leonardo da Vinci).

Problems and Critical Thinking Questions

1. Which protein is responsible for the resting potential in nerve and muscle cells?

2. Which proteins are responsible for action potentials in nerve and muscle cells?

3. Is the resting potential due to passive or active transport of ions? What about the action potential?

4. Distinguish voltage-gated from ligand-gated ion channels. Give examples.

CHAPTER 7

Muscle Contraction

In this chapter, we will examine how skeletal muscle contraction, the most impressive form of movement in living organisms, is performed and controlled. Your appreciation of the processes involved will grow as we explore the exceptional organization of the contractile elements of muscle cells and the plethora of delicate molecular interactions among these elements. Proteins have reserved the leading part in these processes too (justifying their name once more). The most abundant of muscle proteins, myosin, possesses the rare ability to convert the free-energy change of a reaction to kinetic energy. For this reason we often call myosin a *molecular motor*.

7.1   Structure of a Muscle Cell

A muscle cell is extremely specialized. Although it obeys the basic principles of structure and function of all animal cells, it exhibits some unique or uncommon features. To begin with, it is *multinucleated;* that is, it contains many nuclei lying just underneath the plasma membrane. This is the case because each muscle cell is formed by the fusion of many mononucleated *myoblasts* during development. In addition, a muscle cell is *postmitotic,* meaning that it cannot divide by mitosis. A muscle cell looks like an extremely long tube and is therefore better known as a **muscle fiber.** Finally, biochemists and physiologists use special terms to describe two components of a muscle cell: They refer to the plasma membrane as the **sarcolemma** (meaning "wrapping of the flesh" in Greek) and to the cytoplasm as the **sarcoplasm.**

A skeletal muscle fiber presents densely packed striations, perpendicular to its longitudinal axis, under a light microscope (figure 7.1). The striations consist of alternating dark and light areas. Every dark area is called *A band* and every light area *I band.* These striations are the reason that skeletal and cardiac muscles are named *striated muscles.* In contrast, striations are not so evident in smooth muscle because it is less well organized.

Each muscle fiber is filled with about 1,000 parallel rods of contractile material (think of a pack of spaghetti). Each rod has a diameter of 1 to 2 μm and is called a **myofibril.** The myofibrils are embedded in cytosol. The alternation of A and I bands, which, as we saw, characterizes the muscle fiber as a whole, persists in every myofibril. In fact,

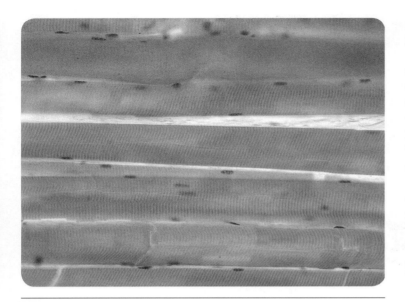

a muscle fiber looks striated because of the alignment of A bands in adjacent myofibrils.

At higher resolution—under an electron microscope—one can see that the A and I bands are not uniform (figure 7.2a). The A band has a stripe in the middle, the *H zone,* which is less dense than the rest of the band. Then, in the middle of the H zone, there is a dense line called the *M line.* The I band too bears a dense line in the middle called the *Z line,* or *Z disc.*

This symmetrical pattern is repeated along the entire myofibril and permits the definition of a minimal complete functional unit, within which we can witness and examine muscle contraction. This unit is called the **sarcomere** and is defined as *the segment of a myofibril between two Z lines.* The sarcomere has a filamentous appearance and is approximately 2.3 μm long in a muscle at rest.

What lies behind this mosaic of bands, zones, and lines of alternating density? Electron micrographs of cross sections of a myofibril (figure 7.2b) reveal that it contains two kinds of filaments: the **thick filaments** with a diameter of about 15

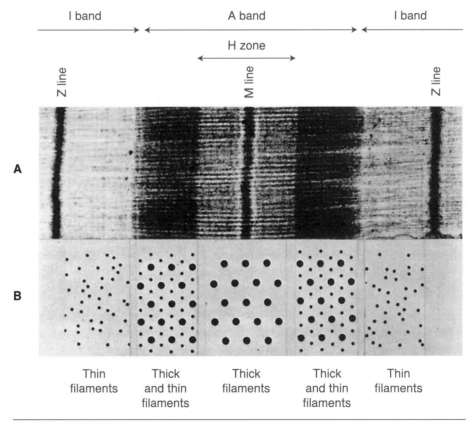

▶ **Figure 7.2**   Electron micrograph of a sarcomere. *(a)* A longitudinal section of a myofibril presents a repetitive symmetrical pattern. The part between two Z lines shown here is called a sarcomere. *(b)* Cross sections at different points along the sarcomere reveal its structural details.

Courtesy of Dr. H. E. Huxley.

nm, and the **thin filaments** with a diameter of 9 nm. The I band consists of thin filaments, while the A band has a mixed makeup. Its H zone consists of thick filaments,

whereas the rest of the band contains both kinds of filaments in a symmetrical arrangement: Every thick filament is surrounded by six thin filaments, and every thin filament is surrounded by three thick filaments. The filaments are composed of proteins. The thick filaments contain mainly **myosin**, while the thin filaments contain mainly **actin, tropomyosin**, and **troponin.** Additionally, the M line contains the proteins *myomesin* and *M-protein*, while the Z line contains the protein *α-actinin*.

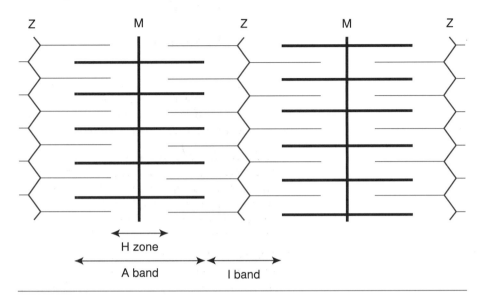

▶ **Figure 7.3** Filament arrangement. Deduced from the electron micrographs in figure 7.2, this diagram depicts two sarcomeres consisting of overlapping thick (black) and thin (color) filaments.

The data just presented show that thick and thin filaments interdigitate at the two ends of the A band (figure 7.3). A wealth of other data convinces one that, in addition, the two kinds of filaments interact through **cross-bridges** that are part of the myosin molecules. Approximately 500 cross-bridges protrude from the surface of each thick filament and point toward the surrounding thin filaments. The gap between thick and thin filaments covered by the cross-bridges is 13 nm. As we will see, the interaction of the myosin cross-bridges with actin is what generates the force of contraction.

7.2   The Sliding-Filament Theory

The shortening of a muscle during contraction is reflected in its elementary functional unit, the sarcomere. Indeed, each sarcomere shortens in proportion to the whole muscle. However, there is no proportional shortening in the parts of the sarcomere. Instead,

as the examination of sarcomeres by electron microscopy and X-ray diffraction has shown, the A band does not change, whereas the I band and the H zone shorten during contraction (figure 7.4). These observations led two independent groups of investigators, one consisting of Andrew Huxley and Ralph Niedergerke, and the other consisting of Hugh Huxley and Jean Hanson, to propose the **sliding-filament model** early in the 1950s.

The model maintains that *when the sarcomere contracts, the lengths of the thick and thin filaments do not change but their overlap increases.* Therefore, *contraction is caused by the active sliding of thick and thin filaments past each other.* The sliding-filament model has been verified repeatedly by the experimental data and is now accepted as

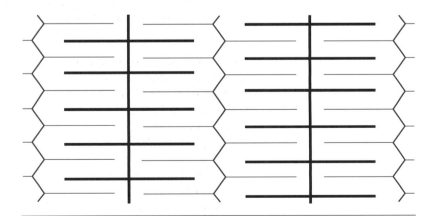

▶ **Figure 7.4** Sarcomere shortening. The length of each sarcomere decreases when a muscle contracts. The decrease is not due to a shortening of the filaments but to an increase in their overlap (compare the length of the filaments and the sarcomeres in this figure to those in figure 7.3).

a theory. But what is it that draws thick and thin filaments past each other? To answer this question we have to explore the major players in force generation, myosin and actin, as well as their interplay.

7.3   The Wondrous Properties of Myosin

Myosin is the leader of the protein pack during contraction. It is the most abundant protein of muscle tissue, accounting for half its protein mass. Because of its abundance, myosin can be purified and studied with relative ease. Decades of research have ascribed three important biological properties to this protein.

- *Self-assembly.* Myosin is soluble in aqueous solutions of high ionic strength. If we start with such a solution of myosin at physiological pH and decrease the ionic strength by adding water until it reaches the physiological level ($0.3 \, \text{mol} \cdot \text{L}^{-1}$), the solution becomes turbid, as the myosin molecules form filaments by themselves. These filaments resemble the thick filaments of intact myofibrils. We say that *myosin is capable of self-assembling.*
- *Enzyme activity.* Apart from constituting the main structural protein of muscle, myosin is an enzyme. As Vladimir Engelhardt and Militsa Lyubimova discovered in 1939, myosin hydrolyzes ATP to ADP and P_i. This action renders myosin an ATPase. *The ΔG of ATP hydrolysis is exactly what drives muscle contraction.*
- *Binding to actin.* Myosin binds to the polymerized form of actin, the main component of the thin filaments (see section 7.5). This binding is necessary for the movement of thick and thin filaments past each other during contraction.

7.4   Structure of Myosin

Myosin's structure is as interesting as its properties. The main features of the myosin molecule are the following:

- It is very large, having a molecular mass of 520 kDa.
- It consists of six subunits (figure 7.5a): two large ones (220 kDa each) named *heavy chains* and four small ones (20 kDa each) named *light chains.*
- It is asymmetric, consisting of a double-headed region linked to a very long tail (170 nm). The tail contains part of each heavy chain in α-helical conformation, coiled around the same part of the other heavy chain. Each of the two heads is formed by the rest of the heavy chain and two light chains. Each head is 16.5 nm long.

The three properties described in the previous section may be attributed to specific regions of the myosin molecule. This can be shown experimentally by **proteolysis,** that is, treatment with enzymes that hydrolyze proteins and are called **proteases.** Proteolysis under mild conditions may yield information on particular functional domains of a protein because such domains are usually separated by stretches of the polypeptide chain exposed to the environment and therefore vulnerable to mild hydrolysis.

Treatment with *trypsin,* a protease secreted by the pancreas, cleaves myosin into two segments named *light meromyosin* and *heavy meromyosin* (figure 7.5b). Heavy meromyosin contains the double-headed region and part of the tail, whereas light meromyosin corresponds to the remaining length of the tail. Another protease, *papain,* derived from the exotic papaya fruit, cleaves heavy meromyosin further to three *subfragments:* two identical heads denoted by S1 and one rod denoted by S2 (figure 7.5c).

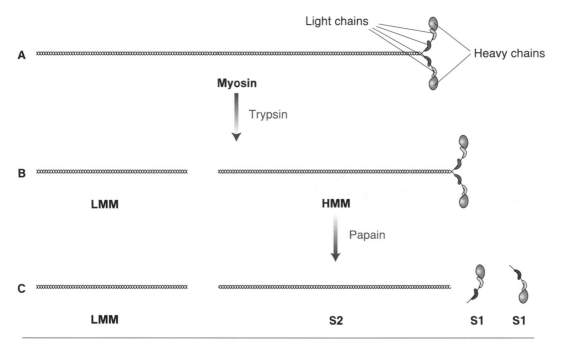

Figure 7.5 Myosin. *(a)* Myosin looks like two intertwined snakes. It consists of six polypeptide chains: two heavy chains accounting for the tail and most of the two heads, and four light chains, which are shared by the heads. *(b)* Myosin can be cut by limited proteolysis. Treatment with trypsin (a protease) splits it into light meromyosin (LMM) and heavy meromyosin (HMM). *(c)* Further treatment of HMM with papain (another protease) divides it into two subfragments S1 and one subfragment S2.

How then are the three properties of myosin distributed among its segments? Light meromyosin has the property of self-assembly (figure 7.6a); that is, it forms filaments, just as intact myosin does. However, whereas cross-bridges protrude from the filaments of intact myosin, the filaments of light meromyosin are smooth. This shows that light meromyosin constitutes the trunk of the thick filaments, whereas heavy meromyosin corresponds to the cross-bridges. Light meromyosin lacks ATPase activity and does not bind to actin.

In contrast to light meromyosin, each S1 hydrolyzes ATP and binds to actin (figure 7.6b) but does not form filaments. Finally, S2 possesses none of myosin's properties; it just serves to link the other segments.

7.5 Actin

Actin is the main component of thin filaments. It consists of a single 42 kDa polypeptide chain and exists in two forms. In solutions of low ionic strength, actin molecules are present as monomers. This form is designated *G-actin* (G stands for globular). If we increase the ionic strength toward the physiological value, actin polymerizes to form fibers of *F-actin* (F stands for fibrous). Actin monomers are placed in such a way

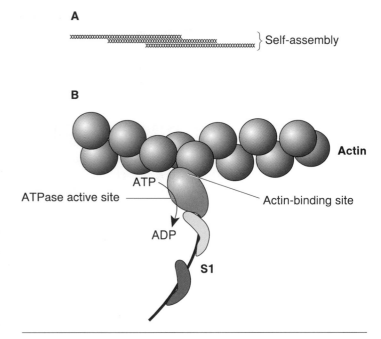

Figure 7.6 Assigning properties to myosin's segments. *(a)* LMM is capable of self-assembling to form smooth filaments corresponding to the trunk of the thick filaments in intact muscle. *(b)* S1 is capable of hydrolyzing ATP and binding to actin. The ATPase active site and actin-binding sites are close to each other but distinct.

Actin polymerization
is accompanied by
the hydrolysis of ATP
to ADP. In the 1940s
and 1950s this misled
several investigators
to propose that the
mechanism of muscle
contraction was based
on actin.

in F-actin that they give the impression of a necklace with two intertwined strands of beads (figure 7.7). F-actin forms the trunk to which tropomyosin and troponin attach in the thin filaments.

F-actin greatly increases the ATPase activity of myosin. To be exact, it does not increase the efficiency of ATP hydrolysis (myosin alone is as good at that) but increases the rate at which the products, ADP and P_i, are released from the active site (this rate is low in the absence of actin). Thus, the active site is emptied of the products sooner and is capable of carrying out more reaction cycles per unit of time. To use a term introduced in section 3.16, actin increases the turnover number of myosin ATPase 200-fold. Actin owes its name to this activation of myosin ATPase.

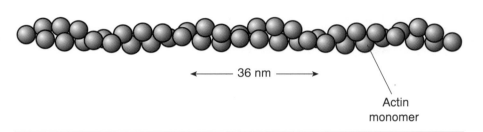

← 36 nm →

Actin
monomer

▶ **Figure 7.7**   Actin. F-actin is formed by the welding of pairs of actin monomers that turn gradually, giving the impression of two intertwined fibers. A full turn of F-actin encompasses 13 monomers and is 36 nm long.

When a solution of actin is added to a solution of myosin, a complex called *actomyosin* forms and the mixture becomes viscous (it thickens). The mixture can become thin by the addition of ATP, which means that *ATP dissociates actin and myosin.* However, because of the presence of actin, myosin hydrolyzes ATP quickly. When ATP is depleted, actomyosin reforms but can be dissolved by another addition of ATP. The description of these interactions by Albert Szent-Györgyi in the 1940s shed the first light on the mechanism of muscle contraction.

For ATP to dissolve actomyosin, magnesium ions have to be present. The reason is that the actual substrate of myosin ATPase, and all ATPases, is the complex of Mg^{2+} with the ATP^{4-} ion shown in figure 2.3.

7.6   Sarcomere Architecture

Now that we have examined the structure and interaction of actin and myosin in detail, we are in a position to explore the organization of the fundamental contractile unit of muscle. As already mentioned, the sarcomere is delimited by two Z lines composed mainly of α-actinin. Thin filaments, approximately 0.9 μm long, sprout from the Z lines (figure 7.8). *Nebulin,* an extremely large (about 600 kDa) and elongated protein running along every thin filament, is believed to aid in its assembly by serving as a template on which the filament grows during myofibril formation.

The thick filaments, 1.5 μm long, lie in the center of the sarcomere, holding on to the M line. Cross-bridges protrude from the surface of the thick filaments in a helical arrangement every 14.3 nm along the filament axis, except for a 150 nm bare zone (with no cross-bridges) in the middle.

The sarcomere owes part of its exquisite architecture to yet another protein, *titin.* This extraordinary elastic protein is the largest polypeptide known (3,000-3,700 kDa). Twelve titin molecules, six on each side of the M line, extend through each thick filament all the way to the Z lines bordering the sarcomere. Between the end of the thick filament and the Z line, titin forms flexible connections, which are the main source of passive elasticity in muscle. These connections center thick filaments in the sarcomere, so that equal forces are developed by myosin in both halves of the sarcomere.

It is important to note that the direction of the myosin molecules is opposite in the two halves of the thick filament: The tails point to the middle of the sarcomere (hence the bare zone), whereas the heads point to the Z lines. The thin filaments are directional too. The direction is the same for all thin filaments in the same half of the sarcomere

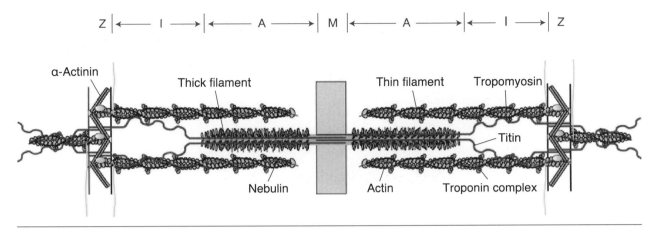

▶ **Figure 7.8**   Sarcomere organization. A sarcomere contains precisely aligned thick and thin filaments anchored to the M and Z lines, respectively. The direction of the myosin molecules is reversed in the middle of the sarcomere. This is also the case for the actin, tropomyosin, and troponin molecules in the thin filaments. These opposite directions ensure the coordinated movement of the Z lines toward the M line during contraction. Both kinds of filaments are held in place by two giant elongated proteins, titin and nebulin. Note that the diagram depicts a longitudinal section of part of a sarcomere. To get the real picture you have to imagine many more filaments in space, with the thin filaments arranged in hexagonal symmetry around the thick filaments.

Reprinted from *Trends in Cardiovascular Medicine*, vol. 13(5), A.S. McElhinny et al., Nebulin: The nebulous, multifunctional giant of striated muscle, pg. 15, Copyright 2003, with permission from Elsevier.

but is reversed in the other half. Thus, we can generalize by saying that *both thick and thin filaments have opposite directions in the two halves of the sarcomere.* This reversal of directions causes the thick filaments to pull the thin filaments toward the M line during contraction, thus decreasing the distance between the Z lines.

The architecture of the sarcomere is such that a muscle can be shortened but not stretched by its own power. This is so because the direction of the myosin molecules permits the active movement of the Z lines toward the M line, not away from it. However, muscle contraction with shortening, known as **concentric contraction**, is not the only type of contraction. When the muscle meets a resistance it cannot overcome, such as when we push against a firm wall, its length does not change, although it spends energy. This is **isometric contraction**. Finally, it is even possible for a muscle to be lengthened despite developing contractile force (and, naturally, spending energy) because a higher opposing force is applied to it. Then it performs **eccentric contraction**, as does the biceps brachii of a person losing a *bras de fer*.

7.7   Mechanism of Force Generation

We have now all that is needed to answer the question posed at the end of section 7.2. Although we do not know in every molecular detail the mechanism by which the force that moves the thick and thin filaments past each other is generated, we do have a fairly good picture of it.

The force of contraction is produced by repeated cycles of attachment, pulling, and detachment between the myosin heads in the thick filaments and F-actin in the thin filaments. Each head in a myosin molecule operates independently of the other. At rest (figure 7.9a), the heads protrude from the thick filaments but are unable to bind to the thin filaments because, as we will see in the next section, tropomyosin covers the binding sites on actin. ATP is hydrolyzed on the heads at a low turnover number because the binding of actin is hindered.

When the muscle is excited, tropomyosin moves to the side (for a reason that we will see in the next section also), allowing the myosin heads to stick to actin monomers on the thin filaments (figure 7.9b). Actin binding pushes P_i out of the ATPase active

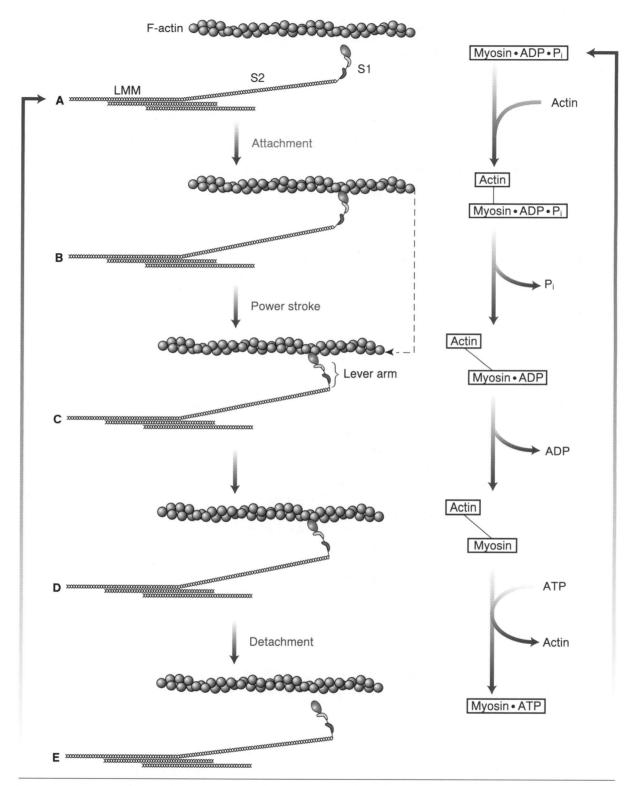

▶ **Figure 7.9** Mechanism of muscle contraction. The most probable mechanism of muscle contraction, depicted schematically at the left and in terms of molecular interactions at the right, is as follows. *(a)* At rest, S1 of myosin (for clarity, only one of the two S1 is shown) hydrolyzes ATP to ADP and P_i but does not liberate the products readily and does not bind to F-actin. *(b)* Excitation of the muscle fiber leads to the attachment of S1 to F-actin. *(c)* P_i leaves the ATPase active site, causing the lever arm to swing and the thin filament to move relative to the thick filament (note the black broken line and arrow). *(d)* ADP leaves the active site next. *(e)* ATP enters the active site, forcing S1 to detach from actin. *(e → a)* S1 hydrolyzes ATP, resumes its initial orientation, and is ready to start all over again.

site. Researchers believe that the departure of P_i triggers an important conformational change in S1, that is, a swinging of the stalk, known as the *lever arm,* to which the light chains are bound. Since S1 is hooked to the thin filament, the swinging of the lever arm results in a movement of the thin filament by approximately 10 nm relative to the thick filament (figure 7.9c). The movement of the myosin head is called the *power stroke,* as this is what generates muscle contraction.

Following the power stroke, ADP departs the ATPase active site (figure 7.9d) and ATP moves in, breaking actomyosin (figure 7.9e). S1 hydrolyzes ATP, returns to its original conformation, and is ready for another cycle of attachment, pulling, and detachment. This cycle is repeated for as long as the binding of actin and myosin is allowed.

Recent evidence suggests that the angle by which the lever arm swings when P_i is released, and hence the displacement of the thin filament, decreases with increasing load on the muscle.

7.8 Myosin Isoforms and Muscle Fiber Types

The genome of an organism often contains several genes, differing little in base sequence, that encode proteins differing little in amino acid sequence and serving essentially the same function. Slightly different forms of a protein can also arise from one gene by alternative splicing of exons, which produces two or more mRNA (section 4.12). Such different forms of a protein are called **isoforms.** Isoforms usually differ quantitatively in biological activity, often serving the particular needs of the cells in which they are found.

If a protein is an enzyme, its different forms are also referred to as *isoenzymes* or *isozymes.*

Myosin is a protein displaying isoforms. Animals and humans have multiple forms of both the heavy and the light chains of the protein, although the ones primarily affecting myosin action are the heavy chain isoforms. Adult humans have three such isoforms, denoted by I, IIa, and IIb. The latter isoform has been renamed IIx in recent years. Investigators separate and measure myosin heavy chain isoforms, and isoforms of other proteins, by *gel electrophoresis,* a technique introduced in the sidebar at the end of chapter 2. A muscle fiber usually contains one isoform and is designated as type I, IIA, or IIX. The latter is slowly replacing the IIB designation in the literature.

How do the three myosin heavy chain isoforms differ? IIa and IIx have a higher ATPase activity than isoform I and endow the respective muscle fibers with a higher maximal shortening velocity. Because of this, type IIA and IIX fibers are characterized as *fast-twitch* and type I as *slow-twitch.* In addition, type IIA and IIX fibers are less economical, as they spend more ATP to develop a given force.

A motor unit contains muscle fibers of the same type, which are all excited when the motor neuron fires action potentials. However, not all motor units in a muscle fire when it contracts. Instead, there seems to be a gradual and additive *recruitment* of motor units as the force of contraction increases: from motor units whose neurons have small cell bodies, that display low excitation thresholds (section 6.3), and that control few muscle fibers, to motor units whose neurons have large cell bodies, that display high excitation thresholds, and that control many muscle fibers. It is believed that the former contain slow-twitch fibers, whereas the latter contain fast-twitch fibers.

A complementary, and finer, means of controlling the force of contraction is the frequency of action potential firing: The higher the frequency, the higher the force output per muscle fiber, motor unit, and muscle. Slow-twitch fibers respond to lower frequency ranges (5-30 Hz) than fast-twitch fibers (30-65 Hz). (Hz is the symbol of the *hertz,* the unit of frequency equal to one event per second.) The electrical activity in a muscle can be studied by *electromyography* through the attachment of electrodes to the skin over the muscle or, better, the insertion of needle electrodes into the muscle.

Differences among fiber types are not confined to contractile behavior but are extended to metabolism. However, before we discuss the metabolic differences of muscle fiber types we need to explore muscle metabolism both at rest and during exercise. Therefore, we will return to the issue of muscle fiber types in chapter 13.

7.9   Control of Muscle Contraction

For over a century (since 1883) it has been known that muscles cannot contract in the absence of calcium ions. However, 80 years had to go by before the exact role of Ca^{2+}

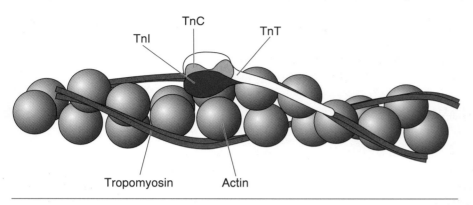

was discovered. Today we know that *Ca^{2+} controls muscle contraction* and that it does so by permitting the binding of myosin to actin. This action of Ca^{2+} is not direct but indirect. As Setsuro Ebashi discovered in the 1960s, control is exerted through tropomyosin and troponin, the two proteins that coexist with actin in the thin filaments, constituting about one-third of the thin-filament mass. Let's meet them.

Tropomyosin has a molecular mass of 70 kDa and consists of two similar stringlike subunits in an α-helical conformation. The two subunits wind around each

▶ **Figure 7.10** Thin-filament proteins. A thin filament consists of an F-actin fiber (the double necklace of figure 7.7), two series of tropomyosin molecules, and troponin (Tn) complexes placed at regular intervals. Every troponin complex consists of TnC, which is the Ca^{2+} acceptor; TnI, which binds to actin; and TnT, which extends by way of a long tail along tropomyosin.

Adapted, by permission, from L.B. Smillie.

> Tropomyosin owes its name to the similarity of its structure and some of its physical properties to myosin (*trópos* means "mode" in Greek).

other just as the myosin heavy chains do in the myosin tail. The extremely elongated tropomyosin molecules join in a row to form fibers. Two such fibers run along each thin filament while following the twisting of the actin monomers (figure 7.10).

Troponin, symbolized as Tn, is a complex of three different subunits: TnC (18 kDa), TnI (24 kDa), and TnT (37 kDa). TnC binds Ca^{2+}, TnI binds to actin, and TnT binds to tropomyosin. Two troponin complexes appear on the two sides of a thin filament every 39 nm, approximately the length of a tropomyosin molecule. One troponin complex attached to one tropomyosin molecule controls approximately seven actin monomers.

When a muscle is at rest (relaxed), the cytosol has a very low $[Ca^{2+}]$, approximately 10^{-7} mol $\cdot$ L^{-1}. Researchers believe that, in this case, interactions among actin, TnI, TnT, and tropomyosin hold the latter close to the sites on the actin monomers where the myosin heads bind. Thus, tropomyosin hinders the interaction of thin and thick filaments. As we will see in the next section, muscle excitation by the nervous system results in the release of Ca^{2+} from an intracellular reservoir called the sarcoplasmic reticulum. This causes a surge in the cytosolic $[Ca^{2+}]$ by 100 times, from 10^{-7} to 10^{-5} mol $\cdot$ L^{-1}. The increased Ca^{2+} ions encounter TnC, bind to it, and elicit a change in its conformation. As a result, TnC detaches TnI from actin. This lets tropomyosin move over the surface of the thin filament, away from the binding sites of myosin on actin. Myosin then binds to actin, and the muscle contracts.

The muscle contracts until Ca^{2+} is sequestered in the sarcoplasmic reticulum (in a manner that we will examine shortly). The actin-TnI-TnT-tropomyosin interaction is then restored; tropomyosin returns to a position hindering cross-bridge formation; and the muscle relaxes. The chain of events through which Ca^{2+} controls muscle contraction is summarized in figure 7.11.

7.10   Excitation–Contraction Coupling

In chapter 6, we examined how a nerve impulse travels from its birthplace to the surface of a muscle fiber. Then, in the present chapter, we considered how the interior of a

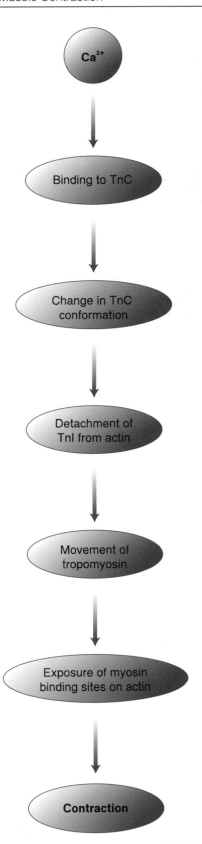

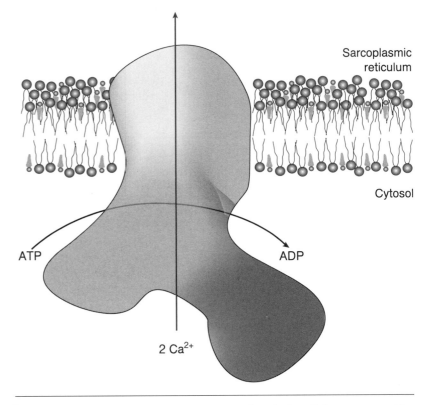

muscle fiber is constructed, how the fiber contracts, and how Ca^{2+} controls contraction. Now we can complete the sequence of events leading from nervous excitation to muscle contraction by providing the missing link: the transmission of the signal from the surface to the interior of a muscle fiber.

We saw that the opening of the acetylcholine receptor at the plasma membrane of a muscle fiber (the sarcolemma) results in its depolarization and the generation of action potentials (section 6.7). The action potentials are then transmitted to the interior of the fiber as they travel along a system of **transverse tubules,** which is also known as the **T system** (refer to the introductory figure of part II). The transverse tubules are an extension of the sarcolemma and are closely apposed to a complex membranous system of delicate sacs surrounding the myofibrils and covering the entire length of the sarcomere. This membranous system is called **sarcoplasmic reticulum** and forms the Ca^{2+} reservoir that responds to nervous excitation.

Ca^{2+} accumulates in the sarcoplasmic reticulum by the action of an integral protein of its membrane, the Ca^{2+} **pump** (figure 7.12). This protein has a

▶ **Figure 7.11** Control of muscle contraction. Ca^{2+} controls muscle contraction through a series of protein interactions involving all four major muscle proteins.

▶ **Figure 7.12** Calcium pump. The sarcoplasmic reticulum Ca^{2+} pump resembles the α subunit of the plasma membrane Na^+-K^+ pump (cf. figure 6.4). The Ca^{2+} pump concentrates Ca^{2+} inside the sarcoplasmic reticulum at the expense of ATP. The protein and membrane are drawn to scale.

single 110 kDa subunit, similar to the α subunit of the Na⁺-K⁺ pump. The Ca²⁺ pump constitutes 80% of the protein mass of the sarcoplasmic reticulum membrane and occupies nearly half its surface! The pump sequesters Ca²⁺ from the cytosol and raises the [Ca²⁺] inside the sarcoplasmic reticulum to above 10^{-3} mol · L⁻¹. As with the Na⁺-K⁺ pump, the energy for the operation of the Ca²⁺ pump comes from the hydrolysis of ATP; thus, the pump is also referred to as **Ca²⁺ ATPase.** For every molecule of ATP that it hydrolyzes, the ATPase transports two Ca²⁺ ions.

The maintenance of such a steep concentration gradient across the sarcoplasmic reticulum membrane ($10^{-3}/10^{-7}$ = 10,000-fold at rest) is facilitated by *calsequestrin.* This protein is located inside the sarcoplasmic reticulum and features approximately 40 low-affinity sites for Ca²⁺, thus acting as a molecular sponge: By binding calcium ions loosely, it lowers their tendency to leak out of the sarcoplasmic reticulum.

The transmission of an action potential from the sarcolemma to the transverse tubules causes the sudden opening of another integral protein of the sarcoplasmic reticulum membrane acting as a Ca²⁺ channel (figure 7.13). This particular channel is known as the **ryanodine receptor.** Exactly how the action potential opens the ryanodine receptor is not known with certainty, but possibly it causes a conformational change in a protein of the transverse tubule membrane called the **dihydropyridine receptor.** This protein is in touch with the ryanodine receptor in the adjacent sarcoplasmic reticulum membrane and may transmit the conformational change to it, resulting in its opening.

The [Ca²⁺] gradient between the sarcoplasmic reticulum and the cytosol is so steep in the resting state that the rate of efflux through the Ca²⁺ channel upon excitation of the muscle fiber exceeds the rate of influx through the Ca²⁺ pump. Thus, the [Ca²⁺] in the cytosol rises 100-fold, although it remains well below the [Ca²⁺] inside the sarcoplasmic reticulum (figure 7.14). By analogy, think of a rich man possessing 10,000 gold coins and a poor man possessing one gold coin. If the rich man gives the poor man 99 coins, the poor man will increase his money 100-fold but the rich man will still be wealthier and will have lost just 1% of his money.

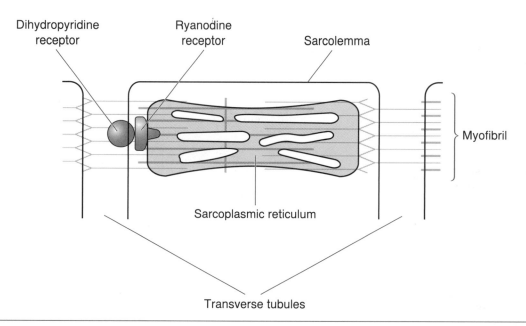

▶ **Figure 7.13** From nervous excitation to muscle contraction. The transverse tubules and the sarcoplasmic reticulum, two delicate membranous systems inside muscle fibers, permit the coupling of excitation to contraction. The transverse tubules carry action potentials from the sarcolemma to the depths of a muscle fiber. A voltage-sensitive protein of the transverse tubule membrane, the dihydropyridine receptor, undergoes a change in shape that opens an adjacent Ca²⁺ channel, called the ryanodine receptor, in the sarcoplasmic reticulum membrane. Ca²⁺ ions then flow from the sarcoplasmic reticulum to the cytosol, where they bind to TnC and permit the interaction of actin with myosin. The sizes of the dihydropyridine receptor and the ryanodine receptor have been exaggerated for diagrammatic purposes.

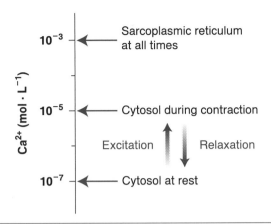

▶ **Figure 7.14** Ca^{2+} concentrations in muscle. A large Ca^{2+} reservoir, the sarcoplasmic reticulum, causes an instantaneous 100-fold increase in the cytosolic $[Ca^{2+}]$ when the muscle is excited by a nerve impulse. Note that the scale is logarithmic.

When the $[Ca^{2+}]$ in the cytosol rises to 10^{-5} mol · L^{-1}, the binding of Ca^{2+} to TnC activates muscle contraction as described in the previous section. When the excitation of the muscle fiber by the motor neuron ceases, the calcium channel closes and prevents the efflux of Ca^{2+}. Calcium ATPase then gets the upper hand: It brings Ca^{2+} back to the sarcoplasmic reticulum, decreasing its concentration in the cytosol from 10^{-5} to 10^{-7} mol · L^{-1} and causing muscle relaxation. This activity is thought to require one-third of the ATP hydrolyzed in a contracting muscle.

Problems and Critical Thinking Questions

1. Place the following terms in a logical order.

   Thick and thin filaments, muscle, myofibril, sarcomere, muscle fiber

2. Consider a muscle having average sarcomere, A band, and H zone lengths of 2.5, 1.5, and 0.7 μm, respectively, at rest. Suppose that the muscle shortens by 20%. Calculate the length of the following parts at rest and after contraction.

   a. Sarcomere

   b. A band

   c. H zone

   d. I band

   e. Thick filaments

   f. Thin filaments

   g. Overlap of thick and thin filaments

3. What is the immediate source of energy for muscle contraction? Which part of the muscle machinery utilizes it?

4. What is the cause of rigor mortis, that is, the stiffening of the body after death?

5. Is muscle contraction feasible if we remove tropomyosin and troponin from the thin filaments by an experimental treatment?

6. Addition of a caffeine solution to a muscle with severed sarcolemma (to permit the entrance of caffeine to the sarcoplasm) causes contraction. Suggest a mode of action for caffeine.

Part II Summary

In chapters 6 and 7, we followed, step by step, the complex route of transmission of a nerve signal to a skeletal muscle and the ensuing contraction. Despite the multitude of molecules and interactions involved, the whole process is very brief, usually less than a second. In fact, some of its individual steps, like the release of neurotransmitter in the synaptic cleft and the cyclic attachment and detachment of myosin and actin, are executed in less than 1 ms.

Let's recapitulate the biochemical processes leading to muscle contraction through an example. A sprinter is standing at the start of a track, ready for a race. His nervous and muscular systems are on alert. The starter fires. The sound waves from the shot vibrate the sprinter's eardrums and excite the auditory nerves. Nerve signals in the form of action potentials run along the axons of the relevant neurons. The action potentials are due to the opening of a sodium channel in the plasma membrane of the axon, which permits Na^+ to enter the cytosol from the extracellular fluid, followed by the opening of a potassium channel, which permits K^+ to exit the cytosol toward the extracellular fluid. Na^+ and K^+ move down their concentration gradients, which are maintained by the Na^+-K^+ pump, or Na^+-K^+ ATPase.

At the synapses between neurons, action potentials cause the opening of a calcium channel in the plasma membrane, through which Ca^{2+} enters the cytosol and forces synaptic vesicles to empty the neurotransmitter they contain into the synaptic cleft. Synaptotagmin 1 appears to be the Ca^{2+} sensor for this process in the vesicle membrane. The released neurotransmitter binds to receptor molecules in the postsynaptic membrane and alters their biological activity.

By employing electrical and chemical transmission, the auditory stimulus reaches the brain, which processes it and converts it to a decision to contract certain muscles. The decision is conveyed through the spinal cord to motor neurons in basically the same way the auditory stimulus was conveyed. Not all motor neurons controlling a muscle are excited. Rather, the number of motor neurons recruited relates to the force of contraction that is to be produced. At the neuromuscular junction, the nerve impulses cause the release of acetylcholine, which diffuses to the surface of muscle fibers and binds to the acetylcholine receptor. The receptor opens, permitting mainly Na^+ ions to enter the muscle fibers. In effect, the sarcolemma is depolarized.

The depolarization is transmitted deep inside the muscle fibers thanks to an extensive network of transverse tubules. A voltage-sensitive protein in the tubule membrane, the dihydropyridine receptor, causes the release of Ca^{2+} ions from the adjacent sarcoplasmic reticulum through the ryanodine receptor, a calcium channel. Ca^{2+} rushes to the cytosol and binds to TnC, which is located in the thin filaments of the myofibrils along with two other troponin subunits (TnI and TnT), F-actin, and tropomyosin. The binding of Ca^{2+} to TnC detaches TnI from actin and lets tropomyosin step aside from between F-actin and the myosin heads that protrude from the thick filaments, ready to form cross-bridges.

Myosin then binds to F-actin, which drives P_i, the product of ATP hydrolysis, away from the heads. The departure of P_i causes a swinging of the lever arm, a flexible stalk in the heads. As a result, the thick filaments pull the thin filaments to the center of the sarcomere, the elementary contractile unit of muscle. The sarcomeres of the excited muscle fibers shorten, and muscles in the legs, the trunk, the hands, and so on contract. The sprinter springs forward and the finish line begins to get closer and closer. . . .

Exercise Metabolism

Our study [of muscle], however, needs light also from another aspect, it requires the skilled labours of the organic and biochemists. [...] To take another analogy, the completion of the drawing will rest with the chemists: we physicists can only provide a sketch; we can indicate the type of machine and its properties, the chemists must describe it in detail.

–Archibald V. Hill (Nobel Prize in Physiology or Medicine 1922), *Nobel Lecture*

In chapter 2, we defined metabolism as the sum of the chemical reactions occurring in a living organism. We introduced some basic knowledge about metabolism, including the energy changes accompanying metabolic reactions, the molecules playing central roles in energy transactions and oxidation–reduction reactions, and the basic processes yielding energy from foodstuffs.

The present part of the book deals with the metabolism of humans and animals during and after exercise. Let me clarify right from the beginning that the metabolic reactions taking place in an organism are the same whether the organism is resting or exercising. What changes spectacularly upon shifting from one state to the other is the rate of reactions: Some speed up, while others slow down, thus letting the body respond to the different demands of each state. This is **metabolic control** and is manifest not only in exercise but also in any state of an organism (be it growth, food consumption, starvation, stress, or disease). In fact, each state is distinguished by the dominance of different reactions; it has what we call a *metabolic profile* of its own. We can therefore say that the subject of part III is the *metabolic profile of exercise*.

III.1   Principles of Exercise Metabolism

Metabolic control (both in exercise and in general) is exerted through biochemical mechanisms that we have only begun to comprehend in recent decades. Although most metabolic pathways have been elucidated, our knowledge about their regulation is incomplete. (In the context of metabolism, I will use the terms control and regulation interchangeably.) The available data, however, show that metabolic control is both efficient and flexible.

Exercise is one of the most powerful modulators of metabolism. Few other stimuli cause such impressive changes. The metabolic changes caused by exercise are the subject of the chapters to come. Through these the following basic principles will emerge:

- Exercise metabolism obeys the need for increased energy supply to the contracting muscles.

- Most kinds of exercise cause increased breakdown of carbohydrates and lipids.

- Exercise changes metabolism not only in the exercising muscles but also in other organs and tissues (such as the liver and adipose tissue).

- Metabolism does not return to the resting state immediately after the end of exercise. Many changes persist for hours or days, while others are manifest during recovery rather than exercise.

- Regular repetition of exercise—what we call training—can change metabolism to an extent great enough to make the metabolic profile of a trained individual (not only during exercise but also during rest) different from that of an untrained individual.

III.2   Exercise Parameters

Not all exercises have the same effect on metabolism. Although a thorough examination of all possible kinds of exercise is beyond the scope of this book, we need to be clear and specific as to which kinds of exercise the metabolic changes that we discuss do apply to.

Three main identifying parameters of exercise are *type, intensity,* and *duration.* Three widely recognized types of exercise are **endurance, resistance,** and **sprint.** Endurance exercise is characterized by prolonged continuous or intermittent periods of contractile activity against low resistance. Jogging is an endurance exercise. In contrast, resistance exercise involves short periods of contractile activity against high resistance. Weightlifting is a resistance exercise. Finally, sprint exercise consists of short periods of maximal contractile activity against low resistance. A competitive 50 m swim is a sprint exercise.

An alternative way of describing exercise type is by the terms aerobic and **anaerobic.** These terms refer to the predominant means of energy production during an exercise. Aerobic exercise draws energy mainly from biochemical processes requiring oxygen either directly or indirectly, whereas anaerobic exercise draws energy from processes not requiring oxygen in any way. (We will explore all these processes in the next chapters.) Endurance exercise is aerobic, whereas resistance and sprint exercises are anaerobic.

Turning to exercise intensity, this parameter can be expressed in a variety of ways. Some are exercise specific (for example, running speed or

weight lifted), while some apply generally (for example, heart rate or oxygen uptake). Additionally, it is possible to express intensity in relative terms, that is, as a percentage of maximal intensity. A widely used measure of relative intensity is percentage $\dot{V}O_2$max. Since the scientific literature on how exercise affects metabolism is replete with % $\dot{V}O_2$max, I will be using this measure a lot in this book. Unless specific values are necessary, I will refer to exercise up to 49% as *light,* 50% to 74% as *moderate intensity,* 75% or more as *hard,* and around 100% as *maximal.* Although the American College of Sports Medicine (1998) has proposed an alternative classification of exercise intensity based on $\dot{V}O_2$ reserve (that is, maximal minus resting oxygen uptake) as more accurate, this has not been adopted in the literature.

Exercise duration is easier to measure, and there is only one way of expressing it (time). Since there is no agreed-upon classification (that is, there is no consent as to what durations should be called short and long), I will try to be as specific as possible when referring to duration.

A final exercise parameter that we need to discuss relates to whether exercise is executed once or is repeated regularly over a period of weeks, months, or years. Exercise executed once is *acute,* or *short-term;* exercise repeated regularly is *chronic,* or *long-term.* The effects of acute exercise on metabolism usually last from a few minutes to a few days after its cessation, while the effects of chronic exercise usually last from several days to several months after its interruption. The chronic effects are also called **adaptations to exercise** in the sense that the body modifies its metabolic profile to meet the requirements of regular exercise with a slighter disturbance of its homeostasis and a lower chance for damage.

> **Homeostasis** is the maintenance of a biological parameter in an organism (such as the concentration of a substance in the blood) at a relatively stable level despite temporary fluctuations. The mechanisms that are accountable for homeostasis are called *homeostatic.*

III.3   Experimental Models Used to Study Exercise Metabolism

How do researchers study the changes in human metabolism brought about by exercise? The most reasonable approach would be to exercise volunteers (or recruit volunteers that have exercised on their own), take biological samples (such as expired air, blood, or muscle tissue) at various time points, and look for changes in the amount of certain substances. This approach, however, has some limitations. One is that it may not favor the discovery of the molecular mechanisms underlying the observed changes because too many parameters alter in an exercising person at the same time. Additionally, certain organs of interest to exercise scientists are very hard or inappropriate to sample in living humans (for example, the liver, heart, and brain).

Because of such limitations, investigators often use experimental approaches, or models, that permit a more elaborate examination of exercise metabolism. We refer to these as *models of modified contractile activity,* in general, in order to include manipulations that modify the contractile activity of muscles but do not qualify as exercise.

What are these models? Booth and Thomason (1991) have divided them into three categories depending on how closely they resemble human exercise.

1. *Animal models that closely mimic human exercise.* Such models include treadmill running, aimed at imitating endurance exercise, and hanging a weight on a limb (such as the leg of a rat or the wing of a chicken) for up to 2 h per day, aimed at imitating resistance exercise. Of course, things could be quite different in another world (figure III.1).

2. *Animal models of increased contractile activity that do not mimic human exercise.* The most frequently used model of this kind is the electrical stimulation of a muscle or a muscle group **in vivo** (in a live animal) or **in situ** (in an organ exposed for manipulation). Stimulation is achieved via the insertion of electrodes and transmittal of electric current (figure III.2). Most of the time, researchers administer currents of low frequency (about 10 Hz) for many hours a day in order to imitate endurance exercise. To imitate resistance exercise they administer currents of high frequency (up to 100 Hz) for a short while or stretch a muscle in vivo. This is achieved either by immobilizing a limb in a stretched position or by implanting a special inflatable material. Alternatively, one may remove synergic muscles by surgery, thus increasing the load on the remaining muscles, or employ other inventive designs (figure III.3).

3. *Models of cultured muscle fibers.* Investigators have used electrical stimulation of muscle fibers **in vitro** (in the test tube) and stretching of muscle fibers in special containers covered with flexible silicone. Obviously, these models are even further removed from human exercise than those cited previously.

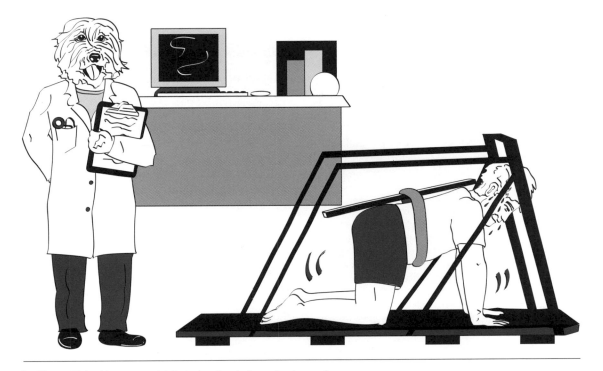

▶ **Figure III.1** Human model that closely mimics animal exercise.

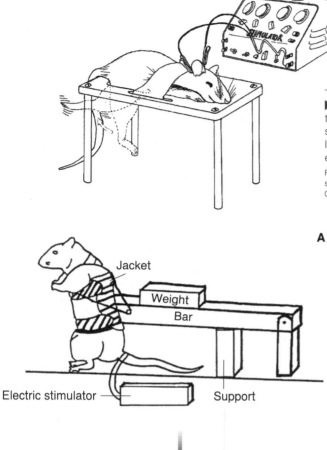

► **Figure III.2** Animal model of increased contractile activity that does not mimic human exercise. Electrodes implanted in the sciatic nerve and pulled up to the neck under the skin enable the leg muscles of a rat to contract when they receive current from an electric stimulator.

From K. Baar and K. Esser, 1999, "Phosphorylation of p70(S6K) correlates with increased skeletal muscle mass following resistance exercise," *American Journal of Physiology* 276: C120-C127. Used by permission of The American Physiological Society.

A

B

► **Figure III.3** Animal model of resistance exercise that does not mimic human exercise. An intelligent arrangement lets researchers study the effect of resistance exercise on metabolism. A rat wears a vest connected with a wooden rod carrying a weight. *(a)* The arrangement forces the animal to stand on its hind limbs without bearing any weight at rest because the bar leans on a support. *(b)* The animal bounces whenever it receives a slight electric shock through electrodes connecting its tail to a stimulator. By changing the weight and the program of administering the electric shocks, one can create a wide variety of resistance exercise programs.

European Journal of Applied Physiology, "Effect of resistance exercise training on mass, strength and turnover of bone in growing rats," T. Notomi et al., vol. 82: 268-274, 2000, © Springer-Verlag.

Note that the categories of models mentioned lie on one side of modified contractile activity, that is, increased contractile activity. There are also models of reduced contractile activity, as this too elicits changes in metabolism. Examples of such models are the interruption of training, the suspension of the hind limbs of an experimental animal (so that the muscles of these limbs do not receive the load of the body), and muscle denervation.

III.4   Five Means of Metabolic Control in Exercise

Mammals (including humans) have developed a multitude of mechanisms to modify their metabolism during exercise. Let's try to group these mechanisms.

Allosteric Regulation

Metabolic reactions are catalyzed by enzymes. A usual way of regulating the rate of enzyme reactions is through allostery (section 3.12). Allosteric regulation applies to exercise metabolism as well: The binding of a compound whose concentration increases with exercise to a site on an enzyme distinct from the active site may modify the enzyme's activity. If the activity increases, the compound is called an **activator**; if the activity decreases, the compound is called an **inhibitor.**

The modification of enzyme activity by an *effector* (that is, an activator or inhibitor) is instantaneous: It takes place within milliseconds. Moreover, it is reversible: If the effector concentration decreases, then its dissociation from the enzyme is favored and the enzyme regains its initial activity. We will encounter many cases of enzyme activation or inhibition by compounds whose concentrations change with exercise.

▶ **Figure III.4** Feedback inhibition. A usual way to control the rate of a metabolic pathway from compound A to compound Z is to inhibit the enzyme catalyzing the first irreversible reaction (which happens to be the second reaction of the pathway in this example) with the end product Z. This inhibition signals the production of Z to slow down when its concentration becomes exceedingly high.

A usual control point in a metabolic pathway is the enzyme catalyzing the first irreversible reaction. This reaction is termed the **committed step** because it commits the compound entering the pathway to a course of no return. A common inhibitor of the enzyme catalyzing the committed step is the product of the pathway (figure III.4). This effect is called **feedback inhibition.**

Covalent Modification

The activity of certain enzymes is controlled by the reversible addition of chemical groups to their molecules. This is **covalent modification** and takes place within seconds. The addition and subsequent removal of chemical groups are catalyzed by specific enzymes, which can be activated or inhibited by biochemical factors affected by exercise. The control of glycogen synthesis and breakdown by phosphorylation of the responsible enzymes, as discussed in chapter 9, is a case in point.

Changing Substrate Concentration

The increase or decrease in the concentration of a reaction's substrate during exercise may cause a similar change in reaction rate (section 3.16). For example, the increase in the amount of glucose entering the exercising muscles from the blood speeds up its use for energy supply.

Changing Enzyme Concentration

The quantity as well as activity of enzymes is controlled. Different kinds of exercise create biochemical stimuli that modify the concentrations of various enzymes. Cells often achieve this by altering the transcription rates of the genes encoding these enzymes. Changing enzyme concentration is the slowest means of regulation—it requires hours. Nevertheless, it is also the most long-lived, constituting the basis of adaptations to exercise. We will discuss the effect of exercise on gene expression in chapter 12.

Nervous and Hormonal Control

Metabolic regulation in animals and humans exceeds the boundaries of individual cells thanks to two large communication systems, the nervous and the hormonal systems. These systems convey signals to which cells bearing the appropriate receptors respond by altering their function. The route from a signal to a cellular response can be quite complicated, involving intricate molecular interactions and conversions, as we saw in part II and will also see in subsequent chapters. These routes are termed **signal transduction pathways.** We already dealt with the nervous system and its signals (action potentials and neurotransmitters) in chapter 6, so now it is proper to present the hormonal system.

Hormones are compounds that coordinate the functioning of different cells in multicellular organisms. Coordination enables an organism to respond to external or internal stimuli in a concerted manner and to avoid conflicting activities by individual tissues or cells. The many hormones known to this day display a great variety of structures and functions. Despite this variety, however, they share a number of features:

- They are synthesized in minuscule quantities in organs or tissues characterized as *endocrine glands.*
- They are secreted in the blood, which disseminates them to the rest of the body.
- They alter specifically the function of tissues, organs, or cells called *target tissues, target organs,* or *target cells.*

The term hormone is derived from the Greek *hormón,* meaning "the one who rushes." It thus implies a mobilization of the body, although some hormones repress bodily functions.

The nervous and hormonal systems intertwine and collaborate to afford the smooth working of a body in health. They also collaborate in the case of exercise and dictate the coordinated modification of metabolism in different tissues and organs such as muscle, the liver, and adipose tissue. In this way, for example, there is an increased flow of glucose from the liver to the exercising muscles and an increased breakdown of body fat. Both of these processes (which we will examine in detail in chapters 9 and 10) result in the supply of external energy sources to the exercising muscles.

III.5   Four Classes of Energy Sources in Exercise

As we saw in section III.1, exercise metabolism obeys the need for increased energy supply to the contracting muscles. Four classes of biological substances serve this need:

- Compounds of high phosphoryl transfer potential
- Carbohydrates
- Lipids
- Proteins

We will examine these classes in the next four chapters. Their metabolism for energy production is often referred to as *energy metabolism*. As we will see, there is almost no kind of exercise deriving its energy from one source only. On the contrary, two or more sources contribute, depending on a number of factors including the exercise parameters and the characteristics of the exercising organism.

Compounds of High Phosphoryl Transfer Potential

Compounds of high phosphoryl transfer potential are a small group of compounds sharing two features:

They bear a phosphoryl group.

Their hydrolysis releases a high amount of energy.

In accordance with what you learned in chapter 2, we can symbolize a compound of high phosphoryl transfer potential as $A \sim P$ and write the following equation:

$A{\sim}P + H_2O \rightarrow A + P_i$ $\qquad\qquad \Delta G^{o\prime} < 0$ $\qquad\qquad$ **(equation 8.1)**

In a variation of this equation, compound $A \sim P$ can transfer its phosphoryl group to a compound B, thus increasing the energy content of the latter.

$A{\sim}P + B \rightarrow A + B\text{-}P$ $\qquad\qquad \Delta G^{o\prime} < 0$ $\qquad\qquad$ **(equation 8.2)**

This justifies calling $A \sim P$ a compound of high phosphoryl transfer potential. The product B-P is usually not such a compound because part of the energy of $\sim P$ is lost to the surroundings. This is the price we pay for equation 8.2 to have a negative $\Delta G^{o\prime}$ and thus be favored. Nevertheless, $A \sim P$ may have such a high phosphoryl transfer potential that it endows B-P with a high potential too (thus producing $B \sim P$). This is the case with the production of ATP from creatine phosphate, as we will see in section 8.3.

We will consider three compounds of high phosphoryl transfer potential: ATP, ADP, and creatine phosphate. (AMP is not a compound of high phosphoryl transfer potential, as it lacks a $\sim P$.)

Compounds of high phosphoryl transfer potential are also known by the shorter but less accurate term **phosphagens.**

8.1 The ATP-ADP Cycle

In chapter 2, we considered ATP as the energy currency of cells. We saw that ATP hydrolysis to ADP feeds anabolism with energy, whereas the energy released during catabolism feeds ATP synthesis from ADP. It is now time to complete the picture of

127

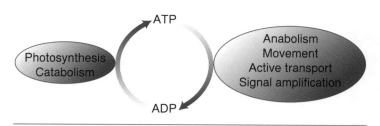

Figure 8.1 Energy exchange. The ATP-ADP cycle is the main route of energy exchange in biological systems. Shown in the ellipses are the processes driving the conversion of ADP to ATP and vice versa. This figure extends figure 2.5.

> "Signal amplifiers" inside us resemble the amplifier in our stereo, which amplifies weak signals such as the one from a laser beam reading a compact disc. This amplifier spends energy too (of course, electric energy, not ATP), as evidenced by the heat you will feel if you touch the device at the vents.

this interconversion by inserting four more biological processes: photosynthesis, movement, active transport, and signal amplification (figure 8.1).

Photosynthesis is a source of ATP for plant cells; it does not operate in animals, which have to rely solely on catabolism for ATP synthesis. Let's then turn our attention to the functions that degrade ATP.

- *Anabolism.* I have already mentioned (in section 2.4) the need to couple the endergonic biosynthetic reactions to the exergonic hydrolysis of ATP in order for anabolism to be thermodynamically favored.

- *Movement.* We have seen (in chapter 7) that movement in the form of muscle contraction is made possible by ATP hydrolysis in the myosin heads. ATP hydrolysis is also needed for other forms of movement in living organisms.

- *Active transport.* The transport of a substance against a concentration gradient requires ATP hydrolysis by proteins such as the Na^+-K^+ pump (section 6.2) and the Ca^{2+} pump (section 7.10).

- *Signal amplification.* Certain biological signals such as those initiated by hormones are despairingly weak because hormones and other signal molecules are synthesized in minuscule amounts. To affect metabolism the signals have to be "amplified." Signal amplification is performed by regulatory mechanisms operating at the expense of ATP. We will encounter such mechanisms when we examine the effect of exercise on carbohydrate metabolism, lipid metabolism, and gene expression.

ATP serves as the immediate source of energy for all the functions in the preceding list. As it is hydrolyzed to ADP and P_i, it is also regenerated mainly thanks to the catabolism of carbohydrates and lipids. Thus, ATP and ADP participate in a reciprocative process known as the **ATP-ADP cycle.** ATP has a very high turnover: An average human hydrolyzes and resynthesizes approximately 40 kg of it daily! Nevertheless, the amount of ATP in the body at any moment is a mere 0.1 kg. This would suffice for only a few minutes of operation were it not for the catabolic pathways that can replenish it. (By analogy, each of us consumes hundreds of kilograms of food each year, but we have only a few kilos stored at home at any moment.)

8.2   The ATP-ADP Cycle in Exercise

Exercise stimulates three of the four processes that require ATP: movement, active transport, and signal amplification. Stimulation of movement is self-evident. Stimulation of active transport consists mainly in the increased operation of the Na^+-K^+ pump in the plasma membrane of neuronal axons and muscle fibers participating in exercise, as well as the Ca^{2+} pump in the membrane of the sarcoplasmic reticulum. The increased operation of the Na^+-K^+ pump is due to the higher frequency of action potentials; this augments the perturbation of the $[Na^+]$ and $[K^+]$ gradients across the plasma membranes. The increased operation of the Ca^{2+} pump is due to the high $[Ca^{2+}]$ in the cytosol of contracting muscle fibers. Finally, signal amplification is stimulated because of enhanced secretion of certain hormones such as epinephrine (section 9.3). In contrast, most anabolic processes are inhibited during exercise, as we will see in chapters 9 through 11.

The net result of these changes is the acceleration of ATP breakdown. An athlete may degrade as many as 70 kg daily, while the rate of ATP hydrolysis during maximal exercise may rise to 0.5 kg per minute! But what is happening on the other side of the ATP-ADP cycle? Naturally, catabolism is activated to meet the increased demand for ATP. Thus, the whole ATP-ADP cycle is accelerated during exercise. This is why the cycle has acquired a permanent place in the logos of the International Biochemistry of Exercise Conference (figure 8.2).

▶ **Figure 8.2** Biochemistry of Exercise Conference. The ATP-ADP cycle adorns the logos of the International Biochemistry of Exercise Conference (IBEC) held around the world every three years. On the left, the logo of the 10th IBEC. Inside the cycle is a drawing of the famous Sydney Harbor Bridge and Opera House. In the middle is the logo of the 11th IBEC held in Little Rock, Arkansas. In this abstract version of the cycle, stars have replaced ATP and ADP. Inside the cycle is a drawing of the State Capitol. On the right is the logo of the 12th IBEC held in Maastricht, the Netherlands. The logo depicts the country outline and the ATP-ADP cycle with Maastricht at its center.

The ATP content of a skeletal muscle at rest, measured in biopsy samples, is about 6 mmol $\cdot$ kg^{-1}. As the muscle begins to contract, the cytosolic [ATP] decreases. It is estimated that ATP would vanish in about 3 s of maximal contraction—and that would be the end of contraction—in the absence of sources and processes ensuring its almost instantaneous replenishment. Obviously, such sources and processes are present, and we will start considering them in the next paragraph. Thanks to them, the decrease in ATP is limited. The largest decreases that have been reported are by approximately 50% after 30 s of maximal exercise. The longer the duration and the lower the intensity of exercise, the smaller the decrease in ATP because the processes that regenerate it have more time to balance its breakdown.

In contrast to the [ATP] that decreases, the [ADP] and [P$_i$] increase in exercising muscles. The [AMP] also increases thanks to the reaction

2 ADP $\leftrightharpoons$ AMP + ATP   **(equation 8.3)**

In this reaction, the terminal phosphoryl group of one ADP is transferred to another ADP, converting the former to AMP and the latter to ATP. Thus, we squeeze some extra energy out of ADP. To put it in numbers, one could write the equation 2 $\cdot$ 2 = 1 + 3 for the phosphoryl groups or 2 $\cdot$ 1 = 0 + 2 for the phosphoanhydride linkages. Reaction 8.3 (equation 8.3) is one of the fastest ways of resynthesizing ATP. It has a maximal rate of 0.9 mmol ATP per kilogram muscle per second, and it is anaerobic.

The enzyme catalyzing reaction 8.3 is **adenylate kinase. Kinases** are enzymes catalyzing the phosphorylation of compounds, with ATP serving as the phosphoryl donor. Kinases are specified by their substrates. Adenylate kinase is so named because if you reverse reaction 8.3 (remember, enzymes catalyze reactions both ways; section 3.15), you will see that ATP phosphorylates adenylate (synonym of AMP, section 2.3). Adenylate kinase is also known as **myokinase,** as it abounds in muscle.

8.3   Creatine Phosphate

Skeletal muscles contain considerable quantities of **creatine** and **creatine phosphate (CP),** or **phosphocreatine.** The constitutional formulas of these compounds appear in figure 8.3. Creatine is an amino acid but not one of the 20 amino acids that constitute proteins, although it is synthesized in the body from three of those amino acids (glycine, arginine, and methionine). Creatine phosphate is derived from the attachment of a phosphoryl group to one of creatine's nitrogens. A muscle at rest contains about 12 mmol creatine and 20 mmol creatine phosphate per kilogram.

Researchers often prefer to report the amount of ATP and other biomolecules in muscle per kilogram of dry muscle (denoted as dm, which happens to stand for both dry muscle and dry mass). Dry muscle is produced in the laboratory by desiccation of the natural, "wet" muscle. Because 1 kg of dry muscle is derived from about 4 kg of wet muscle, we convert the content of dry muscle to content of wet muscle by dividing by 4. Thus 24 mmol ATP $\cdot$ kg^{-1} dry muscle is 6 mmol ATP $\cdot$ kg^{-1} wet muscle. In this book, contents will refer to wet muscle, the natural form.

Although ADP is a compound of high phosphoryl transfer potential, it cannot support muscle contraction directly because myosin hydrolyzes only ATP. All other energy sources that we will consider (creatine phosphate, carbohydrates, lipids, and proteins) are also indirect: One way or another, they regenerate ATP.

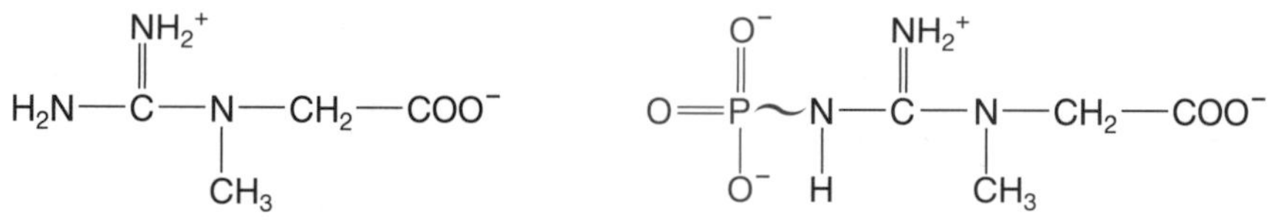

► **Figure 8.3** Creatine and creatine phosphate. These compounds aid ATP homeostasis in muscle through their interconversion.

Creatine phosphate is the fastest source for ATP resynthesis, rendering it a valuable energy source during maximal exercise. The reason is that creatine phosphate regenerates ATP from ADP in a single reaction:

$$CP + ADP + H^+ \rightleftharpoons ATP + C \qquad \Delta G^{\circ\prime} = -3 \text{ kcal} \cdot \text{mol}^{-1} \qquad \text{(equation 8.4)}$$

CP symbolizes creatine phosphate, and C symbolizes creatine.

The negative $\Delta G^{\circ\prime}$ of this reaction means that CP has a higher phosphoryl transfer potential than ATP. This is also clear from the fact that the $\Delta G^{\circ\prime}$ of CP hydrolysis is more negative than that of ATP hydrolysis.

$$CP + H_2O \rightleftharpoons C + P_i \qquad \Delta G^{\circ\prime} = -10.3 \text{ kcal} \cdot \text{mol}^{-1} \qquad \text{(equation 8.5)}$$

As the cytosolic [ATP] decreases and [ADP] increases at the onset of exercise, reaction 8.4 (equation 8.4) is shifted to the right. The reaction rate is high owing to the high concentration of the catalyzing enzyme. It is called **creatine kinase** (CK) because it catalyzes the phosphorylation of creatine by ATP (reverse reaction 8.4). CK in the cytosol consists of two subunits; each can be either of two isoforms denoted by B for brain and M for muscle. This variation yields three combinations:

CK-BB, or CK1, which dominates in the brain and smooth muscle

CK-MB, or CK2, which dominates in the heart

CK-MM, or CK3, which dominates in skeletal muscle

Part of CK-MM is bound to the M line thanks to its high affinity for myomesin and, to a lesser degree, M-protein (section 7.1). Thus, CK-MM is appropriately positioned close to the myosin heads, ensuring the rapid resynthesis of ATP from CP and ADP during contraction.

ATP regeneration by CP is an anaerobic process. Owing to the superiority of CP over ATP in terms of both concentration and phosphoryl transfer potential, the high activity of CK, and its proximity to the myosin heads, ATP is efficiently resynthesized in the first seconds of a maximal effort. The maximal rate of ATP resynthesis in human muscle is estimated to be about 2.6 mmol $\cdot$ kg^{-1} $\cdot$ s^{-1} and to be attained within 1 to 2 s of maximal contraction. Thus, as mentioned in the previous section, ATP suffers a rather small decline compared to its massive hydrolysis to support muscle contraction. By way of contrast, CP is decimated: Its concentration, from three to four times the ATP concentration at rest, may drop below the [ATP] after one-half minute of maximal exercise. An example of such changes is presented in figure 8.4.

Because its amount is rather limited, CP is the main source for ATP resynthesis (that is, it contributes more to ATP than does any other source) in maximal sporting activities lasting only a few seconds. Exactly how many seconds? This is hard to determine because measuring the contribution of all potential sources to energy production during very short periods of exercise is experimentally difficult. At any rate, research findings of recent years indicate that the dominance of CP does not exceed 7 s of maximal exercise.

The term creatine phosphokinase (abbreviated CPK) that is used sometimes in the literature is inaccurate, since the enzyme phosphorylates creatine, not creatine phosphate.

Thus, CP is the main source of energy in sports or events like weightlifting, jumps, throws, and 60 m sprint. It is also the main source of energy during short maximal efforts in intermittent events, such as the sprint of a football player or soccer player, the attack of a boxer, and a spike in volleyball.

Apart from CP breakdown during maximal exercise, of great interest is CP replenishment during and after exercise. Because this process requires first the resynthesis of ATP by aerobic mechanisms that we have not yet examined, we will defer the discussion of CP replenishment until we reach section 13.21, where we will also discuss the hot issue of creatine supplementation.

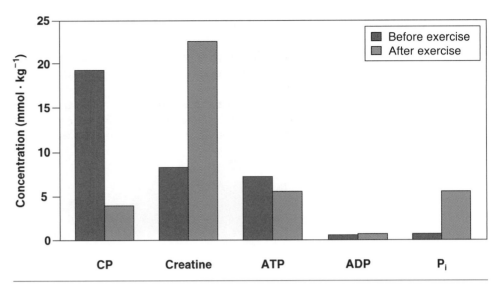

▶ **Figure 8.4** Changes in the concentrations of compounds related to energy production during maximal exercise. Creatine phosphate decreases by 80%, being converted to creatine, in human quadriceps muscle after 30 s of maximal cycling. ATP decreases slightly, while ADP and P_i increase. The graph was constructed based on data from Bogdanis and colleagues (1995).

8.4   A Window Into the Sarcoplasm

Measurement in small biopsy samples is the source of most of our knowledge of changes in the muscle concentrations of compounds related to energy production during exercise. An alternative way of studying these changes is through noninvasive imaging techniques such as *nuclear magnetic resonance spectroscopy*. This method is painless and does not destroy muscle fibers. Moreover, it allows us to watch muscle metabolism as it changes during exercise much as we watch the traffic outside our window.

Nuclear magnetic resonance, or NMR, is based on the magnetic properties of the nuclei of certain chemical elements, such as the nucleus of the natural isotope of phosphorus, [31]P. When these nuclei are placed in a strong magnetic field, they can jump from a low- to a high-energy state by absorbing electromagnetic radiation of an appropriate frequency (in the area of microwaves) emitted by a transmitter. We say then that the nuclei *resonate* with this frequency. The graph of the absorbed energy as a function of frequency is called an *NMR spectrum*.

The frequency with which a nucleus in a compound resonates depends primarily on the intensity of the applied magnetic field. Additionally, it depends on the minuscule magnetic field formed by the surrounding atoms in the compound. Thanks to this property, nuclei of the same element resonate with different frequencies if they belong to different compounds or even if they occupy different positions in the same compound. Thus, we can identify the compounds in a sample by examining the frequencies at which absorption peaks appear in the sample's NMR spectrum. We can even measure a compound from the height of its characteristic peak, height being proportional to concentration.

Our sample can be a solution, a piece of tissue, a small animal, or even a human inside a powerful cylindrical magnet. There is no drug or radioisotope injection, no ionizing radiation, and no blood sampling or muscle biopsy, and the subject feels nothing during the experiment or afterward. The subject may exercise inside the magnet; this lets researchers monitor changes in the concentrations of compounds in the subject's tissues during exercise. However, two drawbacks of the technique are the high cost of the equipment and the fact that it cannot detect abrupt concentration changes like the ones occurring during hard exercise. This is the case because data collection requires several minutes. By contrast, an experienced person can perform muscle biopsy within seconds.

The ^{31}P NMR spectra of a human muscle before and after exercise are presented in figure 8.5. Five major peaks project in this impressively simple reflection of a complex living tissue. Three of the peaks account mainly for the α, β, and γ phosphorus atoms

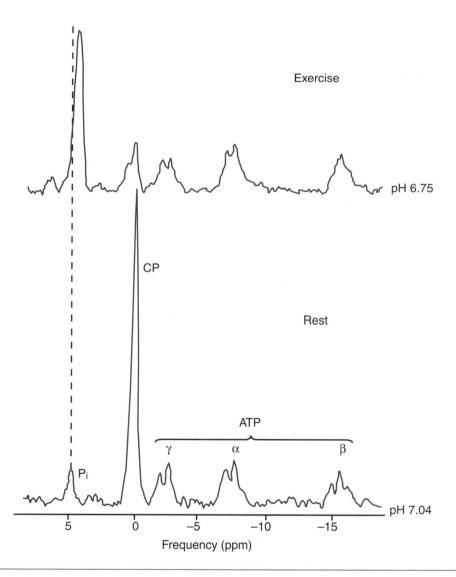

▶ **Figure 8.5** Studying exercise metabolism with NMR. Nuclear magnetic resonance affords the bloodless monitoring of metabolism in a human tissue. The ^{31}P NMR spectrum of the vastus medialis muscle shows spectacular—and inverse—changes in [CP] and [P$_i$] after 15 min of exercise on a horizontal step ergometer at 20%, 35%, and 45% of a person's maximal strength (5 min at each intensity in sequence). In contrast, the peaks corresponding to the α, β, and γ phosphorus atoms of ATP have remained practically unaltered. Note how much more CP there is compared to ATP at rest. The P$_i$ peak has moved slightly to the right after exercise (the dotted line shows its position at rest), which is a sign of a drop in pH. On the basis of the radio frequencies with which P$_i$ resonated at rest and after exercise, the researchers calculated that the pH decreased from 7.04 to 6.75. Frequency is expressed as deviation from the frequency of a reference compound and is measured in millionths (ppm, that is, parts per million) of the instrument's operating frequency.

Reprinted, by permission, from G. Bernus, J. Alonso, J.M. Gonzalez de Suso, and C. Arús, 1992, "^{31}P nuclear magnetic resonance spectra of human vastus medialis muscle at rest and exercise," *Sandoz Sport Research Project, 1988-1995.* Unpublished data.

of ATP (consult figure 2.3). The remaining peaks belong to the phosphorus atoms of CP and P_i.

The spectrum taken after exercise shows a dramatic drop in [CP] and an impressive rise in [P_i]. On the other hand, the [ATP] has not changed. Experiments like this let us watch "live" the movement of phosphoryl groups from one compound to another during exercise.

We can obtain additional information on muscle metabolism during exercise by studying the shift in the P_i peak along the frequency axis. The position of the peak is affected by the cytosolic pH because this determines the dominant ionic form of P_i. At physiological pH, hydrogen phosphate (HPO_4^{2-}, figure 2.4) is the dominant form. However, if protons are produced, formation of dihydrogen phosphate ($H_2PO_4^-$) is favored according to the reaction

$$HPO_4^{2-} + H^+ \leftrightharpoons H_2PO_4^- \qquad \text{(equation 8.6)}$$

Incidentally, the interconversion of the two phosphate ions constitutes an important buffer system in the body, the *phosphate system*. Now, because the P nucleus is surrounded by different atoms in HPO_4^{2-} and $H_2PO_4^-$, its resonance frequency depends on the proportion of the two ions and hence the cytosolic proton concentration.

The main reason for proton production in muscle during exercise is the anaerobic breakdown of carbohydrates to lactate. By using standard solutions of phosphate salts (that is, solutions of defined composition and pH), investigators are able to measure the drop in cytosolic pH and estimate lactate production from the shift in the P_i peak.

8.5   Loss of AMP by Deamination

In section 8.2, we saw that two ADP can yield one AMP and one ATP, thus providing some additional ATP for muscle contraction. AMP can next be converted to *inosinate* or *inosine monophosphate* (IMP) through the reaction

$$AMP + H_2O + H^+ \rightarrow IMP + NH_4^+ \qquad \text{(equation 8.7)}$$

This is a **deamination**, that is, loss of an amino group. The amino group appears in the products as ammonium ion (NH_4^+), which is the conjugate acid of the base ammonia (NH_3) and predominates at physiological pH. Reaction 8.7 (equation 8.7) is catalyzed by **adenylate deaminase**, an enzyme that is activated during hard, and maximal, exercise. The main reason for this activation is the decrease in cytosolic ATP.

ATP controls the activity of adenylate deaminase in an impressive way: The enzyme has a domain that binds ATP and a domain that binds to the myosin heavy chains (precisely, S2; section 7.4), as Hisatome and colleagues (1998) discovered. The presence of ATP in the first domain inhibits the binding to myosin. The enzyme then remains soluble in the cytosol and inactive. Over 90% of adenylate deaminase is in this state at rest. When ATP decreases during hard exercise, 50% to 60% of the enzyme loses its ATP and binds to S2. Binding activates adenylate deaminase and speeds up AMP deamination. Another possible mechanism of activation is the decrease in cytosolic pH as a result of the anaerobic breakdown of carbohydrates, since adenylate deaminase exhibits maximal activity at pH 6.5.

What is the significance of AMP deamination? The answer lies in the shift of reaction 8.3 (equation 8.3) to the right as one of the products (AMP) is eliminated. The shift increases the amount of the other product (ATP), thus speeding up its regeneration. We see here another way in which ATP ensures its homeostasis during hard exercise: When it abounds, adenylate deaminase is relatively inactive; when it decreases, adenylate deaminase is activated and indirectly speeds up ATP resynthesis (figure 8.6).

▶ Inosine monophosphate (IMP).

This benefit comes at a price, though: AMP deamination reduces the pool of adenine ribonucleotides (ATP, ADP, and AMP), thus making ATP regeneration during recovery from exercise more time- and energy-consuming, as we will see in section 13.21. It seems, however, that the gain from the extra supply of ATP under the highly demanding conditions of hard exercise outweighs the extra burden during recovery.

The positive effect of ATP depletion on adenylate deaminase activity causes an increase in IMP and ammonia concentrations in vigorously contracting muscles. Ammonia crosses the sarcolemma, appears in the extracellular fluid, and enters the bloodstream. IMP, in the cytosol of muscle fibers, may lose its phosphoryl group to become *inosine*. Inosine may then lose its ribose to become *hypoxanthine*. Hypoxanthine exits to the bloodstream and is taken up by the liver, where, by the action of *xanthine oxidase,* it suffers two successive oxidations to *xanthine* and *urate*. The latter is the product of purine degradation in humans (remember purines in section 4.3) and is excreted in the urine (hence the name).

Figure 8.7 recaps the series of conversions that we considered in the present chapter.

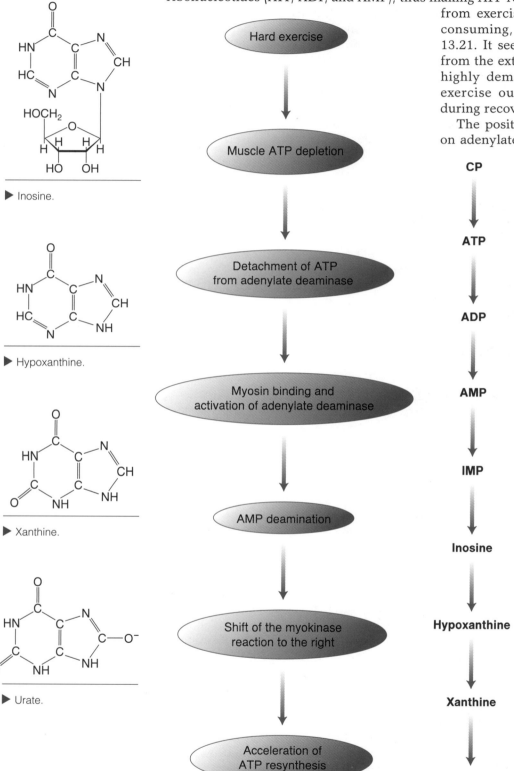

▶ Inosine.

▶ Hypoxanthine.

▶ Xanthine.

▶ Urate.

CP

ATP

ADP

AMP

IMP

Inosine

Hypoxanthine

Xanthine

Urate

▶ **Figure 8.6** A compound taking care of itself. The decrease in ATP during hard to maximal exercise signals the acceleration of its resynthesis by the myokinase reaction through the activation of adenylate deaminase.

▶ **Figure 8.7** Conversions of compounds of high phosphoryl transfer potential and their degradation products during exercise.

Problems and Critical Thinking Questions

1. How can one deduce equation 8.5 from equations 2.5 and 8.4?

2. The concentrations of ATP and CP in a muscle are 6 and 20 mmol · kg⁻¹, respectively. The amount of ATP suffices for 3 s of maximal contraction. Suggest probable concentrations for ATP and CP after 3 s of maximal contraction.

3. Treatment of a muscle with the compound fluoro-2,4-dinitrobenzene causes a rapid decline in [ATP], whereas [CP] does not change during a series of contractions. Propose an explanation.

4. Name three ways in which muscle fibers ensure ATP homeostasis during hard exercise.

5. What changes does hard exercise cause to the concentrations of ATP, ADP, AMP, CP, creatine, P_i, IMP, and ammonia in muscle?

CHAPTER 9

Carbohydrate Metabolism in Exercise

In chapter 5 we explored the variety (in terms of both structure and function) of animal and plant carbohydrates. In the present chapter our spectrum narrows: We will deal only with carbohydrates that are present in animal and human tissues. In particular, we will consider the involvement of glycogen, glucose, and their metabolites in energy production during exercise.

Glycogen is the most abundant animal carbohydrate. It can be easily broken down to glucose, which is present in the body in much smaller quantities. This relationship reflects the general tendency of living organisms to allocate much larger quantities to macromolecules than to their free monomers and readily break down the former to the latter in case of need. Indeed, we will encounter this tendency in lipids and proteins as well (figure 9.1). Getting back to carbohydrates, the degradation of glucose yields ATP in quantities and rates capable of supporting a wide variety of exercise tasks. This is why carbohydrates constitute the most precious energy source in the majority of sporting activities. In the following pages we will explore the different ways in which carbohydrates support an organism during exercise.

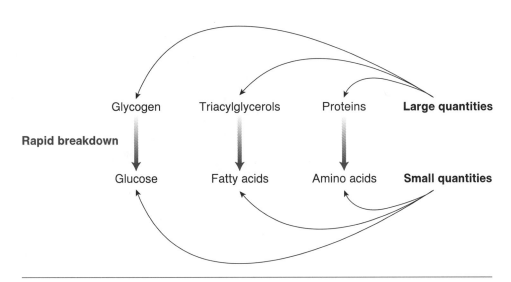

▶ **Figure 9.1** From depots to consumption. Living organisms contain large quantities of storage molecules but small quantities of their building blocks. However, the former can be broken down to the latter through simple reactions that make them available for immediate needs such as energy production for exercise.

9.1   Glycogen Metabolism

In scientific papers the glycogen content of tissues is often expressed as millimoles glucosyl units per kilogram tissue. To convert this to percentage content we must first multiply by the "molecular mass" of the glucosyl unit in glycogen, which depends on how a glucosyl unit is connected to its neighboring ones. Most often a glucosyl unit is connected to two other units, in which case it has a "molecular mass" of 162 Da. Thus, 80 mmol glucosyl units per kilogram muscle would be 12,960 mg or roughly 13 g per kilogram, or 1.3 g per 100 g, or 1.3% by weight.

Glycogen is stored mainly in the liver and muscles. The glycogen content of these organs is influenced considerably by diet and physical activity. If these are not extreme, glycogen ranges between 3% and 7% of the liver mass and between 1% and 1.5% of the muscle mass. How many grams do these percentages translate to? Let our model be a man (because more studies are conducted on men than women) weighing 70 kg (154 lb). His liver weighs roughly 1.8 kg (4 lb) and his muscles weigh roughly 28 kg (62 lb; 40% of body mass). If we adopt 5% as an average glycogen content of the liver and 1.25% as an average glycogen content of the muscles, you can calculate that this man will have 90 g (3.2 oz) in the liver and 350 g (12.3 oz) in the muscles—in all, 440 g (15.5 oz).

Glycogen is located in the cytosol in the form of granules that are 10 to 40 nm in diameter (figure 9.2). In addition to glycogen, the granules contain the enzymes catalyzing the reactions of its synthesis and breakdown, as well as enzymes controlling these processes (we will examine all these enzymes later). There is no glycogen in the blood.

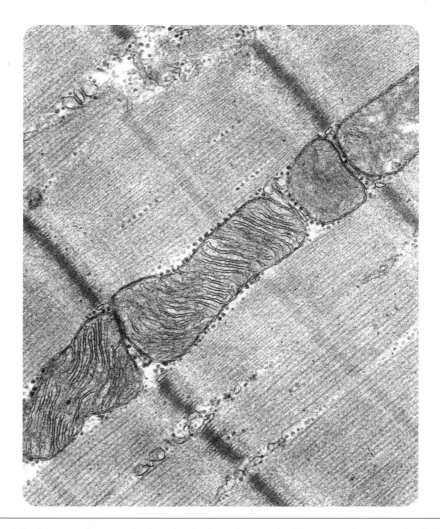

▶ **Figure 9.2**   Glycogen in a muscle fiber. An electron micrograph of the sarcoplasm reveals plenty of glycogen granules seen as dark dots squeezed between myofibrils, between myofibrils and mitochondria, and between mitochondria.

Courtesy of Dr. H. Hoppeler, University of Bern. Photographers H. Claassen and F. Grabe.

Carbohydrate Digestion

Where do the hepatic glycogen and muscle glycogen come from? Their major source is the dietary carbohydrates, that is, starch, sucrose, glucose, and so on. Starch is digested in the mouth and small intestine by the action of *α-amylase*, an enzyme secreted in the saliva and pancreatic juice. *α*-Amylase hydrolyzes the α1 → 4 glycosidic linkages of amylose and amylopectin (section 5.4), producing oligosaccharides like maltose, *maltotriose* (consisting of three glucosyl units in a row), and *α-dextrin* (consisting of several glucosyl units around a branch point in amylopectin). These oligosaccharides are further degraded to glucose by other digestive enzymes.

Sucrose is hydrolyzed to glucose and fructose by *sucrase,* an enzyme located on the outer surface of *microvilli* along with enzymes hydrolyzing other oligosaccharides. Microvilli (figure 9.3) are microscopic fingerlike protrusions of the cells lining the small intestine (the *epithelial cells*), which point toward the interior of the intestine (the *lumen*). Epithelial cells, in turn, cover larger fingerlike protrusions of the intestinal wall called *villi* (figure 9.4).

The monosaccharides—primarily glucose—produced from the digestion of dietary carbohydrates in the small intestine are carried to the cytosol of epithelial cells through transporters located

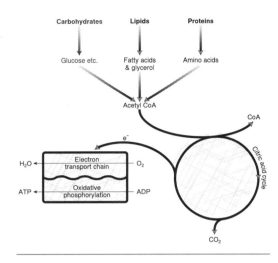

▶ You are here: carbohydrate digestion.

You are here.
A reduction of figure 2.10 (an overview of catabolism) will be displayed throughout part III, a different segment highlighted in color each time, to indicate the specific process discussed and to help you fit it into the overall

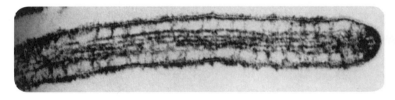

▶ **Figure 9.3** Microvillus. Electron micrograph of a microvillus shows part of the plasma membrane surrounding actin filaments. Microvilli increase the surface of epithelial cells in the intestine, affording a faster breakdown and absorption of food.

Reproduced from *The Journal of Cell Biology*, 1975, vol. 67, pg. 736, by copyright permission of The Rockefeller University Press.

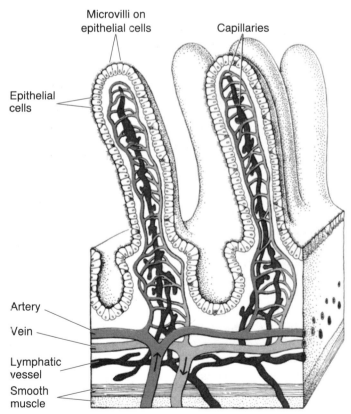

▶ **Figure 9.4** Villi. The inner surface of the intestine is covered by thousands of villi, five of which (two in cross section and three in the background) are depicted here. Villi are covered by a single layer of epithelial cells with tiny processes, the microvilli. The blood capillaries underneath the epithelial cells carry monosaccharides, amino acids, and salts from the food, whereas the lymphatic vessels carry lipids. The smooth muscle surrounding the intestinal tube (as every cavity in the body) contracts slowly to mix and propel the food.

Reprinted, by permission, from A.L. Lehninger, *Principles of biochemistry* (New York: Worth Publishers), 685.

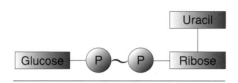

▶ Uridine diphosphate glucose (UDP-glucose).

in their plasma membrane. Finally, monosaccharides exit the epithelial cells toward the capillaries of the villi and are taken to the liver through the portal vein. The liver keeps a portion of the monosaccharides, mainly to synthesize glycogen, and releases the rest to the bloodstream in order to nourish the other organs in the body.

Glycogen Synthesis

Glucose entering the **hepatocytes** (liver cells) and muscle fibers participates in *glycogen synthesis*. This process consists of the successive addition of glucosyl units to a growing chain through the catalytic action of *glycogen synthase*. Being a biosynthetic process, chain elongation requires an input of energy, which comes from the hydrolyses of one ATP and one UTP (uridine triphosphate) per glucosyl unit added. Thanks to these hydrolyses, an activated form of glucose, *uridine diphosphate glucose*, or *UDP-glucose*, is synthesized. This then participates in the reaction of glycogen elongation.

UDP-glucose + glycogen (n glucosyl units) → glycogen (n + 1 glucosyl units) + UDP + H$^+$

(equation 9.1)

In this reaction the UDP unit, acting as a vehicle for glucose, adds it to a nonreducing end (section 5.4) of glycogen. The products are a glycogen molecule, longer by one glucosyl unit, and a UDP molecule.

Glycogen synthase is the leading enzyme in glycogen synthesis but it has a "drawback": It creates only linear chains. However, glycogen is branched. Another enzyme named simply *branching enzyme* detaches segments of the nascent chain from the nonreducing end and transfers them to the interior of the molecule, where it creates branches (figure 9.5).

Glycogenolysis

> The breakdown of a compound by P$_i$ is called *phosphorolysis*, in accordance with hydrolysis.

We turn now to the degradation of glycogen, termed **glycogenolysis.** In this process the $\alpha 1 \to 4$ glycosidic linkages of glycogen are broken down by P$_i$ according to the reaction

glycogen (n glucosyl units) + P$_i$ → glycogen (n − 1 glucosyl units) + glucose 1-phosphate

(equation 9.2)

As the reaction proceeds, glycogen loses, one by one, glucosyl units from its nonreducing ends. Thus molecules of *glucose 1-phosphate* (that is, glucose carrying a phosphoryl group at C1) are produced. The reaction is catalyzed by **glycogen phosphorylase**, or simply **phosphorylase.**

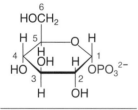

▶ Glucose 1-phosphate.

> In glycogenolysis, glycogen is degraded to glucose 1-phosphate and glucose.

> Glycogen synthesis and glycogenolysis are anaerobic processes.

Phosphorylase stars in glycogenolysis but is not sufficient for the completion of the process because it is unable to remove glucosyl units lying four or fewer places away from a branch point. When a chain becomes this short another enzyme, named simply *debranching enzyme,* comes into play. The debranching enzyme catalyzes two reactions (figure 9.6): the transport of three out of the four glucosyl units of a branch to the end of another branch and the removal of the remaining unit by hydrolysis of its particular ($\alpha 1 \to 6$) glycosidic linkage. The product of the latter reaction is plain glucose.

Thanks to the debranching enzyme, the branch is obliterated, letting phosphorylase take action once again. Molecules of glucose 1-phosphate are produced until the enzyme gets four glucosyl units away from the next branch. The debranching enzyme will eliminate it as already described, and glycogenolysis will continue through the concerted action of the two enzymes.

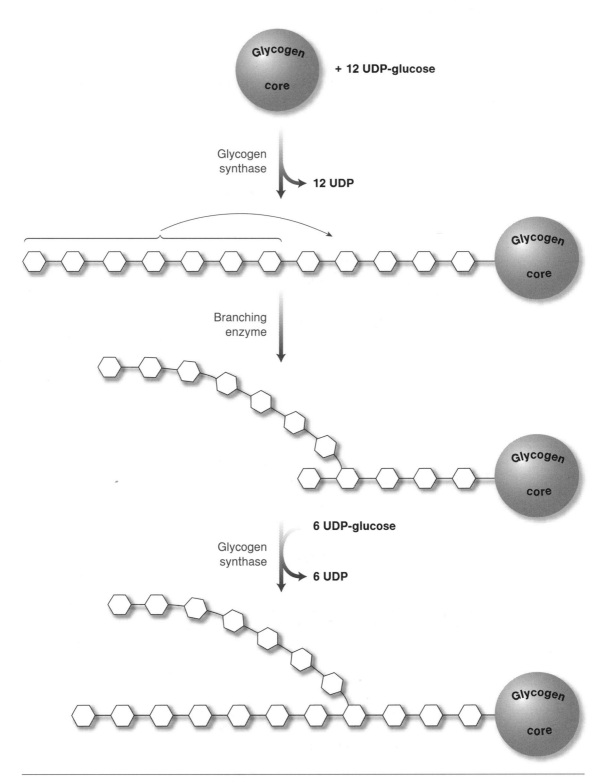

▶ **Figure 9.5** Glycogen synthesis. Two enzymes, glycogen synthase and the branching enzyme, cooperate in glycogen synthesis. The synthase adds glucosyl units (12 in our example) derived from UDP-glucose to a growing chain. The branching enzyme removes seven units and forms a branch connected to the original chain through an α1 → 6 glycosidic linkage. The synthase then resumes work.

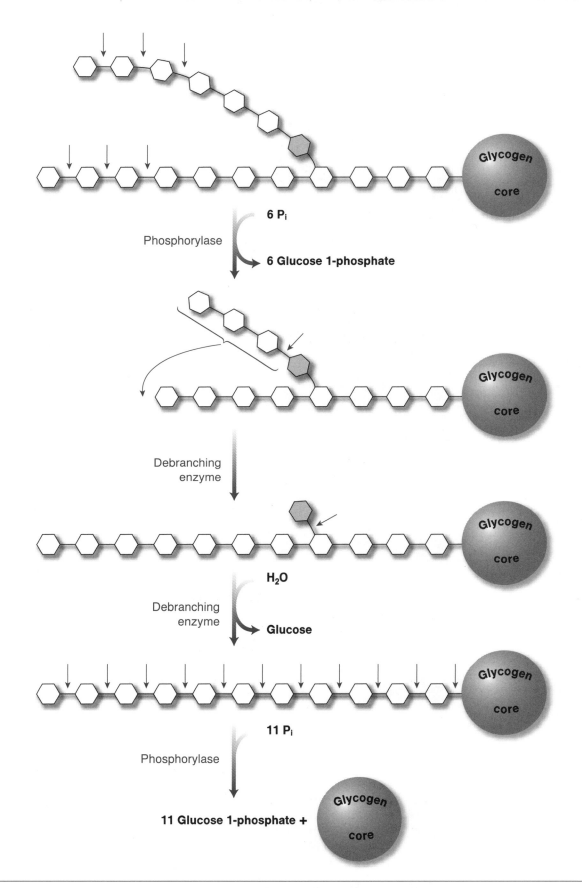

▶ **Figure 9.6** Glycogenolysis. Two enzymes, phosphorylase and the debranching enzyme, are needed to degrade glycogen. Starting with the product of figure 9.5, phosphorylase breaks down glycosidic linkages (vertical thin arrows) and detaches molecules of glucose 1-phosphate (six in our example) until it gets four glucosyl units away from a branch. Then the debranching enzyme (acting at the oblique arrow) carries three glucosyl units from one branch to another. The same enzyme hydrolyzes the glycosidic linkage at the branch point, leaving a linear chain at the disposal of phosphorylase. In the end, of the 18 glucosyl units that constituted the segment of glycogen shown at the top, 17 (the ones in white) are converted to glucose 1-phosphate, and one (in gray) is converted to glucose.

9.2   Exercise Speeds Up Glycogenolysis in Muscle

The rate of glycogenolysis is low in a resting muscle but increases because of changes in the concentration of certain compounds as the muscle begins to contract. The actions of these compounds fit into four of the five means of metabolic control in exercise that we discussed in section III.4: allosteric regulation, covalent modification, changing substrate concentration, and hormonal control. Thus, glycogenolysis offers a first-rate opportunity to apply the knowledge that you acquired in that section.

P_i Increase

The first change that speeds up glycogen breakdown during exercise is the increase in P_i as a result of ATP hydrolysis in the cytosol. Because P_i is a substrate in reaction 9.2 (equation 9.2), its rise speeds up glycogenolysis. In fact, it does so from the very first seconds of exercise.

The other changes that speed up glycogenolysis result in a direct or indirect activation of its leading enzyme, phosphorylase. To understand how this happens we need to consider some of the enzyme's properties. Phosphorylase consists of two identical 97 kDa subunits and exists in two interconvertible forms: a usually inactive *b* form and a usually active *a* form. The latter bears a phosphoryl group covalently attached to the side chain of serine 14 (14th amino acid of the polypeptide chain, counting from the free amino group) in each subunit. The phosphoryl group is derived from ATP and is attached to the hydroxyl group of serine by *phosphorylase kinase*. The resulting *phosphoserine* is a case of amino acid modification after protein synthesis (mentioned in section 3.1). Phosphorylase *b* is thus converted to phosphorylase *a* (figure 9.7).

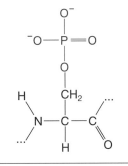

▶ Phosphoserine residue in a protein.

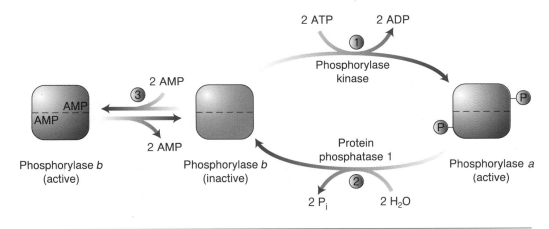

▶ **Figure 9.7**   Regulation of phosphorylase. Phosphorylase activity is controlled by reversible phosphorylation thanks to the action of a specific kinase (1) and a specific phosphatase (2), as well as by AMP binding (3). The active forms of phosphorylase are shown in color.

The reverse conversion (phosphorylase *a* to *b*) is achieved through a **dephosphorylation** (removal of a phosphoryl group) catalyzed by *protein phosphatase 1* (PP1). **Phosphatases** are enzymes catalyzing the dephosphorylation of compounds by hydrolysis. Like kinases, phosphatases are specified by their substrates. PP1 was so named because it was the first to be discovered among a group of enzymes that dephosphorylate proteins.

Following this acquaintance with the two forms of phosphorylase, we are ready to explore the changes that activate it during exercise.

AMP and IMP Increase; ATP Decrease

The first change is an increase in the proportion of AMP to ATP. AMP activates the (usually inactive) phosphorylase b by binding to a site in each subunit that is different from the active site (figure 9.7). Hence, AMP is an allosteric activator of phosphorylase. The binding of ATP to the same site inhibits the activation by AMP. Thus, the increase in AMP and decrease in ATP during exercise lead to an activation of phosphorylase b. Phosphorylase b is also activated by IMP, which increases during hard exercise (section 8.5).

Ca²⁺ Increase

Another change that affects phosphorylase activity is the increase in the cytosolic $[Ca^{2+}]$ because of Ca^{2+} release from the sarcoplasmic reticulum during muscle contraction (section 7.10). Ca^{2+} binds to phosphorylase kinase—in particular, to one of its subunits called **calmodulin**. Calmodulin is a 17 kDa protein resembling TnC, the calcium sensor in the thin filaments (section 7.9). Ca^{2+} binding changes the conformation of calmodulin (as it does with TnC) and activates phosphorylase kinase when the $[Ca^{2+}]$ reaches 10^{-6} mol $\cdot$ L^{-1}, a concentration well within the range of cytosolic $[Ca^{2+}]$ during contraction (figure 7.14). The activated kinase phosphorylates phosphorylase b and converts it to phosphorylase a as already described.

Epinephrine Increase

Finally, and importantly, phosphorylase is activated by the increase in the blood concentration of the hormone epinephrine. This effect is mediated by a series of molecular interactions termed the cyclic-AMP cascade and constituting the culmination of complexity in the control of glycogenolysis. The cascade also controls other processes that will concern us later. For these reasons it warrants treatment in a separate section.

9.3   The Cyclic-AMP Cascade

Epinephrine, or **adrenaline** (figure 9.8), is a relatively small compound belonging to the group of **catecholamines**. **Norepinephrine**, or **noradrenaline**, a closely related compound, belongs to this group too. The two compounds are derived from the amino acid tyrosine and are secreted by the *adrenal medulla*, that is, the interior of the adrenal glands, which are two small organs located one on top of each kidney. Additionally, the two compounds are synthesized in neurons of the sympathetic part of the autonomic nervous system and are released from their terminal branches. Here norepinephrine predominates. Thus, there is a division of labor between the two catecholamines: Epinephrine acts primarily as a hormone, whereas norepinephrine acts primarily as a neurotransmitter. The plasma concentration of norepinephrine is higher than that of epinephrine.

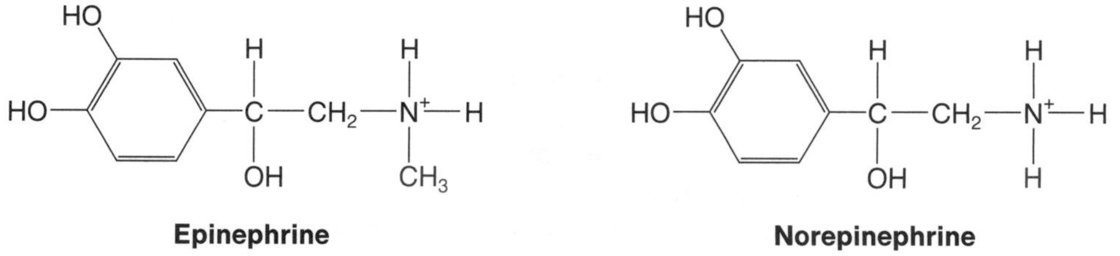

▶ **Figure 9.8** Epinephrine and norepinephrine. The two compounds differ by one methyl group only.

Exercise—or even its anticipation—excites the sympathetic system, which, among other organs, innervates the adrenals. Sympathetic excitation stimulates the adrenals to secrete catecholamines to the bloodstream. In fact, the plasma concentrations of epinephrine and norepinephrine relate to the intensity of exercise. Epinephrine molecules are transported through the circulation to the surface of muscle fibers, where they bind to specific receptors integrated in the sarcolemma (just like the acetylcholine receptor). These receptors are termed **adrenergic** (after adrenaline) and are divided into types of different function and different specificity for epinephrine and norepinephrine. The type participating in the cyclic-AMP cascade is characterized as β.

The β-adrenergic receptor has a molecular mass of 64 kDa and has a binding site for epinephrine on its extracellular surface. Binding of the hormone changes the receptor's conformation and activates the enzyme **adenylate cyclase.** This is another integral protein of the plasma membrane, having a molecular mass of 120 kDa and catalyzing the synthesis of **cyclic adenylate,** or **cyclic AMP** (cAMP, figure 9.9). Cylic AMP is omnipresent in eukaryotic cells and holds a central position in the control of many biological processes. It is synthesized from ATP with the concomitant production of pyrophosphate.

$$ATP \rightleftharpoons cAMP + PP_i \qquad \Delta G°' = 1.6 \text{ kcal} \cdot mol^{-1} \qquad \textbf{(equation 9.3)}$$

The reaction is slightly endergonic, thus not favored. However, it is followed by the hydrolysis of PP_i (equation 2.7), which has a $\Delta G°'$ of –4.6 kcal · mol⁻¹, thus making the overall process exergonic.

The activation of adenylate cyclase by the β-adrenergic receptor is not direct. Rather, it is mediated by a *G protein* belonging to a family of proteins involved in signal transduction. The one mediating the activation of adenylate cyclase is denoted by G_s (s stands for stimulatory). G_s is attached to the cytosolic side of the plasma membrane and consists of three subunits, α, β, and γ. The α subunit has a binding site for the guanine ribonucleotides GTP and GDP (hence the name G protein). When the receptor is vacant, $G_{s\alpha}$ is occupied mainly by GDP and is inactive. However, the binding of epinephrine to the receptor causes the attachment of the receptor to G_s (figure 9.10), the substitution of GTP for GDP in $G_{s\alpha}$, and the separation of $G_{s\alpha}$ from the other subunits, which remain united as $G_{s\beta\gamma}$. The now free $G_{s\alpha}$ diffuses across the membrane, binds to adenylate cyclase, and activates it.

Because of their close relationship, the sympathetic system and the adrenals are often referred to as the *sympathoadrenergic system.*

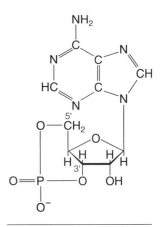

▶ **Figure 9.9** Cyclic AMP. Cyclic AMP differs from AMP (figure 2.3) in that its phosphoryl group is connected with not only the 5' but also the 3' carbon of ribose, thus forming one more ring (hence the term cyclic). This small structural difference endows cAMP with a biological role that is completely different from that of AMP.

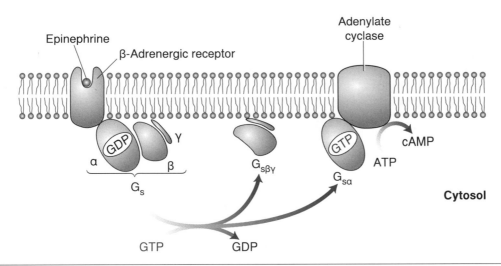

▶ **Figure 9.10** The beginning of the cAMP cascade. The binding of epinephrine to the β-adrenergic receptor in the plasma membrane causes the replacement of GDP by GTP in the α subunit of G_s protein and the detachment of the α from the β and γ subunits. Next, $G_{s\alpha}$ binds to adenylate cyclase and activates it to form cAMP.

Because of these events, cAMP increases in the cytosol of muscle fibers during exercise. Cyclic AMP in turn binds to and activates a *protein kinase* specified by the letter A and abbreviated as PKA for distinction from similar enzymes (protein kinases phophorylate proteins at the expense of ATP). In the absence of bound cAMP, PKA is composed of two pairs of identical subunits. Two subunits (38 kDa each) are catalytic and are denoted by C, while the other two (49 kDa each) are regulatory and are denoted by R. The resulting tetramer, R_2C_2, is inactive. However, when the [cAMP] rises to 10 nmol · L^{-1}, two cAMP molecules bind to each R, dismantling R_2C_2 to R_2 and two C (figure 9.11).

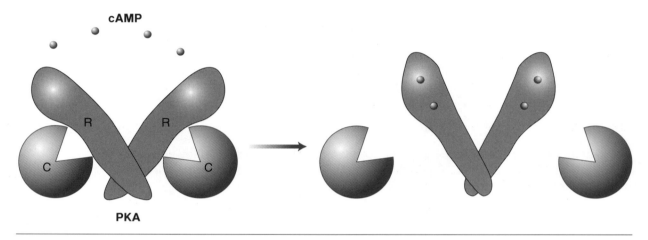

▶ **Figure 9.11**   Activation of protein kinase A. The binding of cAMP to PKA is key to the stimulation of the cAMP cascade. When four cAMP molecules bind to the two regulatory subunits (R), the catalytic subunits (C) are detached and can phosphorylate target proteins.

The liberated catalytic subunits of PKA are now active and phosphorylate cellular proteins of appropriate amino acid sequences. Phosphoryl groups are attached to side chains of serine or threonine. One of the substrates of PKA is phosphorylase kinase. Phosphorylation activates the enzyme; in fact, it acts in an additive manner to the activation caused by Ca^{2+} binding, which we discussed in the previous section. Phosphorylase kinase then phosphorylates and activates glycogen phosphorylase.

This concludes a "cascade" of molecular interactions (figure 9.12) that begins with the binding of epinephrine to the β-adrenergic receptor, involves cAMP as a key molecule, and ends in the acceleration of glycogenolysis. Called the **cAMP cascade**, it accounts for other actions of epinephrine as well, two of which we will explore in sections 9.20 and 10.2. Additionally, the cAMP cascade mediates the action of a large number of other hormones. Discovered half a century ago, it is the best-characterized signal transduction pathway. Cyclic AMP is usually termed a second messenger, the hormone being the first messenger.

Why are there so many steps in this control mechanism? Remember that enzymes are very efficient catalysts. Therefore, one enzyme molecule participating in a step of the cAMP cascade catalyzes the formation of many product molecules, which amplify its effect. Successive amplifications are multiplied, resulting in the release of an enormous number of glucosyl units from glycogen by just a few epinephrine molecules.

On the side of drawbacks, epinephrine does not speed up glycogenolysis as fast as the substances discussed in the previous section do. The main reason is that it takes several minutes for exercise to raise the plasma epinephrine concentration. Additionally, the amplification of the hormone signal is not free of charge, since the cell spends ATP for both cAMP synthesis and enzyme phosphorylation. This corroborates the position of signal amplification in figure 8.1.

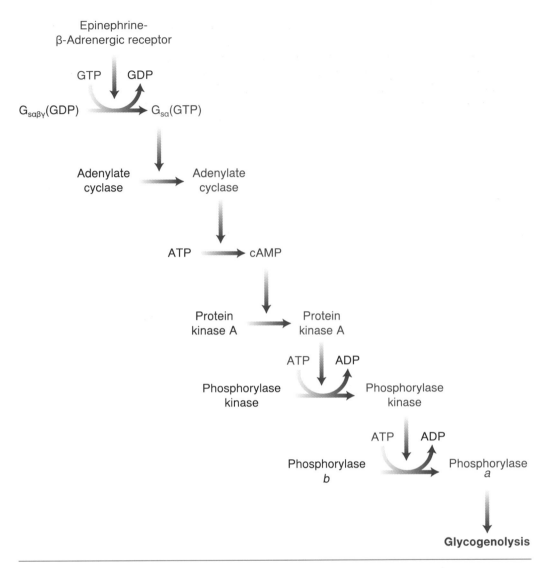

▶ **Figure 9.12**   Control of glycogenolysis by the cAMP cascade. Thanks to a series of sequential interactions, epinephrine leads to a great acceleration of glycogenolysis by binding to the β-adrenergic receptor. The hormone–receptor complex activates $G_{s\alpha}$, which activates adenylate cyclase, which synthesizes cAMP, which activates protein kinase A, which phosphorylates and activates phosphorylase kinase, which, finally, phosphorylates and activates phosphorylase. In this and subsequent signal transduction pathways the more active forms of proteins are depicted in color.

In addition to phosphorylase kinase, PKA phosphorylates glycogen synthase. However, phosphorylation of the enzyme triggers its transition from an active *a* to an inactive *b* form. Thus, glycogen synthesis is inhibited during exercise.

9.4 Recapping the Effect of Exercise on Muscle Glycogen Metabolism

In the last two sections we witnessed the control of a single reaction, the basic one of glycogenolysis (equation 9.2), by a plethora of different factors. The existence of such tight control must have given you an idea of how important glycogenolysis is. Indeed, reaction 9.2 places a valuable energy source at the disposal of the cell while weakening its resources. It is therefore necessary not only to avoid any purposeless breakdown of glycogen but also to ensure its rapid and massive mobilization in case of need. As far as the latter point is concerned, it is interesting to note that the rate of glycogenolysis in human quadriceps muscle during maximal exercise (about 1 mmol glucose per kilogram per second) approaches the V_{max} of phosphorylase.

All these aspects justify the complexity of the regulatory mechanisms that we have examined. Figure 9.13 attempts to moderate this complexity.

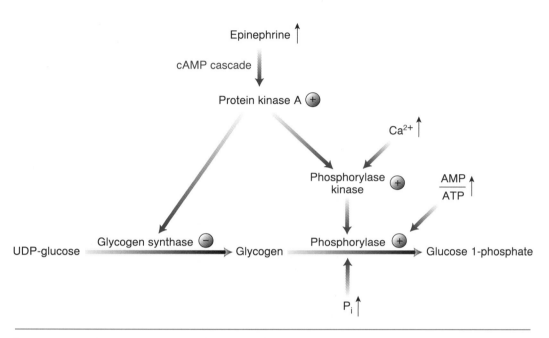

▶ **Figure 9.13** Control of muscle glycogen metabolism in exercise. Exercise speeds up glycogenolysis and slows down glycogen synthesis in muscle through cooperating mechanisms. In this figure and subsequent ones that summarize regulatory mechanisms, the arrows next to compounds denote increase (↑) or decrease (↓) in their concentration during exercise, while the symbols + and – in circles denote enzyme activation and inhibition, respectively.

9.5   Glycolysis

As we saw in section 9.1, glycogenolysis produces glucose 1-phosphate and, to a lesser extent, glucose. Cells also contain glucose originating in the diet or resulting from synthesis from noncarbohydrate sources (see section 9.19). Glucose and glucose 1-phosphate can yield energy through glycolysis. Glycolysis is the breakdown of glucose to **pyruvate;** it is the most common metabolic pathway and operates in all cells. In fact, it is the major energy source for certain cells under certain conditions. Such is the case of intensely contracting muscle fibers. What is more, glycolysis is the sole energy source for erythrocytes.

Glycolysis is an anaerobic process. It consists of 10 reactions taking place in the cytosol and catalyzed by different enzymes. The full sequence of the pathway is shown in figure 9.14. In brief, glucose undergoes two phosphorylations at the expense of two ATP in the first and third reactions, along with an isomerization in between, to become *fructose 1,6-bisphosphate,* a compound of

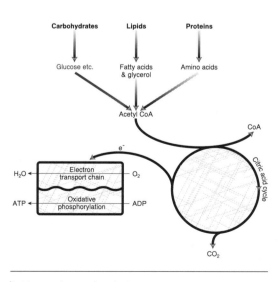

▶ You are here: glycolysis.

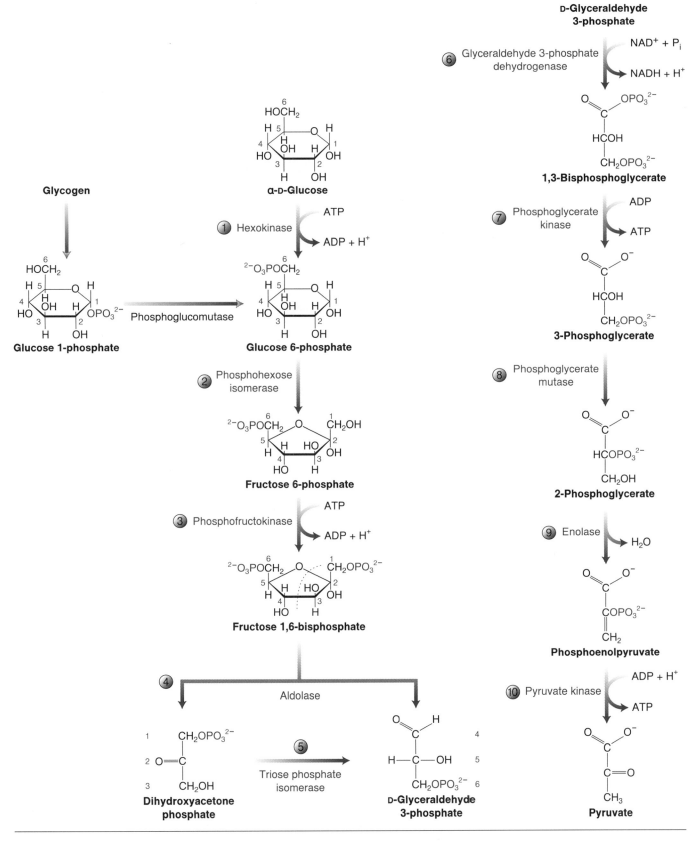

▶ **Figure 9.14** Glycolysis. A pathway of fundamental importance to carbohydrate metabolism, glycolysis transforms glucose and glucose 1-phosphate (the main product of glycogenolysis) to pyruvate while resynthesizing ATP from ADP. The broken line across fructose 1,6-bisphosphate, the product of the third reaction, shows how it is split; and the numbers next to the triose phosphates, the products of the fourth reaction, are the numbers of fructose 1,6-bisphosphate's carbons.

One might wonder why fructose carrying two phosphoryl groups is a *bis*phosphate, whereas adenosine carrying two phosphoryl groups is a *di*phosphate. The prefix bis- is used to describe two groups attached to a molecule at different positions, whereas di- is reserved for two groups attached in a row to a single position of a molecule.

six carbons and two phosphoryl groups. In the fourth reaction this compound is split into *dihydroxyacetone phosphate* and *glyceraldehyde 3-phosphate,* two isomeric triose phosphates (cf. figure 5.1). The former of these is converted to the latter in the fifth reaction; thus, glycolysis continues with two molecules of glyceraldehyde 3-phosphate per molecule of glucose.

In the sixth reaction of glycolysis, glyceraldehyde 3-phosphate accepts a second phosphoryl group, this time from P_i, while being oxidized. NAD^+ (section 2.5) serves as oxidant, being reduced to NADH. This is an important reduction, as we will see in subsequent sections. Finally, *1,3-bisphosphoglycerate,* the product of the sixth reaction, donates its two phosphoryl groups to two ADP, thus converting them to two ATP, in the seventh and (after two more conversions) 10th reactions.

How much energy is produced in glycolysis? Two ATP are spent initially and two ATP are produced toward the end. However, the latter come from one glyceraldehyde 3-phosphate, while the breakdown of glucose yields two triose phosphates, one of which is converted to the other. This means that one glucose yields $2 \cdot 2 = 4$ ATP in the second half of glycolysis. Therefore the net gain from glycolysis is $4 - 2 = 2$ ATP per glucose. Let me point out in advance that this quantity is minimal compared to what pyruvate can yield in the ensuing steps of its catabolism.

Glucose 1-phosphate, the main product of glycogenolysis, enters the glycolytic pathway (figure 9.14) after being isomerized to glucose 6-phosphate (the product of the first glycolytic reaction) through a shift of its phosphoryl group from C1 to C6. In this way glucose 1-phosphate avoids the expense of one ATP in the first glycolytic reaction. Thus, the gain from the conversion of glucose 1-phosphate to pyruvate is three ATP.

9.6   Exercise Speeds Up Glycolysis in Muscle

Exercise can augment the glycolytic rate in a muscle by hundreds of times and by more than one mechanism.

Increased Substrates

First, substrate availability increases. As glycogenolysis is accelerated, there is a rise in the concentration of glucose 6-phosphate, which is substrate to the second glycolytic reaction. Furthermore, the exercising muscles take up more glucose from the blood, mainly because of two factors. The first one is the enhanced blood flow to the active muscles (up to 20 times the flow at rest). This is a characteristic corollary of exercise, although how it happens is not known with certainty.

The second factor responsible for the increased glucose uptake by the exercising muscles is an increase in the number of glucose transporters in the plasma membrane. Let's see what this means and how it happens. Cells take up glucose through integral proteins of the plasma membrane called simply *glucose transporters* and abbreviated as GLUT. Muscle fibers contain several forms of these proteins, which perform passive transport of glucose to the cytosol. **GLUT4,** the most abundant form, is distinct in that it is not always present in the plasma membrane. Instead, it commutes between a population of intracellular vesicles and the sarcolemma or the transverse tubule membrane (figure 9.15), as Ploug and colleagues (1998) showed. Remember that the transverse tubules are an extension of the sarcolemma (section 7.10). For brevity, in the ensuing discussion I will refer to the two membranes as the plasma membrane.

The movement of GLUT4 from the intracellular vesicles to the plasma membrane provides a means of controlling glucose uptake by a muscle fiber: The more GLUT4 molecules present in the plasma membrane, the more glucose will enter the cell.

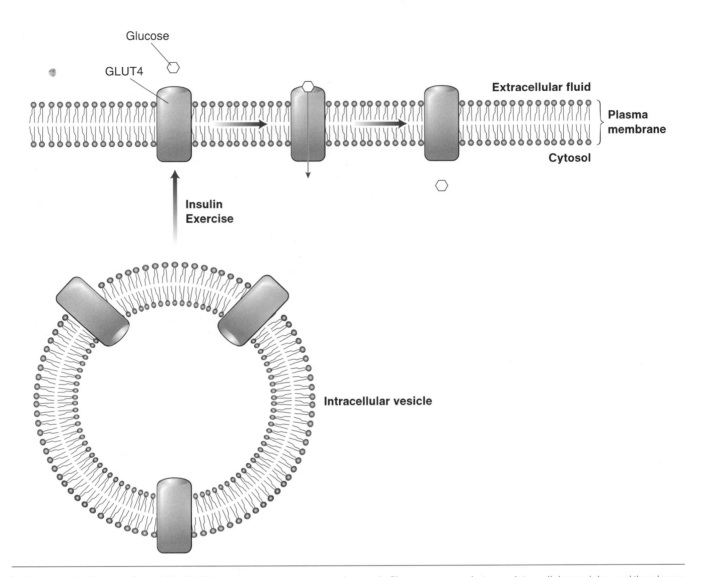

▶ **Figure 9.15** The shuttling of GLUT4. The main glucose transporter in muscle fibers commutes between intracellular vesicles and the plasma membrane. The vesicle membrane fuses with the plasma membrane in much the same way as neurotransmitter-filled vesicles fuse with the plasma membrane during the chemical transmission of nerve signals. Only when present in the plasma membrane can GLUT4 transport glucose from the extracellular fluid to the cytosol. Exercise and the hormone insulin stimulate the movement of the vesicles to the plasma membrane.

Exercise augments the movement of GLUT4 to the plasma membrane, resulting in higher glucose uptake. The biochemical mechanism of this event is not known, but it appears to be triggered by Ca^{2+} release from the sarcoplasmic reticulum. Note that the hormone insulin (which we will know better in section 9.23) also augments glucose uptake by muscle fibers through the movement of GLUT4 to the plasma membrane. However, this movement is by a different mechanism, which is known to some degree.

Phosphofructokinase Activation

The second way in which exercise speeds up glycolysis is through the allosteric regulation of **phosphofructokinase,** the enzyme catalyzing the third reaction. Phosphofructokinase is inhibited by ATP. The inhibition is enhanced by CP and relieved by AMP. This means that the enzyme is relatively inactive in a resting muscle, in which

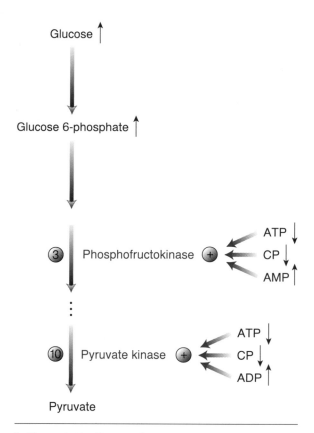

▶ **Figure 9.16** Control of glycolysis. Exercise speeds up glycolysis in muscle because of the concentration changes shown.

the [ATP] and [CP] are high, while the [AMP] is low. However, when the [ATP] and [CP] begin to decrease at the onset of exercise, whereas the [AMP] increases, phosphofructokinase is activated.

Another activator of phosphofructokinase is ammonia, whose muscle concentration increases during exercise because of AMP deamination (section 8.5) and amino acid deamination (as we will see in chapter 11). However, ammonia does not seem to play an important role in the regulation of phosphofructokinase. Finally, the enzyme is inhibited at acidic pH. As shown in figure 8.5 and explained further in section 9.16, the muscle pH becomes acidic during exercise because of the anaerobic breakdown of carbohydrates, most of which is glycolysis. It has been proposed that a decrease in glycolytic rate because of phosphofructokinase inhibition at low pH may protect the muscle fibers and the blood (to which H^+ diffuses) from an excessive and hazardous fall in pH. However, no phosphofructokinase inhibition by acidic pH has been found in vivo. It seems that other compounds produced in an exercising muscle (like AMP and ammonia) compensate for the inhibitory action of acidity.

Pyruvate Kinase Activation

Finally, of interest is the regulation of *pyruvate kinase*, the enzyme catalyzing the last reaction of glycolysis. ATP and CP inhibit pyruvate kinase but ADP activates it. The changes in the concentrations of the three compounds during exercise favor the activation of the enzyme.

The main factors contributing to the acceleration of glycolysis during exercise are summarized in figure 9.16.

9.7   Pyruvate Oxidation

Pyruvate can yield much more ATP than is produced in glycolysis. To do this, it has to pass from the cytosol to the mitochondria, where most of the biological oxidations and ATP resynthesis take place.

Mitochondria are elliptical organelles, approximately 2 by 0.5 μm in size, wrapped in a double membrane (figure 9.17). The outer membrane has pores that let most metabolites pass freely. The inner membrane is the actual permeability barrier separating the mitochondrial interior, termed the *matrix*, from the cytosol. The inner membrane has many folds called *cristae*, which greatly increase its area. This is important, since the electron transport chain and oxidative phosphorylation, the two processes introduced in section 2.6 that account for the synthesis of most of our ATP, take place exactly in the inner mitochondrial membrane.

Upon entering the mitochondrial matrix, pyruvate reacts with coenzyme A (CoA) to yield acetyl coenzyme A. We met the two compounds during the overview of catabolism (section 2.6), but now it is necessary to know more about them.

CoA is a complex compound in whose molecule one can discern adenine, ribose, phosphoryl groups, *pantothenate* (a vita-

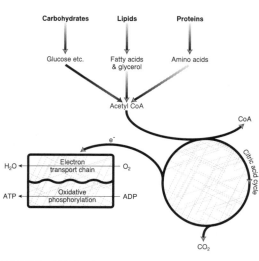

▶ You are here: pyruvate oxidation.

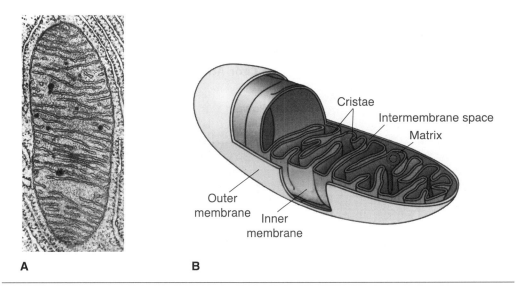

▶ **Figure 9.17** Mitochondria. Mitochondria are units of aerobic energy production in eukaryotic cells. *(a)* Electron micrograph of a longitudinally sliced mitochondrion shows its two membranes and the deep folds of the inner membrane. The organelle is surrounded by endoplasmic reticulum studded with ribosomes. *(b)* A diagram of a cutaway mitochondrion shows the two membranes better. The inner membrane is more important than the outer one because it is less permeable and because it houses the proteins of the electron transport chain and oxidative phosphorylation.

Figure 9.17a Courtesy of Dr. H. Hoppeler, University of Bern. Photographers H. Claassen and F. Grabe.

Figure 9.17b Reprinted, by permission, from J.M. Berg, J.L. Tymoczko, L. Stryer, 2002, *Biochemistry,* 5th ed. (New York: W.H. Freeman/Worth Publishers), 492.

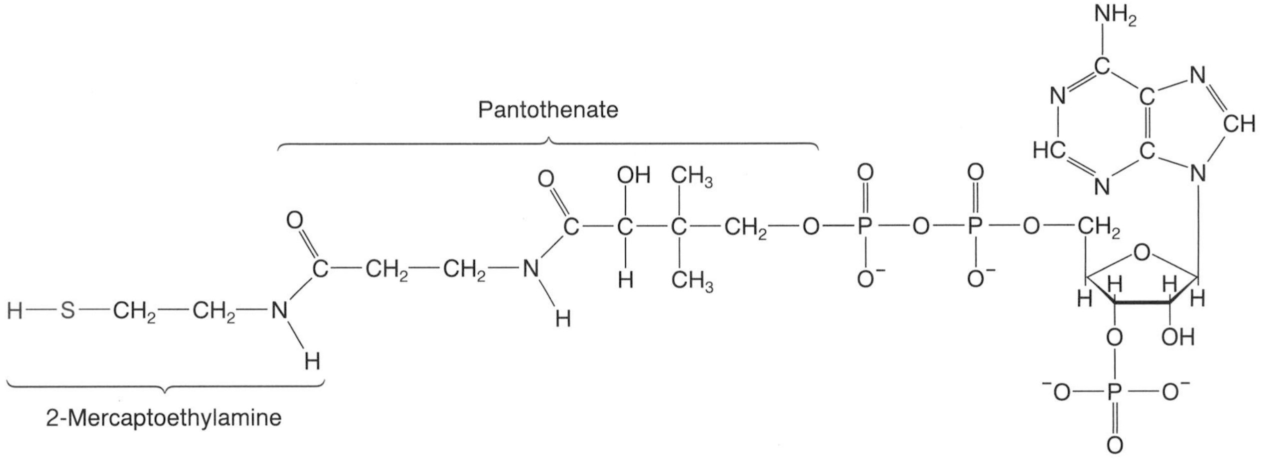

▶ **Figure 9.18** Coenzyme A. CoA stars in carbohydrate, lipid, and protein metabolism thanks to its ability to carry acyl groups. CoA bears a terminal sulfhydryl group (in color), through which it bonds with acyl groups after losing a H.

min), and *2-mercaptoethylamine* (figure 9.18). A *sulfhydryl group* (—SH) at the end of the molecule is the reactive part of CoA. This is why it is symbolized as CoA-SH in biochemical equations. The sulfhydryl group serves as a carrier of the acetyl group or acyl groups in general (section 5.7), which are attached to S through the replacement of H.

Acetyl CoA results from the attachment of an acetyl group to CoA. The acetyl group is what remains from pyruvate after its carboxyl group is removed as carbon dioxide, a process called **decarboxylation.** At the same time pyruvate is oxidized by NAD^+. The whole reaction is

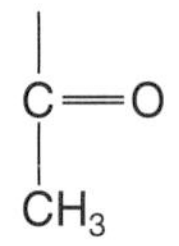

▶ The acetyl group is a very small acyl group.

$$\underset{\textbf{Pyruvate}}{\overset{\displaystyle COO^-}{\underset{\displaystyle CH_3}{C\!=\!O}}} + CoA\text{-}SH + NAD^+ \longrightarrow \underset{\textbf{Acetyl CoA}}{\overset{\displaystyle S\text{-}CoA}{\underset{\displaystyle CH_3}{C\!=\!O}}} + CO_2 + NADH \quad \Delta G^{\circ'} = -8 \text{ kcal} \cdot \text{mol}^{-1}$$

(equation 9.4)

Pyruvate oxidation is irreversible (as evidenced by the very negative $\Delta G^{\circ'}$ value of the reaction); therefore, it is impossible to convert acetyl CoA back to pyruvate. Because acetyl CoA subsequently enters a route of aerobic breakdown, reaction 9.4 (equation 9.4) commits carbohydrates to aerobic catabolism. An indication of the importance of reaction 9.4 is the fact that it is catalyzed by no fewer than three enzymes forming the **pyruvate dehydrogenase complex,** a huge assembly of dozens of polypeptide chains.

9.8 Exercise Speeds Up Pyruvate Oxidation in Muscle

Because it catalyzes such an important reaction, the pyruvate dehydrogenase complex is under strict control. Key to this control is the reversible phosphorylation of pyruvate dehydrogenase, the enzyme that has given its name to the entire complex. Phosphorylation by a specific kinase inhibits pyruvate dehydrogenase, whereas dephosphorylation by a specific phosphatase restores its activity (figure 9.19). At rest, most of pyruvate dehydrogenase is phosphorylated and relatively inactive (b form).

The situation is reversed at the onset of exercise, as pyruvate dehydrogenase phosphatase is activated and pyruvate dehydrogenase kinase is inhibited. The phosphatase is activated by Ca^{2+} released from the sarcoplasmic reticulum. It is also activated by Mg^{2+}, the concentration of which, in free form, increases. This happens because ATP, which binds Mg^{2+} (section 7.5), decreases, whereas the ADP produced does not bind Mg^{2+} that tightly. On the other hand, the kinase is inhibited by pyruvate, which increases because of increased glycolytic rate. In addition, pyruvate speeds up reaction 9.4 as a substrate. The bottom line of all these effects is the increase in the dephosphorylated, active, a form of pyruvate dehydrogenase.

Pyruvate dehydrogenase kinase is under three more regulatory influences: It is activated by high concentration ratios of ATP to ADP, NADH to NAD$^+$, and acetyl CoA to CoA. [ATP]/[ADP] decreases during exercise and contributes to the inhibition of the kinase. [NADH]/[NAD$^+$] does not change in a firm way. The relevant studies show that the ratio increases during the first minutes of hard exercise and then decreases even below resting levels. If exercise is of lower intensity, the ratio does not change appreciably. The rise in the ratio is due to the conversion of NAD$^+$ to NADH in glycolysis, pyruvate oxidation, and (as we will see in the next section) the citric acid cycle. In contrast, the fall in the ratio is due to the oxidation of NADH to NAD$^+$ in the electron transport chain (as we will see in section 9.11). As for [acetyl CoA]/[CoA], it generally

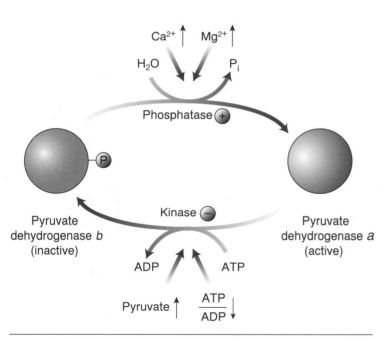

▶ **Figure 9.19** Control of pyruvate dehydrogenase. The enzyme is present in either a phosphorylated or a dephosphorylated form thanks to the action of pyruvate dehydrogenase kinase and pyruvate dehydrogenase phosphatase. During exercise, changes in the concentrations of substances affecting the activity of the kinase or the phosphatase lead to activation of pyruvate dehydrogenase.

increases during exercise because of acetyl CoA production from reaction 9.4 and fatty acid oxidation (section 10.4).

Thus, the changes in the [NADH]/[NAD⁺] ratio during exercise seem to favor the slowing down or speeding up of pyruvate oxidation depending on the exercise conditions, while the changes in the [acetyl CoA]/[CoA] ratio seem to favor the slowing down of pyruvate oxidation. However, in reality, these changes do not seem to affect the activity of pyruvate dehydrogenase very much.

9.9   The Citric Acid Cycle

The citric acid cycle is a series of nine enzyme reactions that process the acetyl group of acetyl CoA derived from the oxidation of carbohydrates, lipids, and proteins. The acetyl group is finally oxidized to carbon dioxide, while the metabolites of the cycle are restored after a full turn. Several of these metabolites serve as precursors for the synthesis of other biomolecules. Finally, a lot of energy is released in the process.

Let us take a closer look at the cycle (figure 9.20). In its first reaction the acetyl group is linked with **oxaloacetate**, a four-carbon compound, to form **citrate**, a six-carbon compound that has given its name to the entire pathway. The pathway is also known as the **Krebs cycle**, after Hans Krebs, who discovered it in 1937. It is also referred to as the **tricarboxylic acid cycle**, since citrate and the products of the second and third reactions, cis-*aconitate* and *isocitrate*, bear three carboxyl groups each.

> "... whatever returns is good, not what passes and is done with. The easiest way to return from where you've been without retracing your steps is to walk in a circle."—Umberto Eco, *Foucault's Pendulum*

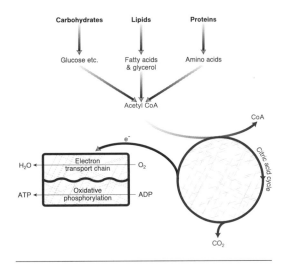

> ▶ You are here: citric acid cycle.

The most interesting events start from the fourth reaction of the cycle. In this and the following reaction, two carbons depart as two CO_2, while isocitrate is oxidized to α-*ketoglutarate* and, next, α-ketoglutarate is oxidized to *succinyl CoA*. NAD⁺ serves as the oxidant in both reactions. In the sixth reaction, GTP is synthesized from GDP and P_i, while in the seventh and ninth reactions, two more oxidations take place, the oxidants being FAD and NAD⁺, respectively. Oxaloacetate is regenerated as the product of the ninth reaction. All reactions of the cycle take place in the mitochondria.

How would we sum up the citric acid cycle? Let's recap the accompanying net changes.

An acetyl group is converted to two CO_2.

Three NAD⁺ are converted to three NADH.

One FAD is converted to $FADH_2$.

One GDP is converted to GTP.

> In the citric acid cycle, the acetyl group of acetyl CoA is oxidized to CO_2.

> The term "cycle" is used metaphorically in biochemistry to denote metabolic pathways that end at the compound they start from. The term does not imply any circular arrangement of metabolites or enzymes.

GTP is equivalent to ATP. In addition, the two compounds are easily interconverted according to the reaction

GTP + ADP ⇌ GDP + ATP $\Delta G^{o\prime} = 0$ **(equation 9.5)**

Is then one ATP the entire energy yield of the cycle? No, the energy produced is much greater, but it is hidden in NADH and $FADH_2$. We will reveal it shortly as we consider the processes of the electron transport chain and oxidative phosphorylation.

No oxygen is directly involved in the citric acid cycle. However, the NAD⁺ and FAD that are consumed in its reactions can be regenerated inside the mitochondria (to be used again as oxidants) only if NADH and $FADH_2$ transfer their electrons to O_2 through the electron transport chain. Thus, oxygen is indirectly required for the operation of the citric acid cycle, which is therefore considered an aerobic pathway.

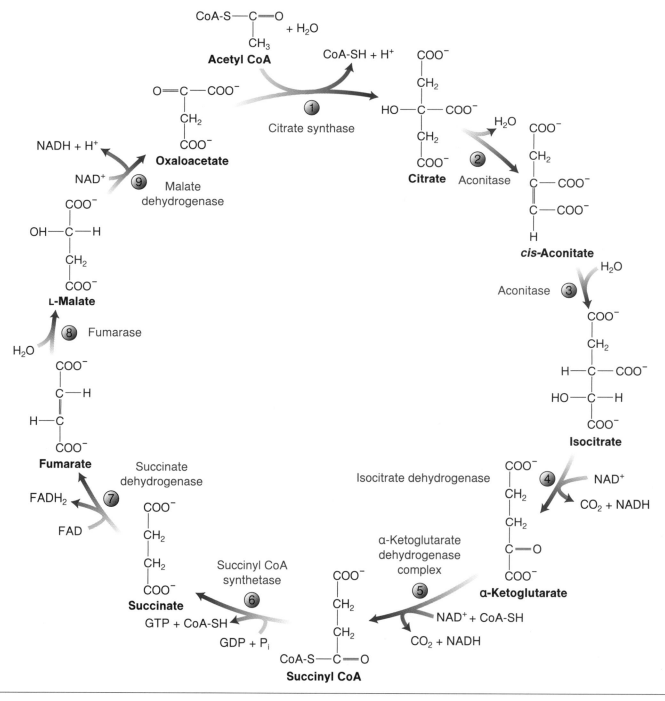

▶ **Figure 9.20** The citric acid cycle. The cycle is a metabolic pathway ending in the very compound it starts from, oxaloacetate. The acetyl group coming from the catabolism of carbohydrates, lipids, and proteins enters the cycle as acetyl CoA and is finally oxidized to carbon dioxide, producing energy.

9.10   Exercise Speeds Up the Citric Acid Cycle in Muscle

The rate of acetyl group oxidation through the citric acid cycle may go up by as much as 100 times in muscle during hard exercise. This increase is primarily due to the rise in acetyl CoA concentration because of the accelerated pyruvate oxidation. In addition, the cycle is accelerated during exercise through the allosteric regulation of three of its

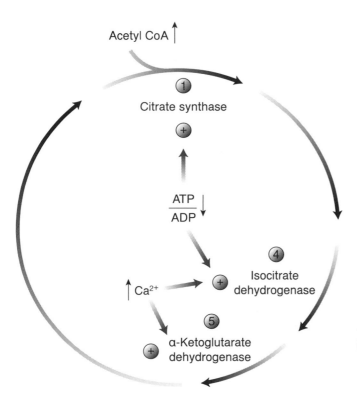

▶ **Figure 9.21** Control of the citric acid cycle. Exercise speeds up the cycle in muscle by increasing the substrate (acetyl CoA) concentration, decreasing the [ATP]/[ADP] ratio, and increasing the Ca^{2+} concentration.

enzymes. These are *citrate synthase, isocitrate dehydrogenase,* and *α-ketoglutarate dehydrogenase* (part of the *α-ketoglutarate dehydrogenase complex*), the enzymes catalyzing the first, fourth, and fifth reactions, respectively. The former two are inhibited by a high [ATP]/[ADP] ratio. The drop in this ratio during exercise activates the enzymes. Isocitrate dehydrogenase and α-ketoglutarate dehydrogenase are activated by Ca^{2+}, whose mitochondrial concentration increases with prolonged exercise. The two enzymes are also inhibited by NADH.

The regulation of the citric acid cycle in muscle during exercise is summarized in figure 9.21.

9.11   The Electron Transport Chain

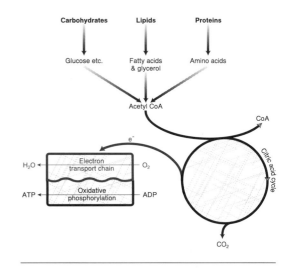

▶ You are here: electron transport chain.

As mentioned in previous sections, the NADH and $FADH_2$ that are produced in the mitochondria during pyruvate oxidation and the citric acid cycle can be oxidized to NAD^+ and FAD by transferring their electrons (along with hydrogens) to oxygen. This is accomplished through the oxidation–reduction reactions

$$NADH + H^+ + 1/2 \ O_2 \rightarrow NAD^+ + H_2O \qquad \Delta G^{\circ\prime} = -52.6 \ \text{kcal} \cdot \text{mol}^{-1} \qquad \textbf{(equation 9.6)}$$

$$FADH_2 + 1/2 \ O_2 \rightarrow FAD + H_2O \qquad \Delta G^{\circ\prime} = -48 \ \text{kcal} \cdot \text{mol}^{-1} \qquad \textbf{(equation 9.7)}$$

In the electron transport chain, electrons flow from NADH and $FADH_2$ to O_2, yielding NAD^+, FAD, and H_2O.

Electrons are not transferred in one step but through an intricate system consisting of three large protein complexes and two mobile electron carriers. The protein complexes are *NADH-Q oxidoreductase, Q-cytochrome c oxidoreductase,* and *cytochrome c oxidase.* The mobile carriers are a lipid-like compound named *ubiquinone,* or *coenzyme Q* or just *Q,* and *cytochrome c,* a small protein. The whole system is called the electron

The electron transport chain is part of the relay race depicted in figure 2.9. It's just that more runners (the components of the chain) intervene between NADH or FADH$_2$ and O$_2$, and an electron pair (rather than hydrogens) is the baton. Be a sportscaster of the event!

transport chain, or **respiratory chain,** since the final destination of the electrons is respiratory oxygen that is carried to the cells by the blood. The components of the electron transport chain are located in the inner mitochondrial membrane. Several of these components contain heme as their prosthetic group, underlining the suitability of this compound for handling oxygen.

A simplified diagram of the electron transport chain is shown in figure 9.22. A pair of electrons from NADH is transferred initially to NADH-Q oxidoreductase, whereas a pair of electrons from FADH$_2$ is transferred initially to ubiquinone. Then the electrons leap from one component of the chain to another until they reach O$_2$. Addition of an electron pair to any one of the components of the chain reduces it, but the subsequent removal of the electron pair restores the component to its oxidized form so that it can accept a new electron pair. In the end, the only net changes are the oxidation of NADH and FADH$_2$ to NAD$^+$ and FAD, as well as the reduction of O$_2$ to two H$_2$O by the addition of four H$^+$ and two electron pairs.

Cells gain two things from the electron transport chain:

- They regenerate NAD$^+$ and FAD.
- They produce high amounts of energy (as evidenced by the extremely negative $\Delta G°'$ values of reactions 9.6 and 9.7 [equations 9.6 and 9.7]), part of which is then utilized to regenerate ATP in oxidative phosphorylation.

Through the electron transport chain, cells can also regenerate the cytosolic NAD$^+$ that is reduced to NADH in the sixth reaction of glycolysis (section 9.5). This regeneration is not as simple as it seems because NADH cannot cross the inner mitochondrial membrane. There are, however, compounds in the cytosol that accept an electron pair from NADH (thus converting it to NAD$^+$), travel to the mitochondria, and transfer the electron pair to either NAD$^+$ or FAD (thus converting them to NADH and FADH$_2$, respectively). The electron carriers themselves, now devoid of the electron pair, return to the cytosol, where they can take part in electron transport again. These compounds are called *molecular shuttles,* as they go back and forth between the cytosol and the mitochondria.

As a side reaction in the final step of the electron transport chain, some oxygen molecules receive one rather than four electrons, thus giving rise to the superoxide radical introduced in section 1.4. O$_2\cdot^-$ and other radicals produced from it or through other routes tend to remove electrons from compounds in their vicinity in order to pair them with their unpaired electrons. In doing so, they may damage DNA, proteins, and lipids, thus causing disease or speeding up aging. On the other hand, they may destroy invading pathogens or trigger signal transduction pathways. Radical production increases during exercise and is therefore attracting the attention of many researchers in exercise science.

9.12 Oxidative Phosphorylation

Eukaryotic cells have developed a way of channeling part of the energy released in the electron transport chain to ATP synthesis. ATP synthesis is carried out according to the familiar endergonic reaction

ADP + P$_i$ + H$^+$ ⇌ ATP + H$_2$O $\Delta G°'$ = 7.3 kcal · mol^{-1} **(equation 9.8)**

Because this process involves the phosphorylation of ADP with the aid of the energy liberated from the oxidation of NADH and FADH$_2$, we call

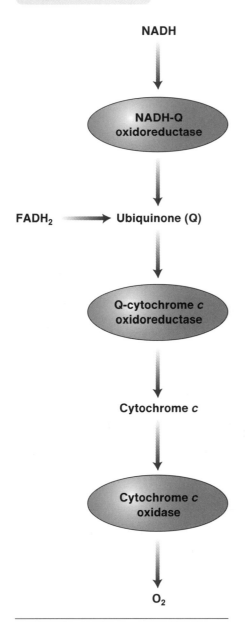

▶ **Figure 9.22** Electron transport chain. The electron transport chain consists of three protein complexes (gray ellipses) and two mobile electron carriers in between. Electrons from NADH and FADH$_2$ are initially transferred to different components of the chain and then flow along it toward their final destination, oxygen.

it oxidative phosphorylation. It is an aerobic process, as it depends on oxygen utilization in the electron transport chain. ATP synthesis is catalyzed by **ATP synthase** (also known as F_0F_1 ATPase), an integral protein of the inner mitochondrial membrane.

The active site of ATP synthase is located in the mitochondrial matrix; therefore, this is where ATP is synthesized. ATP then passes to the intermembrane space with the help of another integral protein of the inner mitochondrial membrane, the *ATP-ADP antiporter*. The protein is so called because for every ATP that it exports, it imports one ADP. This ensures that the ADP produced from ATP hydrolysis in the cytosol is "recharged" in the mitochondria. Both ATP and ADP cross the outer mitochondrial membrane freely. Finally, one more integral protein of the inner mitochondrial membrane, the *phosphate translocase,* imports the phosphate needed for ATP synthesis along with a proton.

> In oxidative phosphorylation, ATP is synthesized from ADP and P_i, powered by the energy of the electron transport chain.

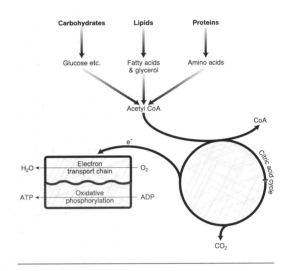

▶ You are here: oxidative phosphorylation.

9.13 Energy Yield of the Electron Transport Chain

Because the electron transport chain and oxidative phosphorylation are distinct processes, the question arises as to how the energy produced by the former is channeled to the latter. This has been one of the most tantalizing problems in biochemistry. The prevalent view today is expressed by the **chemiosmotic hypothesis,** proposed in 1961 by Peter Mitchell, which states that *the link between the electron transport chain and oxidative phosphorylation is protons that are expelled from the mitochondrion as the electron transport chain operates, only to return through ATP synthase.*

Let's take a closer look at the chemiosmotic hypothesis (figure 9.23). As an electron pair from NADH or $FADH_2$ runs along the protein complexes of the electron transport chain, each complex uses part of the energy produced to actively transport H^+ from the mitochondrial matrix to the intermembrane space. Thus, a surplus of H^+ builds up in the intermembrane space, and a deficit of H^+ remains in the mitochondrial matrix. This $[H^+]$ gradient across the inner mitochondrial membrane and the resulting electrical potential difference (negative in the matrix relative to outside) force the protons

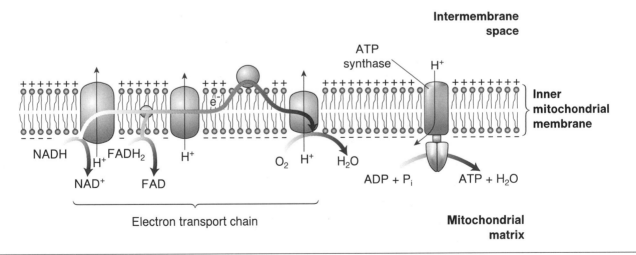

▶ **Figure 9.23** Chemiosmotic hypothesis. This hypothesis explains the coupling of the electron transport chain to oxidative phosphorylation by postulating that protons driven out of the mitochondrial matrix, as electrons flow toward oxygen, return through the ATP synthase and cause the linking of ADP and P_i to form ATP. The five components of the electron transport chain have been placed in the order shown in the previous figure.

back to the matrix. The protons enter through ATP synthase, and the energy of their flow fuels ATP synthesis.

How much ATP is generated by oxidative phosphorylation with the energy of the electron transport chain? Since the motive force for ATP synthesis is the influx of H^+ to the mitochondrion, we need to know two things in order to answer the question.

- How many H^+ does each of the three protein complexes of the electron transport chain drive out of the mitochondrial matrix as an electron pair from NADH or $FADH_2$ flows along the chain?
- How many H^+ need to pass through ATP synthase for one ATP to be formed?

We are currently not certain about these numbers because they are not necessarily constant (they may depend on the energy state of the cell). The best available estimates about the numbers of protons expelled by the three protein complexes of the electron transport chain are four, two, and four in sequence. On the other hand, the synthesis of one ATP requires the passage of approximately three H^+ through ATP synthase. In addition, as mentioned in the previous section, one H^+ enters the mitochondrial matrix with phosphate through the phosphate translocase. This has to be added to the cost of synthesizing ATP. Thus, a total of four H^+ are needed to generate one ATP.

On the basis of the figures just given, we can conclude that the $4 + 2 + 4 = 10\ H^+$ removed from the mitochondrial matrix during the oxidation of one NADH produce about $10/4 = 2.5$ ATP. The gain from $FADH_2$ oxidation is less. Because it transfers its electrons to ubiquinone rather than NADH-Q oxidoreductase, its oxidation results in the removal of only $2 + 4 = 6\ H^+$ from the mitochondrial matrix. Thus, approximately $6/4 = 1.5$ ATP are formed as one $FADH_2$ is oxidized.

9.14   Energy Yield of Carbohydrate Oxidation

Starting anaerobically (through glycogenolysis and glycolysis) and continuing aerobically (through pyruvate oxidation, the citric acid cycle, the electron transport chain, and oxidative phosphorylation), carbohydrate oxidation has come to an end. With the aid of figure 9.24, let's recap what we get per glucose molecule.

- Glycolysis yields two pyruvates, two ATP, and two NADH (one of each per glyceraldehyde 3-phosphate, section 9.5). To yield ATP, the two cytosolic NADH have to transfer their electrons to either mitochondrial NAD^+ or mitochondrial FAD by means of molecular shuttles (section 9.11). When the former happens, one cytosolic NADH gives rise to one mitochondrial NADH; when the latter happens, one cytosolic NADH gives rise to one mitochondrial $FADH_2$. The former seems to prevail in the heart and liver, the latter in skeletal muscle. Thus, two NADH in the cytosol of muscle fibers are equivalent to two $FADH_2$ in the mitochondria.
- Oxidation of the two pyruvates to two acetyl CoA (in the mitochondrial matrix) yields two CO_2 and two NADH.
- Oxidation of the two acetyl CoA in the citric acid cycle yields four CO_2, six NADH, two $FADH_2$, and two GTP. The latter are equivalent to two ATP.

Summing up, the oxidation of one glucose in muscle yields six CO_2, four ATP, eight NADH, and four $FADH_2$. The eight NADH and four $FADH_2$ are oxidized in the electron transport chain by six O_2 (four and two O_2, respectively, according to equations 9.6 and 9.7), thus fueling the synthesis of approximately $8 \cdot 2.5 = 20$ and $4 \cdot 1.5 = 6$ ATP in oxidative phosphorylation. The sum of all the ATP produced is 30.

The yield per glucosyl unit of glycogen, released as glucose 1-phosphate in glycogenolysis, is one ATP more, since glucose 1-phosphate yields three rather than two

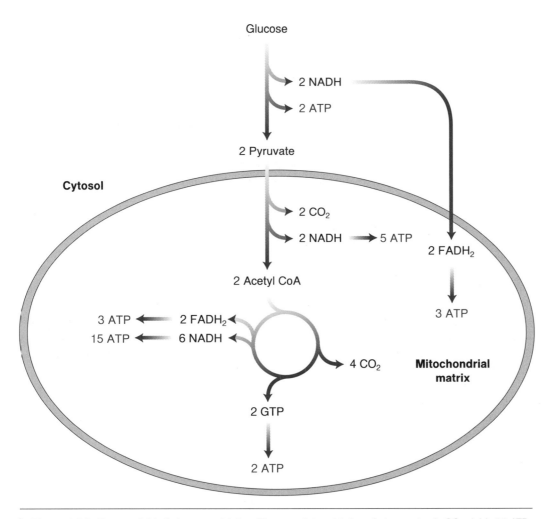

▶ **Figure 9.24** Energy yield of glucose oxidation. The complete oxidation of glucose to six CO_2 yields 30 ATP.

ATP in glycolysis (section 9.5). Thus, the complete oxidation, or combustion, of most glucosyl units of glycogen yields 31 ATP.

9.15 Exercise Speeds Up Oxidative Phosphorylation in Muscle

The most important determinant of the rate of oxidative phosphorylation is the concentration of its substrate, ADP. As ADP increases in a contracting muscle, it speeds up oxidative phosphorylation and ATP resynthesis. Oxidative phosphorylation requires an energy input from the electron transport chain. Because there is usually a tight coupling of the two processes (that is, electrons do not flow along the electron transport chain unless ATP is synthesized at the same time), the acceleration of oxidative phosphorylation drags along the electron transport chain. This, in turn, has three consequences.

- It increases oxygen consumption, which is one of the most evident corollaries of exercise. *Oxygen consumption rises with exercise intensity because so does the rate of ATP breakdown to ADP and P$_i$.*
- It decreases the [NADH], which lifts the brake on pyruvate oxidation and the citric acid cycle.

- It speeds up the regeneration of NAD^+ and FAD, which are indispensable oxidants in glycolysis, pyruvate oxidation, and the citric acid cycle.

Thus, *the last link in the long chain of ATP production from carbohydrate oxidation ensures the supply of materials that are necessary for the operation of the preceding links.*

9.16 Lactate Production in Muscle During Exercise

As we saw in the relevant sections, exercise speeds up glycogenolysis and glycolysis, two anaerobic processes taking place in the cytosol. Their acceleration relates to exercise intensity, since the harder the exercise, the larger the changes—some positive, some negative—in the concentrations of compounds affecting the rates of these processes (CP, ATP, ADP, AMP, P_i, Ca^{2+}, and epinephrine). Now remember that glycolysis is accompanied by the conversion of NAD^+ to NADH. For the glycolytic rate to be maintained, NAD^+ has to be regenerated. If NAD^+ is depleted, glycolysis will be shut down, and the flow of metabolites to subsequent pathways will stop.

The only means of regenerating cytosolic NAD^+ that we have considered so far is the transfer of an electron pair from NADH to a molecular shuttle, entry of the shuttle to the mitochondria, delivery of the electron pair to NAD^+ or FAD, and channeling of the electron pair to O_2 through the electron transport chain. Completion of this process is mandatory for the regeneration, in oxidized form, of all the cellular components mediating the flow of electrons from cytosolic NADH to O_2.

Because of the multitude of the steps involved, the aerobic conversion of cytosolic NADH to NAD^+ is time-consuming. Thus, above a certain exercise intensity, the rate of NADH formation in glycolysis exceeds the rate of aerobic NAD^+ regeneration. This is the point at which the *anaerobic* conversion of pyruvate to **lactate** comes into play.

$$\begin{array}{c} COO^- \\ | \\ C{=}O + NADH + H^+ \rightleftharpoons HO{-}C{-}H + NAD^+ \quad \Delta G^{o\prime} = -6 \ kcal \cdot mol^{-1} \\ | \\ CH_3 \end{array}$$

Pyruvate   **L-Lactate**   (equation 9.9)

The reaction takes place in the cytosol and is catalyzed by **lactate dehydrogenase.** The reaction establishes pyruvate rather than oxygen as the oxidant of NADH; it also establishes the rapid regeneration of NAD^+, which can then go back and help glycolysis out (figure 9.25).

Lactate dehydrogenase is a tetramer composed of a combination of two kinds of similar subunits denoted by H for heart and M for muscle. This results in five combinations, H_4, H_3M, H_2M_2, HM_3, and M_4, also referred to as LD1, LD2, LD3, LD4, and LD5, respectively, or LDH1 through LDH5. The H_4 isoform abounds in type I muscle fibers and the heart, whereas the M_4 isoform abounds in type IIA and IIX muscle fibers, as well as the liver.

> LD and LDH are the new and old abbreviations, respectively, for lactate dehydrogenase.

The muscle lactate concentration can rocket from approximately 1 mmol $\cdot$ kg^{-1} at rest to 30 mmol $\cdot$ kg^{-1} during maximal exercise, indicating a massive anaerobic catabolism of carbohydrates. This is accompanied by an increase in the acidity of the cytosol, since the net conversion of glucose or glycogen to lactate is accompanied by proton production.

$C_6H_{12}O_6 \rightarrow 2\ C_3H_5O_3^- + 2\ H^+$ (equation 9.10)
glucose   lactate

glycogen (*n* glucosyl units) + H_2O → glycogen (*n* − 1 glucosyl units) + 2 lactate + 2 H^+

(equation 9.11)

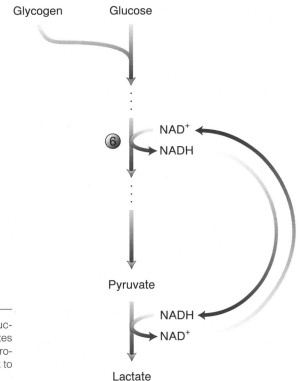

▶ **Figure 9.25** The gain from lactate production. The anaerobic breakdown of carbohydrates to lactate requires no NAD^+, since NADH produced in glycolysis is rapidly converted back to NAD^+ as pyruvate is reduced to lactate.

In contrast, there is no net H^+ production in the aerobic catabolism of carbohydrates to CO_2.

$C_6H_{12}O_6 + 6 O_2 \rightarrow 6 CO_2 + 6 H_2O$ **(equation 9.12)**

glycogen (n glucosyl units) $+ 6 O_2 \rightarrow$ glycogen ($n - 1$ glucosyl units) $+ 6 CO_2 + 5 H_2O$

(equation 9.13)

As a result of proton production in an intensely contracting muscle, the cytosolic pH drops from 7-7.1 at rest to as low as 6.3. Note that the presence of both lactate and H^+ on the right-hand side of equations 9.10 and 9.11 does not mean that lactate is the cause of proton formation during hard exercise, a widely held notion. Rather, proton formation and the drop in cytosolic pH should be viewed as results of the entire process of anaerobic carbohydrate degradation.

The increase in lactate concentration is the most impressive change in the concentration of a metabolite during exercise. Why this happens has puzzled researchers for decades. As you have seen, the explanation that I gave was essentially this: Since NAD^+ is converted to NADH in glycolysis, a reactant increases and a product decreases in reaction 9.9 (equation 9.9). These changes shift the reaction to the right. This explanation, however, is not shared by all researchers, as several of them maintain that the cause of lactate production in a hardworking muscle is a shortage of oxygen. In fact, this is a widely held view in sport science.

A basic reason for this view is that the conversion of carbohydrates to lactate is an anaerobic process. Therefore, the reasoning goes, there has to be a shortage of oxygen for the process to be accelerated. However, the fact that a process is anaerobic does not preclude its acceleration in the presence of abundant oxygen if other factors can speed it up. The majority of the relevant data show that, although the sarcoplasmic oxygen concentration decreases as exercise intensity increases, there is sufficient oxygen to support a maximal rate of ATP resynthesis in the mitochondria even during maximal exercise. Thus, it seems that the true cause of the augmented lactate production, or accumulation, is simply the overproduction of NADH in the cytosol of intensely contracting muscle fibers.

The conversion of glucose (or even glycogen) to lactate is sometimes termed *anaerobic glycolysis*. For the sake of accuracy, I have to note that glycolysis is a specific metabolic pathway (the one leading from glucose to pyruvate) and not just any pathway of carbohydrate breakdown. Glycolysis is solely anaerobic; thus, the term anaerobic glycolysis is redundant. Moreover, it leads one to wonder whether there is also an aerobic glycolysis. To avoid confusion, I believe it is appropriate to call the conversion of carbohydrates to lactate *anaerobic carbohydrate catabolism* (or *degradation,* or *breakdown*).

9.17   Features of the Anaerobic Carbohydrate Catabolism

The anaerobic carbohydrate catabolism is uneconomical: It yields only two ATP per glucose (all that is produced in glycolysis) against around 30 ATP produced in the aerobic breakdown. The corresponding figures for the majority of the glucosyl units in glycogen are three against 31. However, the anaerobic route is rapid. In fact, it is so rapid that despite the low ATP yield, it can produce three times as much ATP as the aerobic route in a given period of time. The maximal rate of ATP resynthesis from the conversion of glycogen to lactate in human muscle is estimated at 1.5 mmol $\cdot$ kg^{-1} $\cdot$ s^{-1} and is reached within 5 s of maximal exercise. In contrast, the maximal rate of ATP resynthesis from the conversion of glycogen to CO_2 is estimated at 0.5 mmol $\cdot$ kg^{-1} $\cdot$ s^{-1} and requires over 1 min of maximal exercise in order to be reached. *The high rate of ATP regeneration is the edge of the anaerobic carbohydrate breakdown.*

When catabolized anaerobically, glycogen is the fastest source for ATP resynthesis, next to CP. In fact, it surpasses CP in the amount of ATP that it can regenerate. While the CP content of a muscle is, as we have said, about 20 mmol $\cdot$ kg^{-1} (yielding an equal amount of ATP), an average glycogen content of 1.25% translates to 77 mmol glucosyl units per kilogram, which can yield close to 3 $\cdot$ 77 = 231 mmol ATP $\cdot$ kg^{-1}. (Note that 1.25% glycogen means 1.25 g per 100 g of muscle, or 12.5 g $\cdot$ kg^{-1}, or 12,500 mg $\cdot$ kg^{-1}. Divide this by the "molecular mass" of the glucosyl unit [162 Da], and you will get 77 mmol $\cdot$ kg^{-1}.)

Thanks to this combination of rate and quantity, the anaerobic glycogen degradation becomes the main source for ATP resynthesis in maximal exercise tasks lasting approximately 7 s to 1 min. Included in this time frame are events such as the 100, 200, and 400 m runs and the 50 and 100 m swims. Maximal, hard, or moderate-intensity exercise tasks lasting longer than 1 min rely on aerobic glycogen degradation as their main energy source. The expansion of the predominance of aerobic metabolism to as short a time as 1 min (compared to the prevalence of anaerobic metabolism extending to 2 or 3 min in older reports) is due to the discovery that the methods used in previous decades to estimate the contribution of anaerobic energy sources were positively biased.

> Exercise relying mainly on ATP resynthesis from anaerobic processes is called anaerobic, while exercise relying mainly on ATP resynthesis from aerobic processes is called aerobic.

The catabolism of carbohydrates to lactate and the creatine kinase reaction are the main anaerobic processes supplying ATP during exercise. To distinguish them while using shorter terms, authors often call the mechanism of energy production by the former process *anaerobic lactic* and the mechanism of energy production by the latter process *anaerobic alactic.*

9.18   Utilizing Lactate

The lactate produced in a hardworking muscle cannot be converted to anything other than pyruvate by a reversal of reaction 9.9 (equation 9.9), since it participates in no other reaction in the body. However, reaction 9.9 cannot be reversed in the cytosol of contracting muscle fibers, since there is a shortage of NAD$^+$. Besides, if NAD$^+$ were abundant, lactate production would not have increased in the first place. Thus, what lactate does is leave the muscle fibers. Their plasma membrane is highly permeable to lactate, which thus diffuses to the extracellular fluid (down its concentration gradient) and from there to the bloodstream, which disperses it all over the body.

The exit of lactate is facilitated by proteins spanning the sarcolemma and transverse tubule membrane, which are called <u>mono</u><u>c</u>arboxylate <u>t</u>ransporters, or MCT, to indicate that, in addition to lactate, they transport other acids with one carboxyl group. MCT carry one lactate and one H^+ in the same direction, thus contributing to the lowering of the cytosolic $[H^+]$. There are several MCT isoforms, two of which, MCT1 and MCT4, are present in human and rat skeletal muscle.

Because it crosses membranes easily, lactate can enter organs in which its concentration is lower than in the blood (again, down its concentration gradient). Such organs are primarily the skeletal muscles that did not participate in the exercise, the heart, and the liver. Small quantities of lactate are also taken up by the brain and kidneys.

It is even possible for fibers in a muscle to take up lactate from other fibers of the same muscle. This can happen because muscles contain a mixture of the three fiber types that we met in section 7.8 and that differ, among other respects, in the ability to resynthesize ATP through anaerobic processes. As we will discuss in section 13.9, type IIA and IIX fibers produce more lactate than type I fibers. Thus, the type I fibers can absorb part of the lactate exiting the type IIA and IIX fibers. Lactate enters the muscle fibers through the same MCT that it uses to exit.

Upon entering a cell, lactate can be oxidized to pyruvate by a reversal of reaction 9.9.

lactate + NAD^+ ⇌ pyruvate + NADH + H^+ **(equation 9.14)**

This conversion is favored by a high $[NAD^+]/[NADH]$ ratio in the cytosol of the cells that take up lactate (unlike that in the cytosol of the active muscle fibers). The pyruvate thus formed follows primarily one of two alternative routes (figure 9.26).

• It is fully oxidized to three CO_2 through the reaction catalyzed by the pyruvate dehydrogenase complex and then through the citric acid cycle. Most of the lactate produced during hard exercise follows this route in the heart and skeletal muscles, releasing a substantial amount of energy. How much energy? Consider the following: Each lactate molecule yields one cytosolic NADH upon conversion to pyruvate. This NADH then transfers an electron pair to a mitochondrial NAD^+ or FAD (section 9.11), thus yielding, respectively, 2.5 or 1.5 ATP through the electron transport chain and oxidative phosphorylation. The conversion of pyruvate to acetyl CoA produces one (mitochondrial) NADH yielding 2.5 ATP. The oxidation of acetyl CoA in the citric acid cycle yields three NADH (7.5 ATP), one $FADH_2$ (1.5 ATP), and one GTP (1 ATP). In all, lactate oxidation produces 15 or 14 ATP. In this way, lactate returns the difference in energy yield between aerobic and anaerobic carbohydrate catabolism (remember that one glucose produces two lactates).

• Pyruvate can be used to resynthesize glucose. This happens primarily in the liver, which hosts the metabolic pathway of gluconeogenesis.

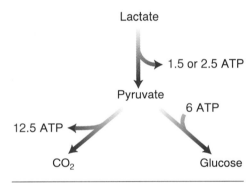

▶ **Figure 9.26** Two fates of lactate. Lactate produced in contracting muscle fibers can be used by other cells in the body either to regenerate ATP through conversion to CO_2 aerobically or to synthesize glucose through gluconeogenesis at the expense of ATP. The former takes place mainly in the heart and skeletal muscle; the latter, in the liver.

9.19   Gluconeogenesis

Gluconeogenesis is the synthesis of glucose from compounds that are not carbohydrates. Such compounds are pyruvate, lactate, glycerol, and most of the amino acids. At present, let's focus our attention on pyruvate and lactate; we will deal with glycerol and the amino acids in chapters 10 and 11.

Glucose synthesis from pyruvate (or from lactate that has been converted to pyruvate, as described in the previous section) involves the reversal of seven of the 10

> Although they are products of carbohydrate breakdown, pyruvate and lactate belong chemically to carboxylic acids rather than carbohydrates.

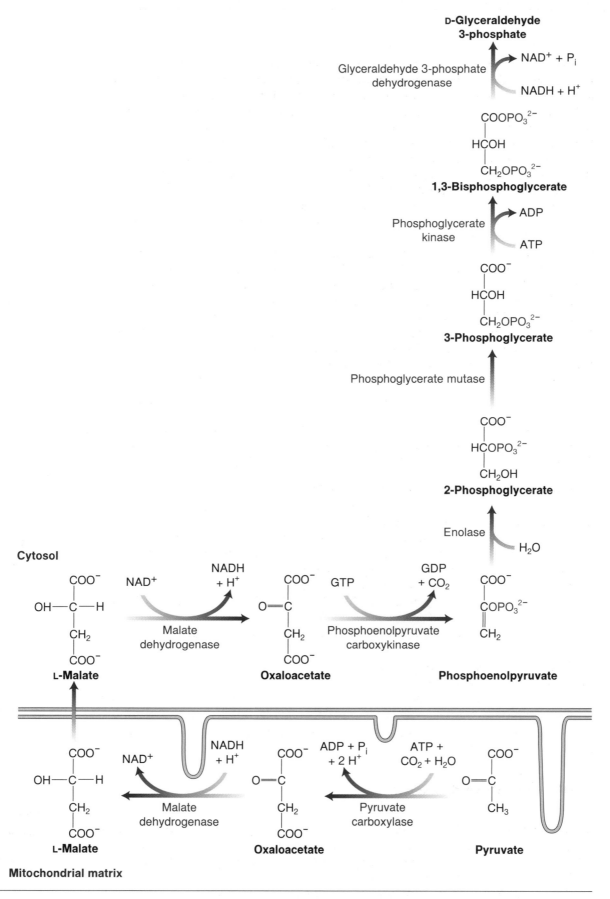

▶ **Figure 9.27** Gluconeogenesis. Gluconeogenesis starts with pyruvate (just above) and ends in glucose. The pathway is presented starting from the bottom right of this page (pyruvate) and ending at the top of the facing page (α-D-glucose) in order to facilitate the comparison with glycolysis in figure 9.14.

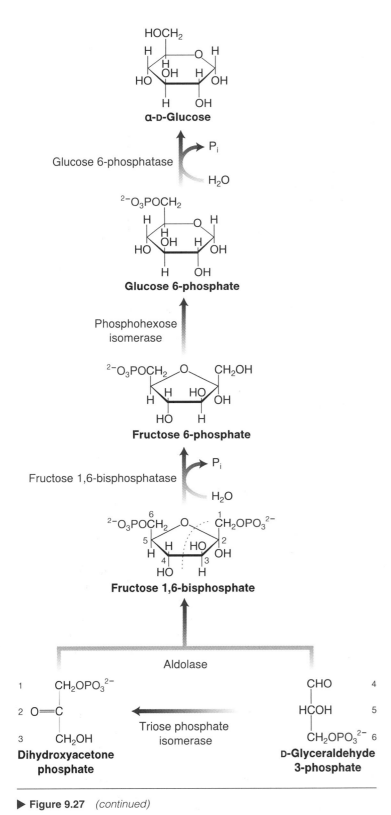

▶ **Figure 9.27** *(continued)*

reactions of glycolysis. As expected, the same enzymes catalyze these reactions in gluconeogenesis as in glycolysis. However, three glycolytic reactions, the first, third, and 10th, are irreversible because they are highly exergonic (therefore, the reverse reactions are highly endergonic and hence not favored). To circumvent this problem, gluconeogenesis lines up different reactions that are exergonic.

Gluconeogenesis takes place largely in the cytosol; however, it starts in the mitochondrial matrix (figure 9.27), where pyruvate is converted to oxaloacetate (a compound of the citric acid cycle) by the addition of one CO_2 and the hydrolysis of one ATP. Oxaloacetate is reduced to *malate* (another compound of the citric acid cycle), which then passes to the cytosol. Malate is oxidized back to oxaloacetate, which is then converted to *phosphoenolpyruvate* (the compound preceding pyruvate in glycolysis) at the expense of one GTP and with the removal of the CO_2 that was added to pyruvate earlier. Phosphoenolpyruvate is then converted to glyceraldehyde 3-phosphate through four reactions involving the hydrolysis of yet another ATP. Part of glyceraldehyde 3-phosphate is isomerized to dihydroxyacetone phosphate, and the two isomers are joined to produce fructose 1,6-bisphosphate. Finally, fructose 1,6-bisphosphate yields glucose through two dephosphorylations and one isomerization.

As expected of an anabolic pathway, gluconeogenesis requires an input of energy. Two ATP and one GTP (in all, three ~ P) are hydrolyzed in the conversion of pyruvate to glyceraldehyde 3-phosphate. However, because we need one glyceraldehyde 3-phosphate and one dihydroxyacetone phosphate to synthesize one glucose, we have to start with two pyruvates and spend six ~ P. Thus, *the energy demand of gluconeogenesis is three times the energy yield of glycolysis.* This confirms the statement, made in section 2.4, that a biosynthetic process is more expensive than the reverse degradation process is lucrative.

The main animal organs in which gluconeogenesis occurs are the liver and kidneys. No gluconeogenesis takes place in muscle because muscle lacks *glucose 6-phosphatase,* the enzyme catalyzing the last reaction of the pathway, that is, the conversion of glucose 6-phosphate to glucose. The absence of glucose 6-phosphatase is not a drawback, as it might seem at first glance. On the contrary, it prevents glucose 6-phosphate, derived from glycogen breakdown by way of glucose 1-phosphate, from becoming glucose and diffusing out of the muscle fibers. Glucose 6-phosphate does not diffuse out of the cells readily. Thus, any glucose 6-phosphate appearing in a muscle fiber is "doomed" to be degraded or stored as glycogen.

In contrast to muscle, the liver has a basic mission to provide the rest of the body with glucose when this is not available through food. The liver accomplishes this partly through gluconeogenesis and partly through glycogenolysis, as we will see in section 9.22.

Although muscle cannot synthesize glucose, it can nevertheless synthesize glucose 6-phosphate from lactate or pyruvate by utilizing all the reactions of gluconeogenesis except the last one. Glucose 6-phosphate can then be isomerized to glucose 1-phosphate, which can be used in glycogen synthesis (figure 9.28). Part of the lactate produced during hard exercise follows this route in the first minutes of recovery, when lactate concentration in the cytosol is still high. Aerobic ATP regeneration during the recovery period provides the energy needed for this process.

> Like glycolysis, gluconeogenesis is an anaerobic process.

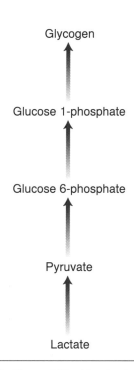

▶ **Figure 9.28** Muscle glycogen from lactate. Lactate produced in a hardworking muscle can be used before it leaves the muscle to synthesize glycogen during recovery.

9.20 Exercise Speeds Up Gluconeogenesis in the Liver

The rate of gluconeogenesis is controlled by a multitude of factors, several of which relate to exercise. To begin with, exercise increases the supply of gluconeogenic precursors. Lactate produced in vigorously exercising muscles and carried to the liver by the blood is one of these. In the following two chapters, we will see that the products of the augmented lipid and protein breakdown during exercise can also be channeled to glucose synthesis.

In terms of enzyme activity, the main means of controlling the rate of gluconeogenesis is the allosteric regulation of *fructose 1,6-bisphosphatase.* This enzyme catalyzes the third reaction before the end, one of the reactions that are not common to gluconeogenesis and glycolysis. The corresponding glycolytic reaction is catalyzed by phosphofructokinase.

The fact that the main control point of gluconeogenesis is a reaction not shared with glycolysis is not chance. If a common reaction were a control point, then an activator of the (common) enzyme catalyzing this reaction would speed up both pathways, whereas an inhibitor would slow both of them down. This would be futile, since the two pathways serve conflicting needs: One serves glucose synthesis at the expense of ATP, whereas the other serves ATP synthesis at the expense of glucose. In contrast, controlling reactions that are not common ensures the regulation of one pathway in a manner that is independent of—and usually opposite to—the regulation of the other.

Fructose 1,6-bisphosphatase and phosphofructokinase are controlled in an opposite fashion in the liver: When one is active, the other is inactive. The key to this opposite control is *fructose 2,6-bisphosphate,* a compound differing from fructose 1,6-bisphosphate, the intermediate of glycolysis and gluconeogenesis, in the position of one of the two phosphoryl groups (it is attached to C2 rather than C1). *Fructose 2,6-bisphosphate activates phosphofructokinase and inhibits fructose 1,6-bisphosphatase.*

Fructose 2,6-bisphosphate is synthesized from fructose 6-phosphate and is degraded to it through two different reactions (figure 9.29) just as fructose 1,6-bisphosphate and fructose 6-phosphate are interconverted in glycolysis and gluconeogenesis. However, different pairs of enzymes catalyze the two interconversions. What is amazing is that the kinase catalyzing the formation of fructose 2,6-bisphosphate (called *phosphofructokinase 2*) and the phosphatase degrading it *(fructose 2,6-bisphosphatase)* reside on the same polypeptide chain, which thus acts as a bifunctional enzyme!

The two enzyme activities are controlled by the hormones epinephrine and glucagon. We have already met epinephrine in section 9.3, where we saw that its increased secretion during exercise contributes to the acceleration of glycogenolysis through the cAMP cascade. **Glucagon,** on the other hand, is a peptide hormone consisting of 29 amino acid residues and having a molecular mass of 3.5 kDa. It is synthesized and secreted by the so-called α *cells* in the pancreas. Its plasma concentration rises when the glucose concentration drops; the concentration also rises during exercise. The target organ of glucagon is the liver; the hormone does not seem to act on any other human organ.

Epinephrine and glucagon affect liver function by binding to particular receptors at the surface of hepatocytes: the former to the β-adrenergic receptor and the latter to the *glucagon receptor.* After binding to the receptors, the two hormones act through the same mechanism: They activate adenylate cyclase (figure 9.10), resulting in a rise of the cytosolic [cAMP] and activation of PKA (figure 9.11). One of the proteins the kinase phosphorylates next is the bifunctional enzyme. Phosphorylation has opposite effects on the two activities of the bifunctional enzyme (figure 9.30): Phosphorylation inhibits phosphofructokinase 2 and activates fructose 2,6-bisphosphatase. Because of this double effect, the concentration of fructose 2,6-bisphosphate drops. Therefore, phosphofructokinase is inhibited (slowing down glycolysis), and fructose 1,6-bisphosphatase is activated (speeding up gluconeogenesis). Glucose production rises, and more glucose exits the hepatocytes to the bloodstream.

Through this complex mechanism, epinephrine and glucagon ensure the increased supply of glucose from the liver to the blood in order to meet the needs of tissues. As has become evident, a key to this mechanism is fructose 2,6-bisphosphate, which can be metaphorically considered *the dimmer of glucose metabolism in the liver.*

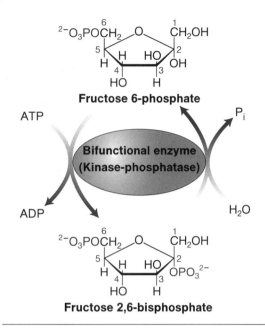

Fructose 6-phosphate

Fructose 2,6-bisphosphate

▶ **Figure 9.29** The dimmer of glucose metabolism in the liver. The synthesis and degradation of fructose 2,6-bisphosphate, the main regulator of glucose metabolism in the liver, are controlled by a bifunctional enzyme possessing both a kinase and a phosphatase activity.

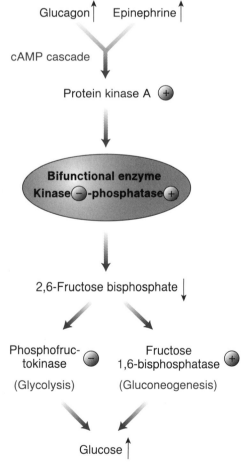

▶ **Figure 9.30** Hormonal control of glycolysis and gluconeogenesis in the liver. Epinephrine and glucagon control the rate of glycolysis and gluconeogenesis in the liver through the cAMP cascade. The activated protein kinase A phosphorylates the bifunctional enzyme. Phosphorylation decreases the kinase activity and increases the phosphatase activity. Thus, the fructose 2,6-bisphosphate concentration drops, resulting in a deceleration of glycolysis and an acceleration of gluconeogenesis. The bottom line is the rise in glucose production in the hepatocytes.

9.21   The Cori Cycle

Now that we have examined the effect of exercise on glucose metabolism in the liver, we can return to lactate utilization. We have seen that, after glucose is anaerobically degraded in a vigorously exercising muscle, the lactate produced diffuses to the blood, and part of it is taken up by the liver, which uses it for glucose synthesis. This glucose is then released to the blood and can be taken up by muscle, which catabolizes it to resynthesize ATP.

This completes a cycle connecting muscle, the blood, and the liver (figure 9.31). It is called the **Cori cycle** after Carl and Gerty Cori, who proposed it in the 1940s. This route recycles lactate to supply additional energy to muscle. The cycle operates at the expense of ATP, since it has a net cost of four $\sim$ P (six spent in gluconeogenesis minus two gained in glycolysis) per glucose. However, one organ pays (the liver) and another gains (muscle). In this way, muscle shifts part of its metabolic burden to the liver, which appears to operate as an "altruistic" organ. In reality, of course, its function obeys the need for survival of the entire organism, as does the function of every other organ in the body.

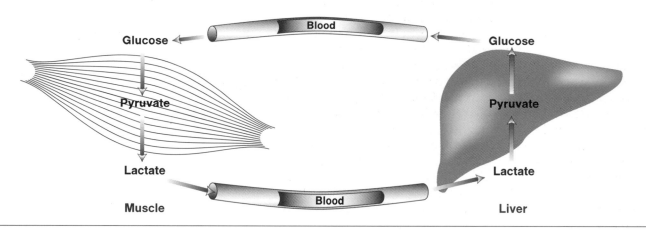

▶ **Figure 9.31**   The Cori cycle. The cycle provides an exercising muscle with glucose that is synthesized in the liver from lactate, the product of the anaerobic operation of muscle.

9.22   Exercise Speeds Up Glycogenolysis in the Liver

Glucose is produced in the liver not only through gluconeogenesis but also through glycogenolysis, since, as we originally discussed in section 9.1, the liver contains substantial quantities of glycogen (figure 9.32). The glycogen content of the liver is greatly influenced by the diet. If the diet is rich in carbohydrates, glycogen can soar to 8%; if it is poor in carbohydrates, glycogen can dwindle to 0.5%. This great variation happens because liver glycogen is used as a donor of glucose to the rest of the body in case of shortage.

The rate of glycogenolysis in the liver is controlled by epinephrine and glucagon. As we have seen in sections 9.3 and 9.20, the two hormones bind to the β-adrenergic receptor and the glucagon receptor, respectively, causing the stimulation of the cAMP

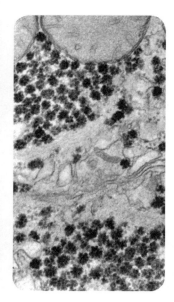

▶ **Figure 9.32**   Glycogen in hepatocytes. Glycogen granules, seen as dark dots under high resolution in this electron micrograph, abound in hepatocytes. Half of the granules are clustered near two mitochondria at the top.

cascade and the activation of PKA. By subsequently phosphorylating enzymes involved in glycogen metabolism (figure 9.13), the kinase speeds up glycogenolysis and slows down glycogen synthesis. In addition, epinephrine binds to another type of receptor in the liver, the α_1-*adrenergic receptor*. This binding stimulates a different cascade of hormone action, the *phosphoinositide cascade* (figure 9.33), which deserves a brief presentation.

When the α_1-adrenergic receptor receives epinephrine in the plasma membrane of a hepatocyte, it stimulates the replacement of GDP by GTP in the α subunit of G_q, a G protein similar to the G_s involved in the cAMP cascade (section 9.3). Following GTP binding, $G_{q\alpha}$ dissociates from $G_{q\beta\gamma}$ and activates *phospholipase C*, one of a group of enzymes that degrade phospholipids. Phospholipase C hydrolyzes a minor phospholipid of the plasma membrane, *phosphatidyl inositol 4,5-bisphosphate* or PIP_2 (a derivative of phosphatidyl inositol, introduced in figure 5.13).

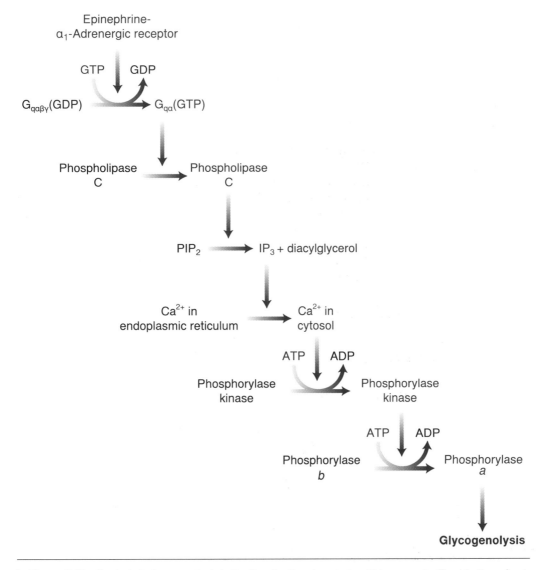

▶ **Figure 9.33** Control of glycogenolysis in the liver by the phosphoinositide cascade. The binding of epinephrine to the α_1-adrenergic receptor in the hepatocytes activates $G_{q\alpha}$, which activates phospholipase C, which generates inositol 1,4,5-trisphosphate (IP_3) from phosphatidyl inositol 4,5-bisphosphate (PIP_2). IP_3 then releases Ca^{2+} from the endoplasmic reticulum. Ca^{2+} activates phosphorylase kinase, which, finally, phosphorylates and activates phosphorylase.

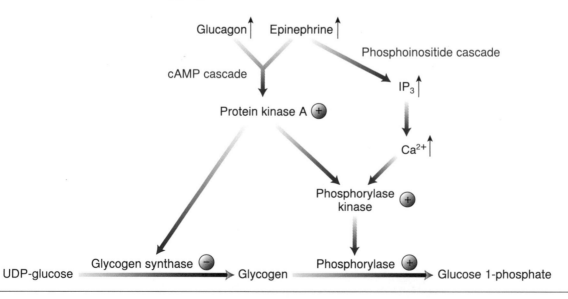

Phosphatidyl inositol 4,5-bisphosphate

(equation 9.15)

Diacylglycerol   **Inositol
1,4,5-trisphosphate**

By analogy to the distinction between the prefixes bis- and bi- made in section 9.5, the prefix tris- denotes three groups attached to a molecule at different positions (as in inositol 1,4,5-trisphosphate), whereas tri- describes three groups attached in a row to a single position of a molecule (as in adenosine triphosphate).

Hydrolysis produces two important second messengers, *inositol 1,4,5-trisphosphate*, or IP$_3$, and *diacylglycerol*. Of interest to the control of glycogenolysis is the former. Being water soluble, IP$_3$ diffuses from the plasma membrane to the cytosol and binds to a ligand-gated Ca^{2+} channel in the endoplasmic reticulum membrane. Binding of IP$_3$ to the channel causes it to open and release Ca^{2+} down its concentration gradient, that is, from the endoplasmic reticulum to the cytosol (figure 9.33). Ca^{2+} then binds to phosphorylase kinase and activates it exactly as it does in muscle, resulting in additional acceleration of glycogenolysis. Figure 9.34 summarizes the ways in which exercise may affect liver glycogen metabolism.

▶ **Figure 9.34**   Control of liver glycogen metabolism in exercise. There are similarities and differences between the liver and muscle with regard to the control of glycogen metabolism (cf. figure 9.13). Glucagon and epinephrine, secreted in increased amounts during exercise, speed up glycogenolysis and slow down glycogen synthesis by stimulating the cAMP cascade. Epinephrine speeds up glycogenolysis even more through the phosphoinositide cascade. In contrast to muscle, P$_i$ and AMP do not affect phosphorylase activity in the liver, since their concentrations do not change much.

The stimulatory actions of epinephrine and glucagon on glycogenolysis in the liver cause a rise in the concentration of glucose 1-phosphate and, subsequently, glucose 6-phosphate. Degradation of the latter is not favored because, as we saw in section 9.20, the two hormones inhibit glycolysis. Thus, the main fate of glucose 1-phosphate in the hepatocytes is conversion to glucose, which is secreted to the bloodstream and sent to tissues for energy supply.

> Because epinephrine and glucagon increase the blood glucose concentration, they are termed *hyperglycemic hormones*.

9.23 Control of the Plasma Glucose Concentration in Exercise

The maintenance of a relatively stable plasma glucose concentration is vital to the functioning of the brain—and hence the entire body—because this organ uses glucose as its almost exclusive fuel and because it practically lacks energy reserves (it contains a minimal amount of glycogen). Thus, the brain requires a continuous supply of glucose from the blood. If the glucose concentration drops below a critical limit, **hypoglycemia** occurs, which is manifest by dizziness, removal from the world, visual disturbances, weakness, unstable gait, and nausea. Glucose is also the exclusive energy source for erythrocytes, which degrade it only anaerobically since they lack mitochondria. On the other hand, excess glucose in the plasma may cause lesions to the vessels in the long run. These lesions increase the risk for cardiovascular disease.

The body has a number of homeostatic mechanisms that protect the plasma glucose concentration against drastic fluctuations. These mechanisms prevent

- an excessive rise in the plasma glucose concentration after a meal, during the so-called *postprandial state;* and

- an excessive fall in the plasma glucose concentration after the absorption of nutrients, during the so-called *postabsorptive state.*

In sections 9.20 and 9.22 we explored how epinephrine and glucagon control carbohydrate metabolism in the liver. The picture would be incomplete without **insulin,** the most important regulatory hormone of fuel metabolism. Insulin is a peptide hormone consisting of 51 amino acid residues and having a molecular mass of 5.8 kDa. It is synthesized and secreted by the β *cells* in the pancreas. Its plasma concentration rises when the glucose concentration rises and decreases during exercise.

> Although insulin and glucagon are synthesized by the same gland, their secretion is affected in an opposite manner by the plasma glucose concentration and by exercise.

Insulin secretion depends on the plasma glucose concentration. When the latter increases after a meal because of carbohydrate digestion, the rate of insulin secretion by the pancreas also increases. The hormone facilitates the entry of glucose and certain other monosaccharides to muscle and adipose tissue cells. When the plasma glucose concentration drops, insulin secretion decreases.

Insulin exerts a plethora of effects not only on carbohydrate but also on lipid and protein metabolism. In subsequent chapters we will consider its effects on lipid and protein metabolism. As far as carbohydrate metabolism is concerned, insulin

- speeds up glycogen synthesis in muscle and the liver,

- slows down glycogenolysis in muscle and the liver,

- speeds up glycolysis in the liver, and

- slows down gluconeogenesis in the liver.

The bottom line regarding the various effects of insulin on carbohydrate metabolism is to decrease the glucose concentration in biological fluids.

In recent years investigators have begun to unravel insulin's mechanisms of action. We know now that the hormone binds to the *insulin receptor,* an integral protein of the plasma membrane of target cells. The insulin receptor is a tetramer of two α (135 kDa)

> Because insulin decreases the blood glucose concentration, it is termed a *hypoglycemic hormone.*

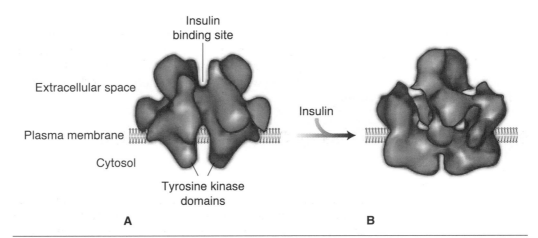

▶ Figure 9.35   Insulin receptor. The insulin receptor has an extracellular binding site for insulin and two intracellular domains possessing tyrosine kinase activity. *(a)* When insulin is free, the receptor has an open conformation that keeps the kinase domains apart. *(b)* When insulin binds, the receptor closes and the kinase domains come close enough to phosphorylate each other. Phosphorylation of each domain at three tyrosine residues stimulates the protein kinase activity of the receptor. The protein and membrane are drawn to scale.

Adapted, by permission, from C.C. Yip and P. Ottensmeyer, 2003, "Three-dimensional structural interactions of insulin and its receptor," *Journal of Biological Chemistry* 278: 27329-27332.

and two β (95 kDa) subunits (figure 9.35). The α subunits are located extracellularly and bind insulin, whereas the β subunits span the membrane and possess protein kinase activity, which is dormant in the absence of insulin. Insulin binding to the α subunits causes a conformational change that brings each β subunit into the active site of the other. The subsequent phosphorylation of three key tyrosine residues in each β subunit activates the receptor, which can now phosphorylate a variety of cytoplasmic proteins.

One of these proteins is *insulin receptor substrate 1* (IRS1), which, when phosphorylated, activates *phosphatidyl inositol 3 kinase,* or PI3K (figure 9.36). This enzyme phosphorylates PIP_2 (see preceding section) to generate *phosphatidyl inositol 3,4,5-trisphosphate* (PIP_3). PIP_3 then activates another protein kinase, *phosphoinositide-dependent kinase 1* (PDK1), which in turn phosphorylates and activates *protein kinase B* (PKB). PKB phosphorylates several proteins, one of which is *glycogen synthase kinase 3* (GSK3). This enzyme phosphorylates and inactivates glycogen synthase (section 9.1), just as PKA does (figures 9.13 and 9.34). However, phosphorylation of GSK3 by PKB inactivates it, thus favoring the dephosphorylated, active, *a* form of glycogen synthase. Through this complex signal transduction pathway, insulin stimulates glycogen synthesis in muscle and the liver.

Insulin promotes glycogen synthesis in two additional ways. First, PKB triggers the movement of GLUT4 to the plasma membrane in muscle (figure 9.15), although we do not know the exact pathway. Thus, insulin promotes the entry of glucose, which is the raw material for glycogen synthesis. Second, it activates protein phosphatase 1, an enzyme introduced in section 9.2 and figure 9.7 as one that dephosphorylates phosphorylase *a*. PP1 also dephosphorylates glycogen synthase *b*, hence activating the enzyme.

In closing the discussion on how insulin affects carbohydrate metabolism, let me note that the activated PP1 in the liver also dephosphorylates the bifunctional enzyme discussed in section 9.20. This reverses the actions that glucagon and epinephrine exert on the enzyme through PKA activation: Now, as a result of insulin secretion and PP1 activation, glycolysis is accelerated and gluconeogenesis is decelerated. To appreciate this better, refer to figure 9.30 and try redrawing it to show the control of the two pathways by insulin.

Insulin action is impaired in **diabetes mellitus,** a disease caused by inadequate insulin production or by **insulin resistance** (malfunctioning of the hormone's signal transduction pathway). When inadequate insulin production is the cause, diabetes is

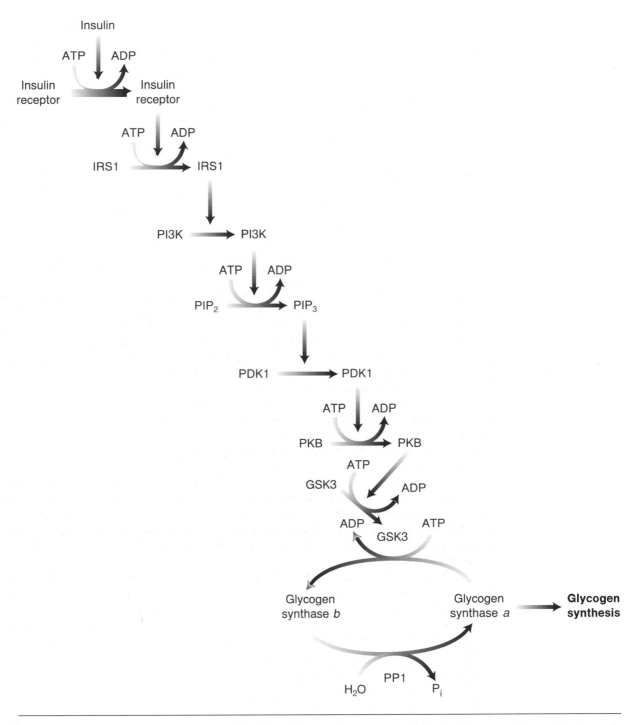

▶ **Figure 9.36** Control of glycogen synthesis by the PI3K cascade. Once activated by insulin, the insulin receptor in muscle and the liver phosphorylates IRS1 (insulin receptor substrate 1), which activates PI3K (phosphatidyl inositol 3 kinase), which phosphorylates PIP_2 to form phosphatidyl inositol 3,4,5-trisphosphate (PIP_3). PIP_3 activates phosphoinositide-dependent kinase 1 (PDK1), which phosphorylates and activates protein kinase B (PKB), which phosphorylates and inactivates glycogen synthase kinase 3 (GSK3). This, combined with the activation of protein phosphatase 1 (PP1) by insulin, promotes the dephosphorylation and activation of glycogen synthase and, hence, the stimulation of glycogen synthesis. Which compound does this cascade share with the phosphoinositide cascade (figure 9.33)?

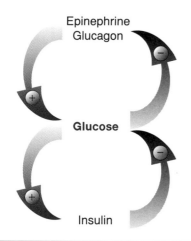

Epinephrine
Glucagon

Glucose

Insulin

▶ **Figure 9.37** Plasma glucose homeostasis. The plasma glucose concentration is maintained within a narrow range mainly thanks to the opposing effects of the hormones epinephrine, glucagon, and insulin. Epinephrine and glucagon signal an increase in plasma glucose, whereas insulin signals a decrease in plasma glucose. Conversely, glucose signals a decrease in plasma epinephrine and glucagon, and an increase in plasma insulin.

classified as *type 1;* when insulin resistance is the cause, it is classified as *type 2.* Uncontrolled diabetes of both types is characterized by exceedingly high plasma glucose concentrations because of limited glucose uptake by muscle and adipose tissue, as well as excess glucose production by the liver, since glycogenolysis and gluconeogenesis are not restrained. Because exercise increases glucose uptake by muscle through an insulin-independent mechanism (section 9.6), it helps diabetic patients control their plasma glucose concentration.

Thanks to the opposite effects of epinephrine and glucagon on one hand, and insulin on the other (figure 9.37), the glucose concentration in the plasma of a healthy person at rest is maintained at 4 to 6 mmol · L^{-1}. This condition is termed **euglycemia** (meaning "good blood glucose" in Greek). In addition to hormonal control, an important euglycemic mechanism is the direct effect of glucose on liver phosphorylase (but not muscle phosphorylase): Glucose binding to an allosteric site on phosphorylase *a,* the phosphorylated and usually active form of the enzyme, deactivates the enzyme. In addition, glucose binding to phosphorylase *a* exposes the phosphorylation site to PP1, which then dephosphorylates and deactivates the phosphorylase. Thus *glucose controls the rate of glycogenolysis in the liver by exerting feedback inhibition on its own production.*

Exercise poses a serious threat to euglycemia: As already discussed (section 9.6), it boosts the rate of glucose uptake by muscle. This threat is not left unanswered, though. The lowering of the plasma glucose concentration is reflected in the cytosol of the hepatocytes, thus lifting the deactivation of phosphorylase according to our earlier discussion. This speeds up glycogenolysis. In addition, the increased secretion of epinephrine and glucagon (because of both exercise and the drop in plasma glucose concentration) activates glycogenolysis and gluconeogenesis in the liver. Thus, more glucose enters the bloodstream to be taken up primarily (by about 80%) by the exercising muscles.

Another factor protecting plasma glucose against drastic decreases during exercise is the slowing down of insulin secretion (in relationship with exercise intensity), which results in a moderation of its hypoglycemic action. Moreover, the attenuation of all the effects of insulin on carbohydrate metabolism, as presented earlier in this section, permits the direct (in muscle) and indirect (by the liver) supply of glucose for the energy demands of exercise.

So, what happens to the plasma glucose concentration during exercise? The answer is not simple. Plasma glucose can increase or decrease depending on the kind of exercise and the size of the liver glycogen depot. It is even customary for plasma glucose to decrease and then increase, or vice versa, during prolonged exercise, even if exercise is performed at a steady pace. In general, light exercise does not affect plasma glucose too much, whereas moderate-intensity or hard exercise tends to increase it initially and then decrease it even to levels below baseline (the value at rest) if exercise is very prolonged.

Because of the multitude of factors fighting over it during exercise, the plasma glucose concentration can reach extreme values. On one end, it can drop to as low as 2.5 mmol · L^{-1} after exhaustive exercise lasting a few hours, which depletes most of the liver glycogen. On the other end, it can exceed 10 mmol · L^{-1} after hard exercise of short duration.

9.24 Blood Lactate Accumulation

The concentration of lactate (or any substance) in the blood is determined by its rates of appearance and disappearance, or removal. The blood lactate concentration at rest is approximately 1 mmol · L^{-1} and reflects a so-called *steady state,* in which the rate of appearance (mainly from muscle and the erythrocytes) equals the rate of disappearance toward the other tissues.

When exercise starts, the steady state of lactate is disturbed because its rate of appearance in the blood rises. The rate of appearance relates to the concentration gradient between the cytosol of the contracting muscle fibers and the plasma. It also relates to the blood flow through the exercising muscles. Exercise increases both of these parameters. As a result, lactate begins to accumulate in the blood. One can measure a rise in its concentration as early as 1 min after the onset of hard exercise.

If exercise is continued at a steady pace, the lactate concentration gradient between the cytosol of the contracting muscle fibers and the plasma starts to diminish because lactate begins to accumulate in the plasma. The gradient diminishes even further when exercise stops, since the rate of lactate production in muscle will return to baseline. In effect, the rate of lactate appearance in the blood gradually declines.

Let's consider now the rate of lactate disappearance from the blood. By analogy to the rate of appearance, the rate of disappearance relates to the concentration gradient between the plasma and the cytosol of the cells that absorb it, such as the cells in nonexercising muscles, the heart, and the liver. It also relates to the blood flow through these organs. The concentration gradient increases gradually during exercise, as the blood lactate concentration rises. The blood flow as a whole does not change appreciably. In effect, the rate of lactate disappearance from the blood gradually rises during exercise.

When the gradually declining rate of lactate appearance meets the gradually rising rate of disappearance, the blood lactate concentration stops rising and the *lactate–time plot* (figure 9.38) peaks. The peak is usually observed 4 to 10 min after the onset of exercise.

The peak lactate concentration in the blood is at best equal to, but usually lower than, the peak lactate concentration in the cytosol of the contracting muscle fibers. This is the case because by the time lactate peaks in the blood, other organs have absorbed a considerable amount of it. Thus, it is not right to consider that, by measuring the peak lactate concentration in the blood, we determine the peak concentration in the exercising muscles. However, the peak concentration in the blood relates to the peak concentration in muscle; and since it is easier to sample a little blood than a little muscle, the former is used widely as a simple marker of anaerobic carbohydrate catabolism. (Chapter 15 has more detail on the utility of measuring blood lactate.)

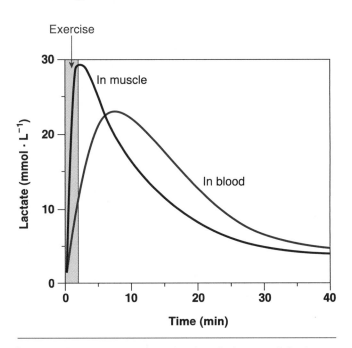

▶ **Figure 9.38** Lactate–time plot. A typical curve of the lactate concentration in an athlete's blood after short maximal exercise exhibits a peak within a few minutes after the start of exercise and then slowly returns to baseline. The lactate concentration in the exercising muscle peaks at the end of exercise, reaching a higher value than in the blood.

9.25 Blood Lactate Decline

When exercise stops, the muscle lactate concentration begins to fall (figure 9.38). The blood lactate concentration follows suit a few minutes later. Now the rate of lactate disappearance from the blood begins to decline, as the concentration gradient between the blood and the cytosol of the cells that absorb lactate gradually decreases. The lactate–time plot then converges slowly to the lactate concentration at rest. The *half-life* of blood lactate (the time it takes a given concentration to halve) is at least 12 min.

The rapid removal of lactate from the muscles after the end of exercise is considered desirable in order for the muscles to regain full contractile capacity. To speed up lactate removal, athletes often continue to exercise lightly (at around 20-40% of $\dot{V}O_2$max) rather than resting after their main exercise bout. This is *active recovery* as opposed to *passive recovery*. Research has shown that active recovery speeds up the decline in blood lactate. To explain this, researchers hypothesize that, compared to rest, more of the existing lactate is oxidized in the muscles under the aerobic conditions of light exercise and thus less is released in the blood. However, there is no agreement among studies as to whether active recovery speeds up the disappearance of muscle lactate. Equally questionable is whether the mode of recovery (active vs. passive) has an effect on performance during subsequent exercise bouts.

9.26 "Thresholds"

In the previous two sections, we examined the blood lactate concentration as a function of the time elapsed from the onset of exercise. Let's consider now the blood lactate concentration as a function of exercise intensity by performing an experiment. Say a person produces a certain amount of work (for example, he or she runs 1 km [0.6 mile]) at a constant low intensity. At the end, we determine the peak lactate concentration in the blood by sequential sampling over a period of about 10 min (see section 15.3 for details). After resting for a while, the person repeats the same task over and over again at an ever-increasing intensity. At the end of each bout, we determine the peak lactate concentration in the blood. Finally, we plot the peak values obtained against exercise intensity (expressed in either one of the ways listed in section III.2). We will then get a *lactate–intensity plot* that is upward and concave (figure 9.39).

Exercise scientists and coaches have strived over the past decades to draw information on exercise metabolism and sport performance from the lactate–intensity plot. Many efforts have focused on identifying the points along the plot with the most useful coordinates (intensity and lactate concentration). To this end, a number of "thresholds" have been proposed, such as the following:

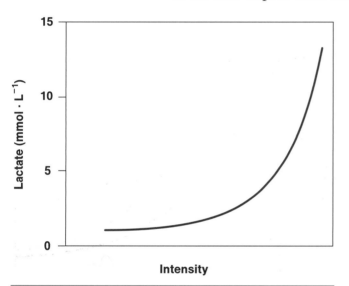

▶ **Figure 9.39** Lactate–intensity plot. The blood lactate concentration increases exponentially with exercise intensity.

- The *lactate threshold*, defined as the exercise intensity above which blood lactate begins to rise; or the exercise intensity corresponding to a blood lactate concentration of 1 mmol · L^{-1} above baseline; or the exercise intensity corresponding to a blood lactate concentration of 2.5 mmol · L^{-1}

- The *anaerobic threshold*, defined as the exercise intensity above which blood lactate begins to rise (same as the first definition in the preceding) or the exercise intensity corresponding to a blood lactate concentration of 4 mmol · L^{-1}

- The *onset of blood lactate accumulation* (OBLA), defined as the exercise intensity corresponding to a blood lactate concentration of 4 mmol · L^{-1} (same as the second definition of the anaerobic threshold)

- The *individual anaerobic threshold*, defined as the maximal exercise intensity that can be maintained for some time (usually 20-30 min) without a continuous rise in the blood lactate concentration (also called the *maximal lactate steady state*) or determined by drawing tangents to the lactate–intensity plot

This is a rather confusing situation: A term is defined in different ways, and different terms mean the same thing. The reason may be the questionable relevance of the terms listed. Take, for example, the anaerobic threshold, by far the most popular of the terms. The anaerobic threshold is often thought of as the point of transit from aerobic to anaerobic metabolism. However, there is no state of exclusively aerobic or exclusively anaerobic muscle function. Even in the resting state, lactate is being constantly produced in the muscles at a slow rate, while, at the other extreme, the exercise that elicits maximal lactate concentrations in the muscles and blood is accompanied by a burgeoning oxygen consumption to support the increased demand for aerobic ATP resynthesis. Thus, it is a mistake to seek the point of passage from aerobic to anaerobic muscle function. Unfortunately, this misunderstanding is encouraged by the term threshold (denoting the border between two states).

How, then, should a lactate–intensity plot be interpreted? As exercise intensity rises, there is a gradual increase in the flow toward lactate and a gradual decrease in the flow toward acetyl CoA at the "pyruvate crossroads" in the exercising muscles (figure 9.40). This happens because of the enhanced conversion of NAD^+ to NADH in glycolysis and because the conversion of pyruvate to lactate absorbs NADH, whereas the conversion of pyruvate to acetyl CoA requires additional NAD^+. Since, as mentioned in section 9.24, the peak lactate concentration in muscle relates to the peak lactate concentration in the blood, the latter increases with increasing exercise intensity. However, the gradual nature of the shift from aerobic to anaerobic function does not justify any threshold terminology—this being even more the case because the blood lactate curve reflects lactate removal as well as production. Even the OBLA, as defined earlier, is not justified, since lactate begins to accumulate in the blood as soon as it departs from the resting value of about 1 mmol $\cdot$ L^{-1}, that is, at intensities that are considerably lower than the one corresponding to 4 mmol $\cdot$ L^{-1}.

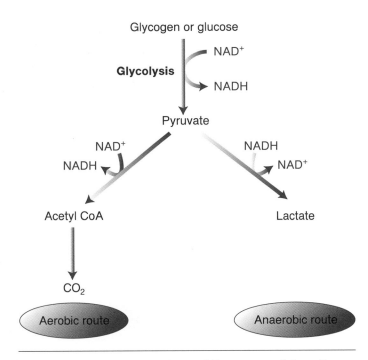

▶ **Figure 9.40** Pyruvate's "dilemma." Pyruvate can follow either an aerobic or an anaerobic metabolic route in muscle. The anaerobic route is gradually favored as exercise intensity increases, but there is no threshold of lactate production or accumulation.

Is the curve useless? No! As analyzed in chapter 15, it can be used to evaluate the aerobic capacity and aerobic training programs. Certain points on it, especially the one at the 4 mmol $\cdot$ L^{-1} level, can be helpful for reference. It's just that it would be preferable to call things by their—admittedly lengthier—real names, such as intensity corresponding to a blood lactate concentration of 4 mmol $\cdot$ L^{-1}, or to use abbreviations such as V4 (velocity corresponding to 4 mmol $\cdot$ L^{-1}), rather than terms of limited, if any, physiological significance.

Problems and Critical Thinking Questions

1. Which substances can speed up glycogenolysis in muscle during exercise through an increase in their concentrations?

2. Can glycolysis begin in vitro if there is no ATP present?

3. Which substances can speed up glycolysis in muscle during exercise through an increase in their concentrations, and which through a decrease in their concentrations?

4. Baker's yeast contains the aerobic microorganism *Saccharomyces cerevisiae*. How does the dough rise?

5. Compare the aerobic to the anaerobic carbohydrate breakdown in muscle in terms of amount of ATP produced, rate of ATP resynthesis, and exercise tasks in which each of the two predominates.

6. In the movie *A Few Good Men*, starring Tom Cruise, a marine dies of "lactic acidosis," and two fellow soldiers are accused of murder. Asked before the court what lactic acidosis is, a physician says, "If the muscles and all the cells in the body burn sugar instead of oxygen, lactic acid is produced." Correct him!

7. Which branch of glucose metabolism and which branch of glycogen metabolism are stimulated in the liver during exercise? Through which mechanisms?

8. Which events tend to decrease and which events tend to increase the blood glucose concentration during exercise?

CHAPTER 10

Lipid Metabolism in Exercise

In chapter 5 we saw that lipids are a diverse class of biological compounds with low solubility in water. Three of the lipid categories that we examined are of interest in exercise metabolism; these are fatty acids, triacylglycerols, and steroids.

Fatty acids and triacylglycerols are of interest as important energy sources in exercise. They relate to each other in the same way that glucose relates to glycogen; that is, fatty acids are present in the body in much smaller quantities than triacylglycerols but can be derived easily from triacylglycerols when called for by the situation (refer to figure 9.1). The subsequent breakdown of fatty acids yields high amounts of ATP, albeit at a low rate. Additionally, we will deal with triacylglycerols and cholesterol (the parent steroid) as compounds whose concentrations in the blood relate to cardiovascular disease and are modified by exercise.

10.1 Triacylglycerol Metabolism in Adipose Tissue

Triacylglycerols are the largest energy depot in the body. An average man has approximately 10 kg (22 lb) and an average woman approximately 14 kg (31 lb) of triacylglycerols, or fat. Most of the fat is concentrated in adipose tissue, a tissue distributed all over the body: under the skin (subcutaneous adipose tissue), around the blood vessels, and in the viscera. Approximately 82% of the adipose tissue mass is triacylglycerols, making this the most amazing accumulation of a substance in a tissue.

The main cells in adipose tissue are the **adipocytes,** or fat cells (figure 10.1). Their cytoplasm is full of triacylglycerol-containing droplets surrounded by a single layer of phospholipids (guess what the orientation of the polar head groups and the hydrophobic tails is; then look at figure 10.2). In addition, the droplets are coated with **perilipins** (from the Greek words *perí,* meaning "around," and *lípos,* meaning "fat"), a family of proteins regulating access to the droplets, as we will see in the next section.

The main metabolic activities of adipocytes are triacylglycerol synthesis and triacylglycerol breakdown. Synthesis prevails during the postprandial state, whereas breakdown occurs mainly during the postabsorptive state and exercise. If the two processes are balanced over the day, the fat mass in the body remains constant. If synthesis outweighs breakdown, we get fat; if breakdown outweighs synthesis, we lose fat. The amount of

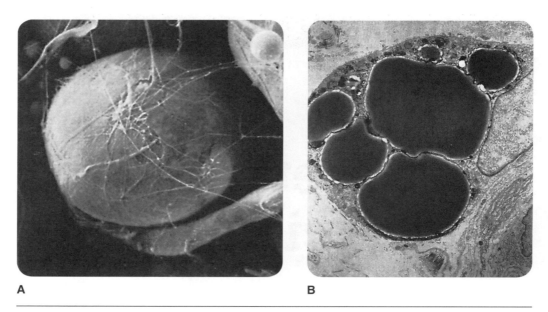

A   **B**

▶ **Figure 10.1** Adipocytes. *(a)* Scanning electron micrograph shows a spherical adipocyte flanked by a capillary blood vessel (the tube at the bottom) and collagen fibers (in the foreground) forming a supporting mesh. The diameter of the adipocyte is about 50 μm. *(b)* Electron micrograph of a cross section of an adipocyte reveals that it is full of lipid droplets (the dark spots) containing mainly triacylglycerols. The long light-colored organelle on the right is the nucleus.

Figure 10.1a © R.G. Kessel and R.H. Kardon, *Tissues and Organs: A Text-Atlas of Scanning Electron Microscopy,* W.H. Freeman, 1979.

Figure 10.1b Reprinted from *The Cell: Its Organelles and Inclusions,* D.W. Fawcett, pg. 308, Copyright 1967, with permission from Elsevier Inc.

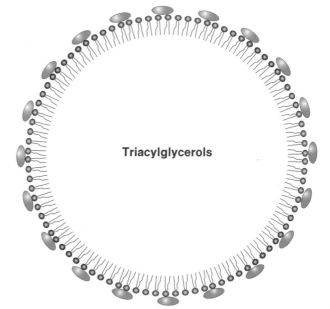

Triacylglycerols

▶ **Figure 10.2** Lipid droplet. Lipid droplets in adipocytes are like fuel tanks. They are full of triacylglycerols and are bound by a phospholipid monolayer (the wall of the tank). The hydrophobic tails of the phospholipids are in contact with the triacylglycerols, while the polar head groups face the cytosol. Perilipins (the gray ellipses on the outside) are proteins that act like guards restricting access to the tanks.

fat deposited or lost in a day, if the two processes are not balanced, is minuscule—usually a few grams or tens of grams at the most. However, if the imbalance remains over weeks, months, or years, a measurable fat—and weight—gain or loss will ensue. Of course, we are more concerned about fat and weight gain, as this may result in *obesity,* a health-threatening condition that has become increasingly prevalent in recent years. Since exercise promotes the breakdown of fat, it is an essential weapon in the fight against obesity.

Triacylglycerol Synthesis

Triacylglycerols are synthesized from activated forms of glycerol and fatty acids. The activated form of glycerol is *glycerol 3-phosphate* and is mainly derived from dietary glucose entering the adipocytes. Through the four first reactions of glycolysis, glucose is converted to dihydroxyacetone phosphate (consult figure 9.14), which is then reduced to glycerol 3-phosphate through a reaction catalyzed by *glycerol phosphate dehydrogenase*.

$$
\begin{array}{c}
\text{H}_2\text{C}-\text{OH} \\
| \\
\text{O}=\text{C} \qquad + \text{NADH} + \text{H}^+ \; \rightleftharpoons \; \text{HO}-\text{CH} \qquad + \text{NAD}^+ \\
| \\
\text{H}_2\text{C}-\text{OPO}_3^{2-}
\end{array}
\qquad \text{(equation 10.1)}
$$

Dihydroxyacetone   **Glycerol**
phosphate   **3-phosphate**

The activated form of a fatty acid is *acyl CoA* and is produced from the joining of an acyl group to CoA. For example, the joining of a palmitoyl group to CoA yields *palmitoyl CoA*. Acyl CoA synthesis requires an input of energy derived from the hydrolysis of ATP to AMP and PP_i.

$$
\text{R}-\overset{\overset{\displaystyle O}{\|}}{\text{C}}-\text{O}^- + \text{ATP} + \text{CoA-SH} \; \rightleftharpoons \; \text{R}-\overset{\overset{\displaystyle O}{\|}}{\text{C}}-\text{S-CoA} + \text{AMP} + \text{PP}_i
\qquad \text{(equation 10.2)}
$$

Fatty acid   **Acyl CoA**

The reaction is shifted to the right by the subsequent exergonic hydrolysis of PP_i to two P_i according to equation 2.7. Acyl CoA synthesis is catalyzed by *acyl CoA synthetase,* an enzyme located in the plasma membrane, the endoplasmic reticulum membrane, and the outer mitochondrial membrane.

Triacylglycerol synthesis from the activated substrates proceeds with the attachment of two acyl groups, borne by two acyl CoA, to positions 1 and 2 of glycerol 3-phosphate.

$$
\begin{array}{c}
\text{H}_2\text{C}-\text{OH} \\
| \\
\text{HO}-\text{CH} \qquad + \text{R}_1-\overset{\overset{\displaystyle O}{\|}}{\text{C}}-\text{S-CoA} + \text{R}_2-\overset{\overset{\displaystyle O}{\|}}{\text{C}}-\text{S-CoA} \\
| \\
\text{H}_2\text{C}-\text{OPO}_3^{2-}
\end{array}
$$

$$\downarrow \qquad\qquad \text{(equation 10.3)}$$

$$
\begin{array}{c}
\qquad\qquad \text{H}_2\text{C}-\text{O}-\overset{\overset{\displaystyle O}{\|}}{\text{C}}-\text{R}_1 \\
\text{R}_2-\overset{\overset{\displaystyle O}{\|}}{\text{C}}-\text{O}-\text{CH} \qquad + 2\ \text{CoA-SH} \\
| \\
\text{H}_2\text{C}-\text{OPO}_3^{2-}
\end{array}
$$

The reaction is catalyzed by *glycerol phosphate acyltransferase*. The product is phosphatidate, familiar from section 5.8 and figure 5.12, which is finally converted to triacylglycerol after losing its phosphoryl group by hydrolysis according to equation 10.4 and accepting a third acyl group derived from yet another acyl CoA according to equation 10.5.

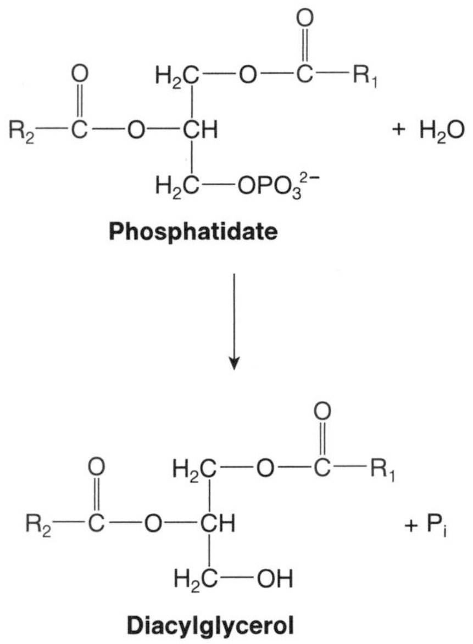

Phosphatide $+ H_2O$

Diacylglycerol $+ R_3-C-S\text{-}CoA$

(equation 10.4)

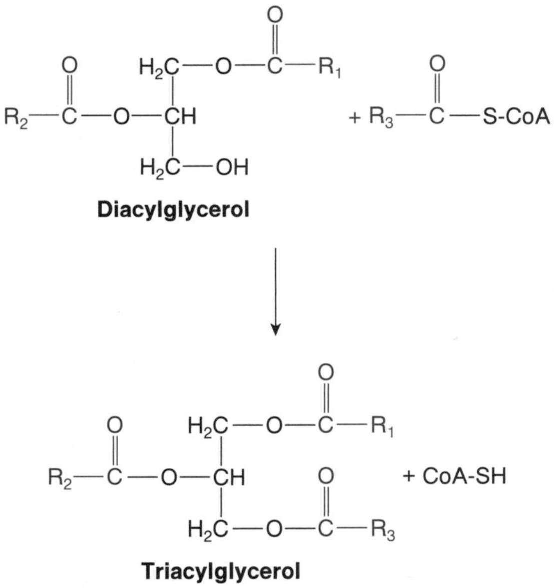

(equation 10.5)

Diacylglycerol $+ P_i$

Triacylglycerol $+ CoA\text{-}SH$

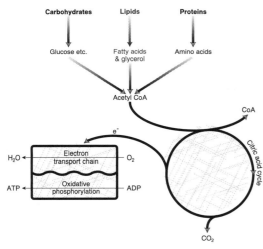

▶ You are here: lipolysis.

The two reactions are catalyzed by the *triacylglycerol synthase complex,* which is bound to the endoplasmic reticulum membrane.

Lipolysis

Triacylglycerol breakdown is called **lipolysis.** Lipolysis takes place in the cytosol with the aid of two **lipases,** *triacylglycerol lipase* and *monoacylglycerol lipase.* The former catalyzes the hydrolysis of the ester linkages at positions 1 and 3 of the glycerol unit to yield *2-monoacylglycerol* and two fatty acids.

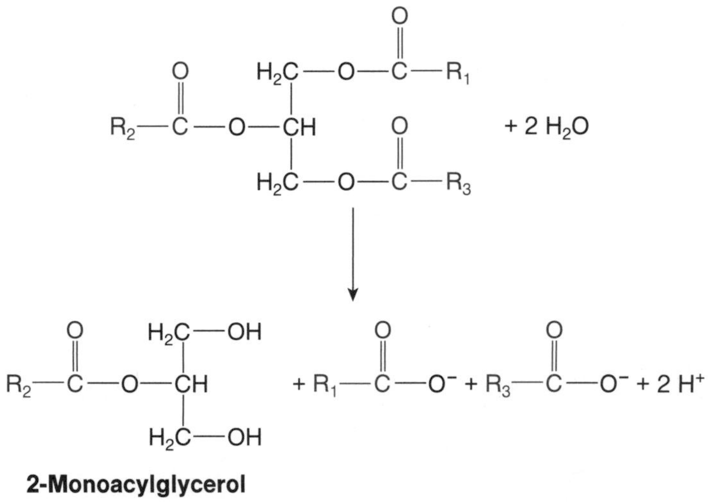

$+ 2 H_2O$

(equation 10.6)

2-Monoacylglycerol $+ R_1-C-O^- + R_3-C-O^- + 2 H^+$

Monoacylglycerol lipase then completes the dismantling of triacylglycerol by hydrolyzing the ester linkage at position 2 of the glycerol unit.

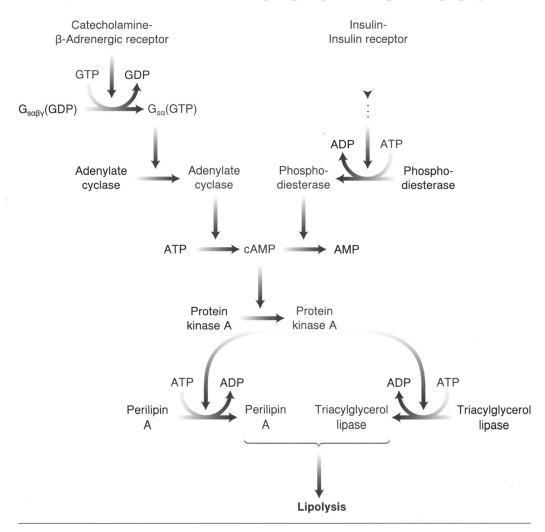

(equation 10.7)

The end products of lipolysis are three fatty acids and one glycerol per triacylglycerol.

In lipolysis, a triacylglycerol is hydrolyzed to three fatty acids and glycerol.

10.2   Exercise Speeds Up Lipolysis

Exercise speeds up lipolysis in adipose tissue through the augmented release of epinephrine from the adrenals and norepinephrine from the endings of sympathetic nerves. Epinephrine and norepinephrine bind to molecules of the β-adrenergic receptor located in the plasma membrane of adipocytes and stimulate the cAMP cascade (figure 10.3), resulting in the activation of PKA. PKA phosphorylates two proteins playing crucial

Catecholamines are *lipolytic*.

▶ **Figure 10.3**   Hormonal control of lipolysis. Epinephrine or norepinephrine binding to the β-adrenergic receptor activates adenylate cyclase, resulting in a rise of the [cAMP]. This activates PKA, which phosphorylates perilipin A and triacylglycerol lipase, thus speeding up lipolysis. Insulin, in contrast, slows down lipolysis by promoting the hydrolysis of cAMP to AMP.

roles in the acceleration of lipolysis. One is perilipin A, the major perilipin isoform: Phosphorylation attracts triacylglycerol lipase to the lipid droplets. The other is triacylglycerol lipase itself: Phosphorylation activates it. As a result, the rate of triacylglycerol hydrolysis to 2-monoacylglycerol and two fatty acids is greatly increased. 2-Monoacylglycerol is then rapidly hydrolyzed to glycerol and fatty acid.

Unlike catecholamines, insulin slows down lipolysis by stimulating the phosphorylation and activation of a *phosphodiesterase.* This enzyme hydrolyzes cAMP to AMP, thus deactivating the cAMP cascade.

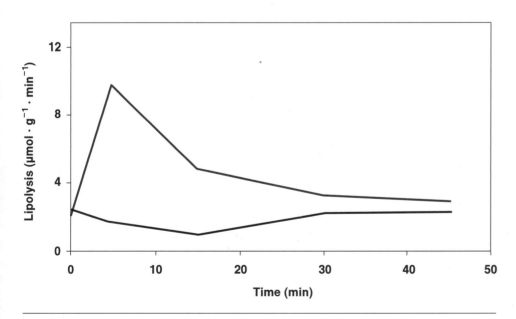

$$\textbf{cAMP} + H_2O \rightleftharpoons \textbf{AMP} + H^+ \qquad \text{(equation 10.8)}$$

In addition, insulin promotes the activation of protein phosphatases that dephosphorylate perilipin A and triacylglycerol lipase. Through these effects, insulin inhibits the lipolytic action of catecholamines. Because of the dependence of its activity on catecholamines and insulin, triacylglycerol lipase is also known as **hormone-sensitive lipase.**

Insulin is *antilipolytic.*

In addition to being antilipolytic, insulin promotes triacylglycerol synthesis in adipose tissue. However, as we have seen, insulin secretion decreases during exercise, which favors lipolysis over triacylglycerol synthesis.

How fast is lipolysis stimulation during exercise? Because of its dependence on the cAMP cascade, stimulation takes some time. In vitro experiments have shown that triacylglycerol lipase can be activated within 1 to 2 min, while an in vivo study showed a sixfold increase in the lipolytic rate in human adipose tissue 5 min after the onset of light exercise (figure 10.4). The lipolytic rate then decreased although exercise continued, which was probably due to *desensitization* of the β-adrenergic receptor, that is, loss of its responsiveness to catecholamines.

▶ **Figure 10.4**   Lipolysis during exercise. The lipolytic rate (expressed as micromoles of triacylglycerols hydrolyzed per gram of adipose tissue per minute) peaks at the fifth minute of light bicycling lasting 30 min (colored line). The lipolytic rate is considerably lower at the 15th and 30th minute of exercise, as well as at the 15th minute of passive recovery. The black line depicts the lipolytic rate in a control group that did not exercise.

Adapted, by permission, from A. Petridou and V. Mougios, 2002, "Acute changes in triacylglycerol lipase activity of human adipose tissue during exercise," *Journal of Lipid Research* 43: 1333.

The Control Group

In addition to the *study group,* which receives an intervention (for example, exercise), many scientific studies employ a *control group,* which receives no intervention but is subjected to the same measurements as the study group. The two groups must be comparable as to the characteristics that may affect the outcome (for example, sex). The researcher conducting such a **controlled study** discovers the effect of the intervention not only by examining the changes in the study group but also by taking into account the changes in the control group. In this way, the researcher reduces unforeseen factors that might affect the validity of the results. In the study shown in figure 10.4, a control group was included to eliminate the possibility that the observed increase in the lipolytic rate with exercise was due to excitation of the adipose tissue by the stress of the biopsy performed before the onset of exercise (at 0 time).

The desensitization of the β-adrenergic receptor is due to phosphorylation by a kinase that is normally located in the cytosol but is attracted to the plasma membrane by $G_{s\beta\gamma}$ (figure 10.5), one of the two parts into which G_s is split after a catecholamine binds to the receptor (see section 9.3). The kinase phosphorylates the receptor where it interacts with $G_{s\alpha}$. Phosphorylation creates a binding site for another protein, β-*arrestin,* which prevents the receptor from interacting with other $G_{s\alpha}$ molecules, thus blocking the cAMP cascade. What is more, β-arrestin binding promotes the sequestering of the receptor from the plasma membrane to the cell interior by **endocytosis,** that is, the formation of intracellular vesicles by invagination of the plasma membrane. The receptor molecules in these vesicles are dephosphorylated and, rejuvenated, return to the

Sensitivity of a cell or organ or tissue to a hormone is the hormone concentration required to produce a half-maximal response. For example, we can measure the sensitivity of adipose tissue to catecholamines by determining the catecholamine concentration that causes 50% of the maximal rate of appearance of glycerol, one of the products of lipolysis, in the extracellular fluid.

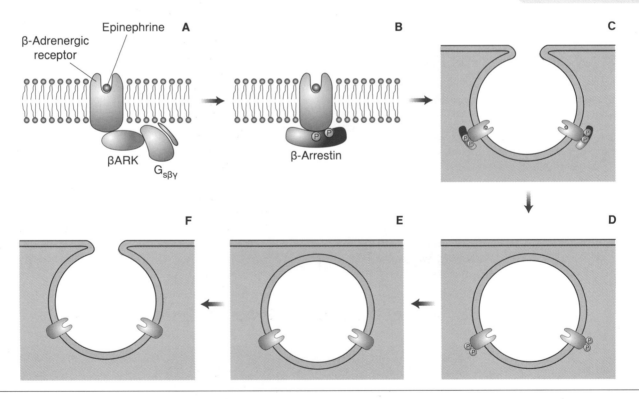

▶ **Figure 10.5** Desensitization of the β-adrenergic receptor. *(a)* $G_{s\beta\gamma}$, produced by the splitting of G_s after a catecholamine binds to the β-adrenergic receptor (refer to figure 9.10), draws β-adrenergic receptor kinase (βARK) to the plasma membrane. *(b)* βARK phosphorylates the receptor, and the added phosphoryl groups attract β-arrestin, which blocks stimulation of the cAMP cascade by catecholamines. *(c)* In addition, β-arrestin causes invagination of the plasma membrane. *(d)* Receptor molecules are removed from the plasma membrane in endocytotic vesicles. *(e)* The receptor is dephosphorylated in the vesicles. *(f)* The vesicles fuse with the plasma membrane, and the functional receptor returns to its original position.

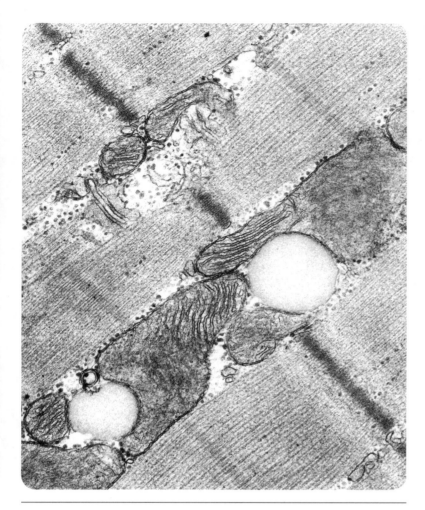

plasma membrane by exocytosis, restoring the sensitivity of the cell to catecholamines after cessation of the hormonal stimulus.

Desensitization is a phenomenon observed in many biological responses. It appears to be important in protecting cells against over-stimulation and causing them to respond to changes in the amount of a stimulus rather than to constantly high values of it. It looks as if cells get tired of receiving the same stimulus continuously. Desensitization is boredom at the cellular level.

What effect does exercise intensity have on the lipolytic rate? Apparently, little. Investigators have measured the lipolytic rate in adipose tissue or in the whole body during exercise tasks ranging in intensity from 25% to 85% of $\dot{V}O_2$max. The results show that even light exercise can elicit maximal lipolytic rate, probably because the amounts of catecholamines secreted are sufficient to maximally stimulate the cAMP cascade. Note, however, that similar lipolytic rates at different exercise intensities do not mean similar slimming effects. For the body to lose fat, the lipolytic products have to be utilized for energy production, and this does depend on exercise intensity. If the lipolytic products are not consumed, the body uses them to resynthesize fat.

Skeletal muscles too contain triacylglycerols either in adipocytes scattered among muscle fibers or inside the muscle fibers themselves. We call the latter *myocellular triacylglycerols*. The triacylglycerol content of muscle fibers generally ranges from 0.2% to 0.8%. Myocellular triacylglycerols are gathered in lipid droplets close to the mitochon-

Figure 10.6 Triacylglycerols in a muscle fiber. Two lipid droplets lie in touch with mitochondria and myofibrils in this electron micrograph of a muscle fiber. The proximity with the mitochondria minimizes the distance that the fatty acids produced from lipolysis need to cross in order to yield energy. As in figure 9.2, you can see glycogen granules scattered in the sarcoplasm.

Courtesy of Dr. H. Hoppeler, University of Bern. Photographers H. Claassen and F. Grabe.

dria (figure 10.6). These droplets are much smaller than the lipid droplets in adipocytes and are not covered with perilipin. The breakdown of myocellular triacylglycerols is also accelerated during exercise, since muscle fibers contain hormone-sensitive lipase. However, we know less about the control of lipolysis in muscle fibers compared to adipocytes.

10.3   Fate of the Lipolytic Products During Exercise

A large portion of the fatty acids and glycerol produced by lipolysis in adipocytes exits to the extracellular space and then to the circulation. Thus, the acceleration of lipolysis during exercise leads to an increase in the rate of appearance of fatty acids and glycerol in the plasma (figure 10.7). Glycerol is soluble in water and circulates easily, but fatty acids are poorly soluble. For this reason they are carried by **albumin,** the most abundant plasma protein (at a concentration of around 40 g · L^{-1}). Albumin binds up to 10 fatty acids; it also serves to carry other substances of low solubility in water.

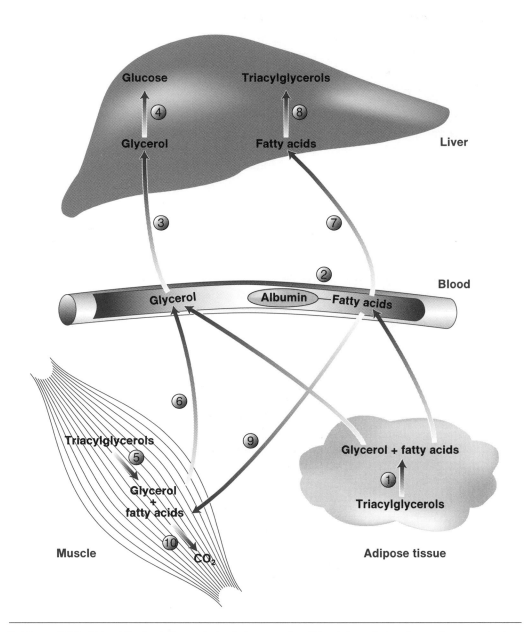

▶ **Figure 10.7** Fate of the lipolytic products during exercise. Triacylglycerols in adipose tissue are broken down to fatty acids and glycerol (1), which are released into the bloodstream and are taken up by other tissues. Fatty acids circulate bound to albumin (2). Glycerol enters the liver (3), which uses it to make glucose (4). Muscle glycerol, derived from the hydrolysis of myocellular triacylglycerols (5), has the same fate (6). Blood-borne fatty acids can either enter the liver (7), which utilizes them to produce triacylglycerols (8), or enter the muscles (9). There they join the fatty acids produced from the breakdown of myocellular triacylglycerols and get oxidized to CO_2 (10).

Plasma glycerol is taken up mainly by the liver and is converted to glycerol 3-phosphate by the action of *glycerol kinase.*

glycerol + ATP $\rightleftharpoons$ glycerol 3-phosphate + ADP + H⁺   **(equation 10.9)**

Glycerol 3-phosphate is then converted to dihydroxyacetone phosphate by a reversal of reaction 10.1 (equation 10.1). As an intermediate of both glycolysis and gluconeogenesis, dihydroxyacetone phosphate can go either way. However, most of it goes in the direction of gluconeogenesis, since this pathway is accelerated, whereas glycolysis is decelerated, in the liver during exercise (section 9.20). Another considerable source of plasma glycerol is the hydrolysis of myocellular triacylglycerols; this glycerol seems to share the fate of glycerol originating from adipose tissue. Thus, triacylglycerols offer a small part of their structure to the support of gluconeogenesis during exercise.

Plasma fatty acids originating from adipose tissue can be taken up by the liver and used for triacylglycerol synthesis. However, most of them are taken up by the exercising muscles and are degraded to regenerate ATP, as we will see in the next section. The entry of fatty acids into cells is facilitated by membrane proteins, which are not well characterized. One of them is *fatty acid-binding protein* (FABP); another is *fatty acid translocase* (FAT), or CD36.

In addition to the entrance of fatty acids into muscles from the plasma, fatty acids derived from the hydrolysis of myocellular triacylglycerols are also degraded to produce energy. The two sources are thought to contribute similarly to the energy demands of moderate-intensity exercise.

10.4   Fatty Acid Degradation

Fatty acids are degraded through the pathway of β oxidation, which occurs in the mitochondria. To get there, fatty acids that have entered the cytosol from outside the cell or have been generated within the cell must be activated and then cross the two mitochondrial membranes.

Fatty acids are activated in the manner discussed in section 10.1, that is, by being converted to acyl CoA (equation 10.2). When their acyl groups have 14 carbons or more (which is the case for almost all acyl groups in our cells), acyl CoA readily cross the outer mitochondrial membrane but not the inner one. **Carnitine,** a compound derived from the amino acid lysine, carries these acyl groups into the mitochondria. Carnitine detaches the acyl group from acyl CoA, producing CoA and *acyl carnitine.*

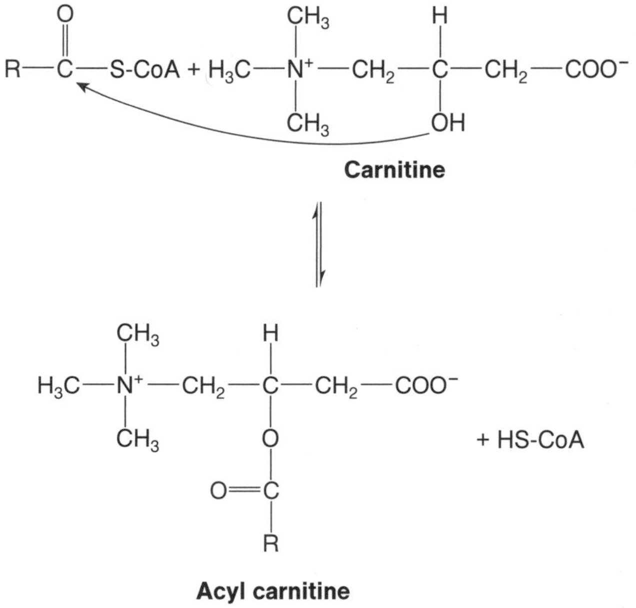

Carnitine

(equation 10.10)

Acyl carnitine

Carnitine acyltransferase I or II is also called *carnitine palmitoyltransferase I or II* because researchers measure its activity by using palmitoyl CoA as substrate.

Carnitine acyltransferase I, an integral protein of the outer mitochondrial membrane, catalyzes the reaction. The acyl carnitine produced crosses the inner mitochondrial membrane with the help of a transmembrane protein called *translocase* (figure 10.8). In the mitochondrial matrix, reaction 10.10 (equation 10.10) is reversed, and the acyl group is linked to CoA again. *Carnitine acyltransferase II,* an enzyme attached to the matrix side of the membrane, catalyzes this reversal. Thus, acyl CoA emerges in the

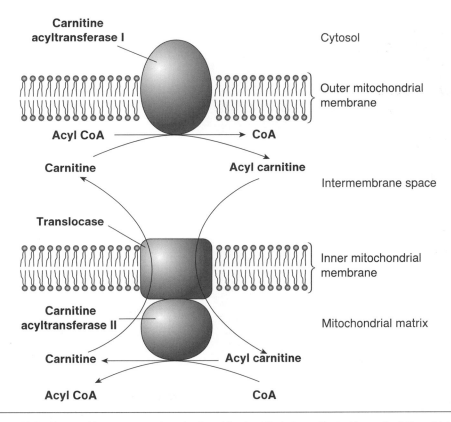

▶ **Figure 10.8** Fatty acid transport to the mitochondria. Acyl CoA, the activated form of a fatty acid, is transported to the mitochondrial matrix "disguised" as acyl carnitine. Carnitine acyltransferases I and II catalyze the interconversion of acyl CoA and acyl carnitine, while translocase transports acyl carnitine and carnitine in opposite directions across the inner mitochondrial membrane.

mitochondrial matrix. Finally, carnitine returns to the intermembrane space through the translocase in exchange for the entry of an acyl carnitine.

Once in the mitochondrial matrix, acyl CoA enters the pathway of β oxidation. This is a series of four reactions ending in the detachment of two carbons in the form of acetyl CoA from the carboxyl end of a fatty acid. The β oxidation reactions are, in sequence, a dehydrogenation (FAD being the hydrogen acceptor), a hydration (addition of a water molecule), another dehydrogenation (NAD^+ being the hydrogen acceptor), and a splitting with the help of CoA (figure 10.9). The products of one "round" of β oxidation are one acyl CoA that is shorter than the initial one by two C, one acetyl CoA, one $FADH_2$, and one NADH.

A new round of β oxidation follows, resulting in the shortening of acyl CoA by another two C. This is repeated as many times as it takes to fully degrade a fatty acid. As an example, follow the fate of palmitate in figure 10.10. The vast majority of fatty acids have even numbers of carbons; thus a fatty acid having n C is degraded to $n/2$ acetyl CoA. These are produced in $n/2 - 1$ rounds of β oxidation because the final round yields two acetyl CoA.

The end products of β oxidation (acetyl CoA, $FADH_2$, and NADH) are funneled to pathways that are known to you from carbohydrate oxidation, without even leaving the mitochondria. Acetyl CoA enters the citric acid cycle, where it is oxidized to two

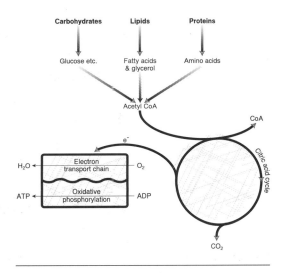

▶ You are here: β oxidation.

In β oxidation, a fatty acid is degraded to a number of acetyl CoA equaling half the number of its carbon atoms.

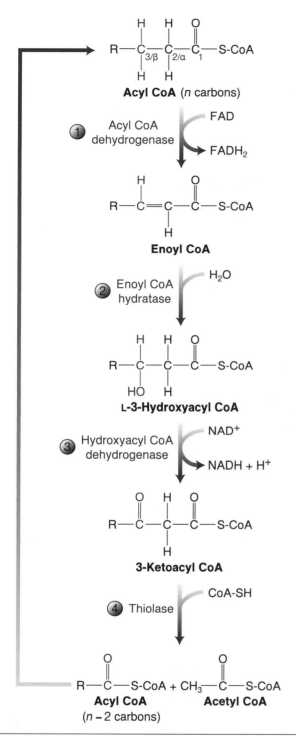

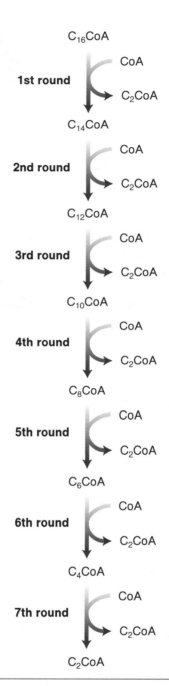

▶ **Figure 10.9**   β Oxidation of fatty acids. Through a series of four reactions, an acyl CoA is broken down to another acyl CoA lacking two C and to acetyl CoA. Only the three terminal carbons on the side of the carboxyl group are presented, since only these take part in the reactions. The acyl CoA produced follows the fate of the initial one through a new round of β oxidation (upward arrow at left). β Oxidation owes its name to the fact that it results in the oxidation of C3 (in color), which, according to older nomenclature, is the β carbon (C2, next to the carboxyl group, is the α carbon). Indeed, compare the initial CoA to the product of the third reaction, 3-ketoacyl CoA, and see that the two H of C3/β have been replaced by an O.

▶ **Figure 10.10**   Palmitate degradation. Starting with the activated form of palmitate (palmitoyl CoA, symbolized as C_{16}CoA), seven rounds of β oxidation detach acetyl CoA (C_2CoA) molecules one after the other until nothing but C_2CoA is left. Each pair of arrows (straight and curved) in this scheme is equivalent to all the downward arrows in the previous figure.

CO$_2$ while yielding three NADH, one FADH$_2$, and one GTP (section 9.9). NADH and FADH$_2$ (from β oxidation and from the citric acid cycle) are oxidized in the electron transport chain to produce large amounts of ATP through oxidative phosphorylation (sections 9.11 to 9.13).

As in the citric acid cycle, oxygen is not directly involved in β oxidation. However, the FAD and NAD$^+$ that are consumed in the process can be regenerated in the mitochondria only through the electron transport chain. Thus, β oxidation requires O$_2$ to operate and must be considered an aerobic process. In addition, the processes through which the products of β oxidation yield energy (that is, the citric acid cycle, the electron transport chain, and oxidative phosphorylation) are also aerobic. Therefore, fatty acid degradation as a whole is aerobic. In fact, there is no anaerobic alternative for energy production from fatty acids as there is with carbohydrates.

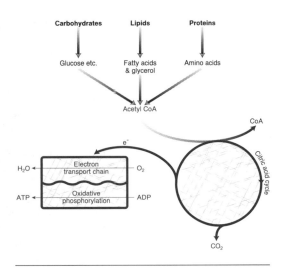

▶ You are here: citric acid cycle, electron transport chain, and oxidative phosphorylation.

10.5 Energy Yield of Fatty Acid Oxidation

Let's wrap up fatty acid oxidation by first keeping in mind the example of palmitate and then generalizing. As you can deduce from figure 10.10 and from the discussion in the previous section, eight acetyl CoA are produced after seven rounds of β oxidation. In the citric acid cycle, these acetyl CoA yield 8 · 2 = 16 CO$_2$, 8 · 3 = 24 NADH, eight FADH$_2$, and eight GTP. Add seven NADH and seven FADH$_2$ produced in the seven rounds of β oxidation, and we are ready to count ATP.

24 + 7 = 31 NADH yield approximately 31 · 2.5 = 77.5 ATP.

8 + 7 = 15 FADH$_2$ yield approximately 15 · 1.5 = 22.5 ATP.

8 GTP are equivalent to 8 ATP.

The result is a sum of about 108 ATP. This value has to be lowered by two ATP that are consumed to activate a fatty acid, although only one ATP seems to be consumed in reaction 10.2 (equation 10.2). The reason is that ATP is hydrolyzed to AMP and PP$_i$, which is further hydrolyzed to two P$_i$. Thus, two phosphoanhydride linkages are broken down, which is equivalent to the loss of two ATP. Therefore the net yield of palmitate oxidation is approximately 108 – 2 = 106 ATP.

Now you can easily show that, in general, a fatty acid of n carbons yields $7n - 6$ ATP. This holds for saturated fatty acids such as palmitate and stearate. If a fatty acid is unsaturated, it follows a slightly different route of degradation in a number of rounds of β oxidation that is equal to its number of double bonds. As a result, it yields, on average, two ATP per double bond fewer than the corresponding saturated fatty acid.

10.6 Fatty Acid Synthesis

Let's consider now how our cells synthesize fatty acids. The starting material is the product of β oxidation, acetyl CoA; and fatty acids grow in two-carbon steps just as they wither. However, the path to synthesis is not the reverse of the path to degradation, as we have seen with the other anabolic pathways that we have encountered so far (glycogen synthesis, gluconeogenesis, and triacylglycerol synthesis). In fact, the differences between fatty acid synthesis and degradation are more pronounced: The two processes take place in different intracellular compartments, and all the reactions and enzymes in the two processes are different.

Preparing the Substrates

Fatty acids are synthesized in the cytosol, but acetyl CoA is produced in the mitochondrial matrix (either from pyruvate oxidation or from β oxidation or, as we will see in the next chapter, from the breakdown of some amino acids). Because acetyl CoA cannot cross the mitochondrial membranes, citrate, formed in the mitochondria from acetyl CoA and oxaloacetate in the first reaction of the citric acid cycle, is transported to the cytosol, where it is broken down to these compounds by *citrate lyase*.

citrate + CoA + ATP → acetyl CoA + oxaloacetate + ADP + P$_i$ **(equation 10.11)**

Thus, acetyl CoA appears in the cytosol at the expense of ATP.

The linking of acetyl groups from acetyl CoA to form a fatty acid is not favored thermodynamically. An even more activated form of the acetyl group is required, and that is *malonyl CoA*. This compound is synthesized from acetyl CoA and bicarbonate by *acetyl CoA carboxylase* at the expense of another ATP.

$$\text{CH}_3 - \overset{\overset{\displaystyle O}{\|}}{\text{C}} - \text{S-CoA} + \text{HCO}_3^- + \text{ATP}$$

Acetyl CoA

> Although the malonyl group has three carbons, only two are used in fatty acid synthesis.

(equation 10.12)

$$^-\text{OOC} - \text{CH}_2 - \overset{\overset{\displaystyle O}{\|}}{\text{C}} - \text{S-CoA} + \text{ADP} + \text{P}_i + \text{H}^+$$

Malonyl CoA

Palmitate Synthesis

Fatty acid synthesis from malonyl CoA is catalyzed by *fatty acid synthase*. This remarkable *multienzyme* catalyzes all the reactions required for synthesis at different active sites. In fact, a fatty acid does not leave the enzyme until synthesis is completed.

In short, fatty acid synthesis begins with the covalent linking of an acetyl group from acetyl CoA and a malonyl group from malonyl CoA to two distinct sulfhydryl groups in fatty acid synthase (figure 10.11). Then the acetyl group is transferred to the middle carbon of the malonyl group, and the carboxyl group of the latter is detached as CO_2. This results in the formation of a four-carbon group, which is subsequently hydrogenated (NADPH, introduced in figure 2.7, being the hydrogen donor), dehydrated (it loses a water molecule), and hydrogenated again (with NADPH as the hydrogen donor again). In the end of this first round of fatty acid synthesis, an acyl group of four C (called *butyryl group*) is produced.

For the second round to begin, the butyryl group shifts to the site where the acetyl group had been. This permits a second malonyl group to be linked to the enzyme and two of its carbons to be added to the butyryl group. Thus, every round results in the extension of the aliphatic chain by two C (figure 10.12). Elongation usually stops at the end of the seventh round. The palmitoyl group formed is detached from the enzyme by hydrolysis, resulting in the release of palmitate.

Energy Demand

How much energy does the anabolic pathway of palmitate synthesis require? As described earlier and shown in figure 10.12, palmitate is synthesized from one acetyl CoA and seven malonyl CoA. The latter are, in turn, synthesized from seven acetyl CoA at the expense of seven ATP (equation 10.12). In addition, eight ATP are consumed to get

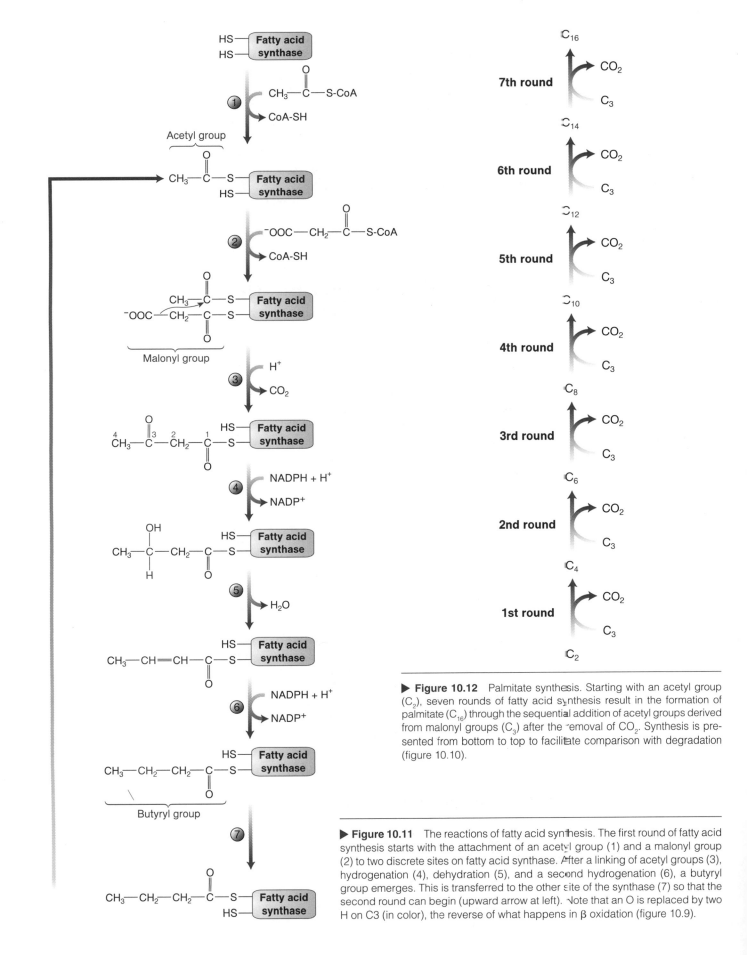

▶ **Figure 10.12** Palmitate synthesis. Starting with an acetyl group (C_2), seven rounds of fatty acid synthesis result in the formation of palmitate (C_{16}) through the sequential addition of acetyl groups derived from malonyl groups (C_3) after the removal of CO_2. Synthesis is presented from bottom to top to facilitate comparison with degradation (figure 10.10).

▶ **Figure 10.11** The reactions of fatty acid synthesis. The first round of fatty acid synthesis starts with the attachment of an acetyl group (1) and a malonyl group (2) to two discrete sites on fatty acid synthase. After a linking of acetyl groups (3), hydrogenation (4), dehydration (5), and a second hydrogenation (6), a butyryl group emerges. This is transferred to the other site of the synthase (7) so that the second round can begin (upward arrow at left). Note that an O is replaced by two H on C3 (in color), the reverse of what happens in β oxidation (figure 10.9).

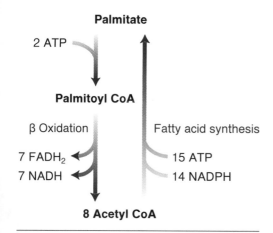

Palmitate

2 ATP

Palmitoyl CoA

β Oxidation   Fatty acid synthesis

7 FADH$_2$ 15 ATP
7 NADH   14 NADPH

8 Acetyl CoA

▶ **Figure 10.13** Comparing β oxidation to fatty acid synthesis. The conversion of palmitate to acetyl CoA requires two ATP and yields seven NADH and seven FADH$_2$, whereas the conversion of mitochondrial acetyl CoA to palmitate in the cytosol requires 15 ATP and 14 NADPH.

Don't be misled into comparing palmitate synthesis to its complete oxidation (which yields about 106 ATP), as the end product of this process is CO$_2$, not acetyl CoA (the starting material of fatty acid synthesis).

the eight acetyl CoA needed for palmitate synthesis from the mitochondria to the cytosol (equation 10.11). Thus, the synthesis of palmitate from acetyl CoA requires 15 ATP, as opposed to two ATP required to activate palmitate and get β oxidation started (figure 10.13).

In addition, palmitate synthesis requires the reducing power of 14 NADPH (two in each of the seven rounds of fatty acid synthesis), which could otherwise direct their electrons to the electron transport chain and yield ATP through oxidative phosphorylation. The demand for NADPH is counterbalanced by the production of seven NADH and seven FADH$_2$ in the seven rounds of β oxidation, although the production of NADPH (which we will not consider) costs more than that of either NADH or FADH$_2$. Once more, we see that a biosynthetic process is more expensive than the reverse degradation process is profitable.

Synthesis of Other Fatty Acids

How do our cells synthesize fatty acids other than palmitate? Shorter fatty acids (composing a small percentage of the total) can be formed when an acyl group is detached from fatty acid synthase at the end of fewer than seven rounds. Longer fatty acids are synthesized by *elongases* lying on the cytosolic side of the endoplasmic reticulum membrane. These enzymes extend palmitate and other fatty acids by catalyzing the same reactions that fatty acid synthase does.

The endoplasmic reticulum membrane also contains *desaturases,* that is, enzymes introducing double bonds and converting saturated fatty acids to unsaturated ones. Desaturases may act before or after elongases. Desaturases remove two hydrogens from adjacent carbons in an acyl CoA. The hydrogen acceptor in this case is one oxygen atom from O$_2$, the other O reacting with NADH to yield NAD$^+$. Two H$_2$O are thus formed.

$$\cdots\!\!-\!\!\overset{\displaystyle H}{\underset{\displaystyle H}{C}}\!\!-\!\!\overset{\displaystyle H}{\underset{\displaystyle H}{C}}\!\!-\!\!\cdots + NADH + H^+ + O_2 \longrightarrow \quad \overset{H}{C}\!\!=\!\!\overset{H}{C} \quad + NAD^+ + 2\,H_2O \qquad \textbf{(equation 10.13)}$$

Essential Fatty Acids

Humans and other mammals have no enzymes capable of introducing double bonds beyond C10 of a fatty acid. Thus we are unable to synthesize, for example, *linoleate* and *α-linolenate* (table 5.1). To obtain these fatty acids we must eat organisms (such as plants and fish) that synthesize them. Linoleate and α-linolenate are important components or precursors of important components of membrane phospholipids and are termed **essential fatty acids.** They are also known as representatives of the *n* – 6 and *n* – 3 fatty acids, respectively (*n* being the number of carbons). This denotes that the ultimate double bond of linoleate is between carbon 18 – 6 = 12 and carbon 13, while that of α-linolenate is between carbon 18 – 3 = 15 and carbon 16. An equivalent nomenclature is ω6 (omega 6) and ω3, ω being the ultimate (methyl) carbon.

10.7   Exercise Speeds Up Fatty Acid Oxidation in Muscle

Exercise speeds up the oxidation of fatty acids in muscle mainly by increasing their concentration. This increase is due to

    the stimulation of lipolysis in adipose tissue,

    the augmented blood flow to the exercising muscles (resulting in augmented delivery of fatty acids), possibly the enhanced translocation of FAT/CD36 from an intracellular reservoir to the plasma membrane (in a manner similar to the translocation of GLUT4 in response to exercise or insulin), and

the stimulation of lipolysis in the muscles themselves.

Thanks to these changes, more fatty acids are converted to acyl CoA and enter the mitochondria, where they are broken down to acetyl CoA through β oxidation. The latter is accelerated as the concentrations of the substrates for its reactions are increased one after another. Additionally, β oxidation is accelerated during exercise as its products, acetyl CoA, $FADH_2$, and NADH, are avidly consumed in the accelerated citric acid cycle, electron transport chain, and oxidative phosphorylation (see sections 9.10 and 9.15).

While speeding up fatty acid oxidation, exercise may slow down fatty acid synthesis in muscle by affecting acetyl CoA carboxylase. The enzyme is inhibited by phosphorylation, catalyzed by an *AMP-dependent protein kinase.* The increase in [AMP] during exercise activates the kinase, which phosphorylates and deactivates acetyl CoA carboxylase. Thus, less malonyl CoA may be synthesized and less substrate may be available for fatty acid synthesis. A decrease in malonyl CoA might also speed up fatty acid oxidation, since malonyl CoA inhibits carnitine acyltransferase I, the enzyme that lets fatty acids enter the mitochondria. Although attractive, the hypothesis that exercise affects fatty acid metabolism through a decrease in malonyl CoA has not been verified experimentally.

> The AMP-dependent protein kinase is different from the cAMP-dependent protein kinase A that participates in the cAMP cascade.

The maximal rate of ATP resynthesis through the oxidation of fatty acids derived from the hydrolysis of myocellular triacylglycerols is only 0.3 mmol per kilogram of muscle per second. The corresponding rate for the fatty acids derived from adipose tissue is even lower, just $0.2 \; mmol \cdot kg^{-1} \cdot s^{-1}$, because the entry of these fatty acids to the muscle fibers is rather slow. As a result, fatty acids cannot support hard exercise. In contrast, they are the major source or a significant source of energy in light and moderate-intensity exercises.

Because fatty acid oxidation makes an important contribution to energy supply during endurance exercise, several efforts to increase endurance performance have targeted increasing fatty acid availability in the mitochondria. One of the means employed is carnitine supplementation, the rationale being that boosting the carnitine concentration in the exercising muscles would increase the rate of fatty acid transport into the mitochondria. Although many athletes involved in endurance activities apparently take carnitine supplements as **ergogenic aids,** controlled studies have shown no effect on performance. Unfortunately, carnitine supplementation does not increase the muscle carnitine content. A few studies have also examined whether carnitine supplementation lowers body fat (another purported effect), again with no encouraging results.

10.8 Changes in the Plasma Fatty Acid Concentration and Profile During Exercise

The concentration of fatty acids in the plasma depends on their rate of appearance, primarily from adipose tissue, and their rate of disappearance toward other tissues, notably muscle. Contrary to the glucose concentration, the fatty acid concentration in the plasma is not subject to strict control. Resting values range from 0.3 to $0.9 \; mmol \cdot L^{-1}$.

Exercise increases the demand for fatty acids in muscle fibers. If the increased demand is not met by an increased hydrolysis of myocellular triacylglycerols, the fatty acid concentration in the cytosol will decrease. This makes the concentration gradient across the sarcolemma (the gradient is always inward) steeper and tends to draw in more fatty acids from the plasma, thus decreasing their concentration there. On the other hand, the acceleration of lipolysis in adipose tissue tends to increase the fatty acid concentration in the plasma.

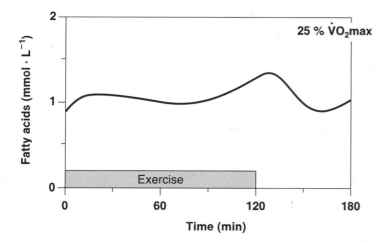

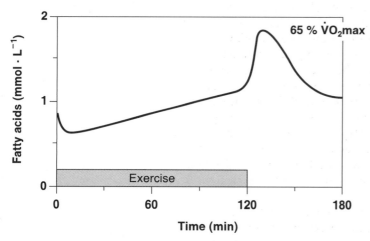

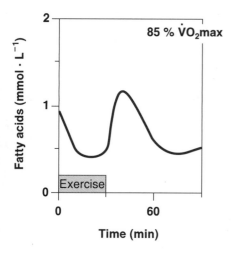

▶ **Figure 10.14** Plasma fatty acid concentration during exercise. Exercise intensity affects how the plasma fatty acid concentration changes. The graphs show average values for endurance cyclists who exercised at 25%, 65%, and 85% of V̇O₂max. Exercise lasted 120 min at the first two intensities but only 30 min at the third intensity because of exhaustion.

Adapted, by permission, from J.A. Romijn et al., 1993, "Regulation of endogenous fat and carbohydrate metabolism in relation to exercise intensity and duration," *American Journal of Physiology* 265: E380-E391.

Thus, the direction of the concentration of plasma fatty acids during exercise depends on which of the factors listed earlier will prevail. This, in turn, depends on exercise intensity. During light exercise, the concentration rises (figure 10.14) because the demand for fatty acids in muscle is low and the lipolytic rate in adipose tissue is high. Moderate-intensity exercise is characterized by an initial drop because the increase in fatty acid degradation is faster than the hormonal stimulation of lipolysis. However, as lipolysis catches up with fatty acid degradation, the concentration rises and can exceed 2 mmol · L⁻¹ in very prolonged exercise tasks.

Finally, hard exercise drives the fatty acid concentration in the plasma below baseline and keeps it there. The main reason is a serious decrease in the blood flow to adipose tissue because of vasoconstriction, which prevents fatty acids from appearing in the circulation. The decrease in the blood flow to adipose tissue, along with the decrease in blood flow to the viscera, counterbalances the increase in the blood flow to the exercising muscles.

Soon after the end of exercise, the concentration of fatty acids in the plasma rises because the decrease in their use by the muscles is almost instantaneous, whereas the cessation of hormonal stimulation of lipolysis takes some minutes. (In section 13.22 I describe how the cAMP cascade is silenced after exercise.) Thus, the adipose tissue continues to supply the blood with increased amounts of fatty acids, while the demand has returned to the resting level. The jump in the plasma fatty acid concentration is higher after hard exercise because, in addition, the blood flow to adipose tissue returns to normal through vasodilation, which permits the fatty acids previously trapped in the extracellular fluid to appear in the bloodstream.

How the fatty acid concentration in the plasma responds to exercise may depend on training state also. Trained individuals experience a slightly lower increase compared to untrained individuals at a given moderate-intensity exercise.

Although all individual fatty acids increase in the plasma during moderate-intensity exercise, they do not increase proportionately. Instead, unsaturated fatty acids increase more than saturated fatty acids, resulting in an elevation of the proportion of the former to the latter. This shift in the profile of plasma fatty acids is in the direction of their major source, the triacylglycerols of adipose tissue, which have a higher proportion of unsaturated to saturated acyl groups. We do not know why this gap between the composition of adipose tissue triacylglycerols and plasma fatty acids exists, but exercise serves to bridge it partially and temporarily.

10.9   Interconversion of Lipids and Carbohydrates

From our discussion so far regarding carbohydrate and lipid metabolism, it should have become obvious that carbohydrates are preferable as an energy source during exercise, since either aerobically or, to a greater extent anaerobically, they resynthesize ATP faster than lipids do. However, the carbohydrate depots in the body are much smaller than the lipid depots. The question then arises: Can lipids be converted to carbohydrates?

Let's explore this question. The most abundant of lipids, triacylglycerols, are hydrolyzed to glycerol and fatty acids, as we have seen. Glycerol can be converted to glucose (section 10.3), but this does not have much of an impact because glycerol is only a small part of a triacylglycerol. Fatty acids, on the other hand, which compose the bulk of a triacylglycerol, are unable to produce glucose. Why? Fatty acid metabolism meets carbohydrate metabolism at acetyl CoA (figure 10.15). However, acetyl CoA cannot be converted to pyruvate, hence glucose, because the reaction catalyzed by the pyruvate dehydrogenase complex (equation 9.4) is irreversible. But

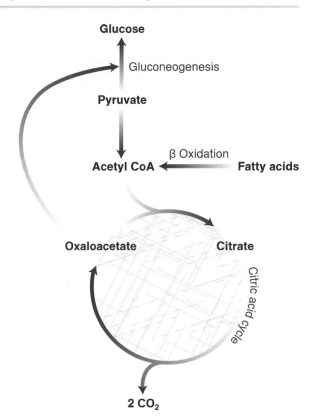

Ten grams of triacyl–glycerols gives rise to only 1 g of glucose.

▶ **Figure 10.15** Fatty acids do not form glucose. Fatty acids are broken down to acetyl CoA, which cannot be converted to pyruvate and then to glucose. Acetyl CoA can reach oxaloacetate, a precursor of glucose, through the citric acid cycle, but this does not represent a net conversion, since the carbons entering the cycle are lost as carbon dioxide.

what if acetyl CoA went all the way around the citric acid cycle to oxaloacetate, which can give rise to glucose through gluconeogenesis (refer to figure 9.28)? Although this seems a reasonable course, there is no net conversion of acetyl CoA to oxaloacetate, since the same number of carbon atoms that enter the cycle exit as CO_2.

Thus, we are unable to convert fatty acids to glucose. In contrast, we can convert glucose to fatty acids through acetyl CoA. Moreover, glucose yields glycerol 3-phosphate for triacylglycerol synthesis, as described in section 10.1. Thus, glucose *can* form triacylglycerols (figure 10.16), although a minimal amount of it is converted to fatty acids under physiological conditions.

It follows from this discussion that lipids cannot replenish carbohydrates adequately. Therefore, one has to ensure carbohydrate sufficiency in the body through proper nutrition.

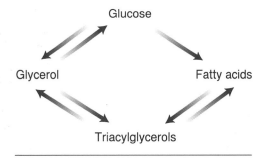

▶ **Figure 10.16** Interconversion of lipids and carbohydrates. Animals and humans can convert glucose to triacylglycerols, but they cannot convert the largest part of triacylglycerols (fatty acids) to glucose.

10.10 Plasma Lipoproteins

In addition to fatty acids, the plasma contains triacylglycerols. However, because they are insoluble in water they do not circulate in free form. Instead they are incorporated in **lipoproteins,** which carry other lipid classes as well. Let's see what lipoproteins are.

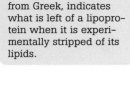

Lipoproteins are globular aggregates of lipids and proteins, as the name indicates. A lipoprotein particle contains thousands of lipid molecules and a few large protein molecules. The protein molecules hold the lipid molecules in place and serve as recognition sites for enzymes and receptors.

The arrangement of lipids in a lipoprotein particle is not random. Hydrophobic lipids such as triacylglycerols and cholesterol esters are sequestered in the interior, whereas amphipathic lipids such as cholesterol and phospholipids stay at the surface (figure 10.17). In fact, amphipathic lipids have a specific orientation: Their polar head groups point toward the aqueous environment, while their hydrophobic tails point to the core of the particle. Remember that phospholipids have the same orientation in the lipid droplets inside adipocytes (figure 10.2). Proteins are placed at the surface of lipoproteins and are called *apolipoproteins* or *apoproteins.*

▶ **Figure 10.17**   Lipoprotein structure. Lipoproteins have a specific arrangement of lipids and apoproteins in the core and shell, which is dictated by the affinity of these components for water.

Lipoproteins are not all alike; rather, they can be separated into classes by *centrifugation,* that is, spinning of a liquid mixture at high speed so as to separate its components according to their densities. If we centrifuge an amount of plasma, the lipoproteins contained in it will be arranged in zones of different densities (figure 10.18), reflecting differences in structure. The higher the proportion of apoproteins to lipids in a lipoprotein particle, the higher its density will be. The major lipoprotein classes, in order of increasing density, are **chylomicrons, very low-density lipoproteins (VLDL), low-density lipoproteins (LDL),** and **high-density lipoproteins (HDL).** Table 10.1 summarizes their features.

I must clarify that each lipoprotein particle is a mixture of chemical compounds and not a specific compound. For this reason it does not have an exact composition. It just happens that each of the four lipoprotein classes contains members of a rather uniform lipid composition and characteristic apoproteins. Thus, the proportions of lipoprotein components in table 10.1 are indicative and may differ slightly among bibliographic sources. To distinguish the apoproteins we use the letters A, B, C, and E, followed, in some cases, by an Arabic or Latin numeral. For example, we have apoprotein B-100.

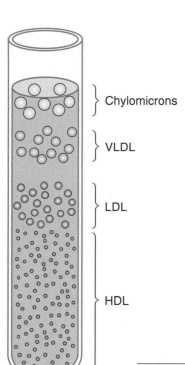

▶ **Figure 10.18**   Lipoprotein separation. By centrifuging a small volume of plasma in a test tube at high speed for many hours, we achieve the separation of plasma lipoproteins. Lipoproteins of high density move to the bottom, whereas lipoproteins of low density stay at the top.

▶ **Table 10.1** The Main Plasma Lipoproteins

Class	Density (g · mL⁻¹)	Percentage composition by weight				Main apoproteins
		Triacylglycerols	Cholesterol and cholesterol esters	Phospholipids	Proteins	
Chylomicrons	0.92-0.94	90	4	4	2	B-48, C, E
VLDL	0.96-1.00	60	15	15	10	B-100, C, E
LDL	1.02-1.06	6	48	23	23	B-100
HDL	1.06-1.21	5	20	25	50	A

10.11 A Lipoprotein Odyssey

Each lipoprotein class has a distinct biological role.

Chylomicrons

Chylomicrons carry dietary fat, which is about 95% triacylglycerols, to the tissues. Dietary triacylglycerols are initially hydrolyzed in the small intestine by **pancreatic lipase,** an enzyme secreted from the pancreas in the duodenum. Pancreatic lipase acts like hormone-sensitive lipase, catalyzing the hydrolysis of a triacylglycerol to 2-mono-acylglycerol and two fatty acids. Hydrolysis is mandatory for fat absorption because triacylglycerols cannot cross the plasma membrane of the epithelial cells. In contrast, 2-monoacylglycerols and fatty acids can. Then, once inside the epithelial cells, they are reunited to form triacylglycerols.

Triacylglycerol synthesis in the epithelial cells differs from triacylglycerol synthesis in the adipocytes (section 10.1). Here two acyl groups are added to 2-monoacylglycerol through the catalytic action of *acyl transferases.*

(equation 10.14)

In the epithelial cells, acyl CoA are synthesized by acyl CoA synthetase as in the adipocytes. In the epithelial cells, however, acyl CoA synthetase and the acyl transferases are organized in a multienzyme system, another *triacylglycerol synthase complex,* also bound to the endoplasmic reticulum membrane.

Once synthesized, triacylglycerols (along with other dietary lipids) are incorporated into chylomicrons and exit the epithelial cells to the extracellular space (figure 10.19). They then enter the lymphatic vessels of the villi (consult figure 9.4), which pour their contents into the left subclavian vein through the thoracic duct. This process goes on for several hours after a meal.

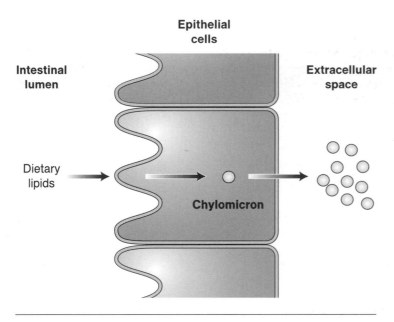

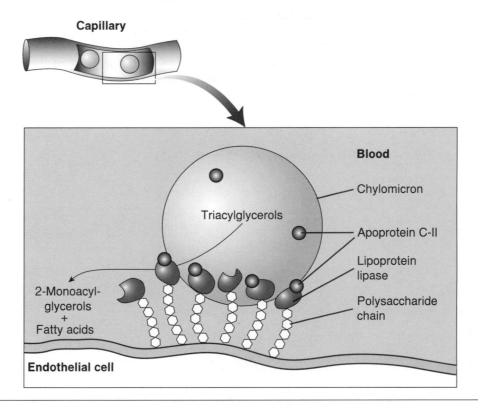

Figure 10.19 Chylomicron forms. Dietary lipids pass from the intestinal lumen to the epithelial cells (cf. figure 9.4), where they are packaged in chylomicrons. Then they are exported to the extracellular space of the villi, from there to the lymph, and from the lymph to the blood.

Within a few minutes of entering the bloodstream, a chylomicron encounters **lipoprotein lipase** on the surface of the capillaries in muscle, adipose tissue, and other extrahepatic tissues, that is, tissues other than the liver (figure 10.20). Lipoprotein lipase too hydrolyzes triacylglycerols to 2-monoacylglycerols and fatty acids. It is synthesized in muscle fibers, adipocytes, and other cells, which export it to their capillaries.

Triacylglycerols in chylomicrons need to be broken down because, again, they are unable to cross cell membranes. Thus, thanks to lipoprotein lipase, tissues take up the components of dietary triacylglycerols. 2-Monoacylglycerols are then hydrolyzed to glycerol and fatty acids by monoacylglycerol lipase (section 10.1). Glycerol usually exits to the bloodstream to be taken up by the liver (section 10.3), whereas fatty acids are either oxidized or used for the synthesis of triacylglycerols, phospholipids, cholesterol esters, and other biomolecules.

The particles remaining after chylomicrons have lost their triacylglycerols are called *chy-*

Figure 10.20 Chylomicron vanishes. A chylomicron travels to the depths of a tissue to meet its own demise in the face of lipoprotein lipase. Molecules of the enzyme stick out from the surface of endothelial cells (the cells forming the walls of the capillaries) toward the blood, tied to the endothelial cells with polysaccharide chains. Apoprotein C-II, a component of the chylomicron, activates the lipase, which begins to hydrolyze triacylglycerols and release the products for pickup by the cells.

lomicron remnants. They contain cholesterol, cholesterol esters, phospholipids, and apoproteins. The liver takes up chylomicron remnants by endocytosis and dismantles them. The liver has a profound biosynthetic and secretory activity, which extends to lipids. It is the main organ of fatty acid and cholesterol synthesis in humans (we will not consider cholesterol synthesis), as well as an important site of triacylglycerol synthesis. Thus, fatty acids derived from food or synthesized in the liver (*endogenous,* as we call the latter) are incorporated in triacylglycerols, packaged in VLDL particles along with cholesterol and cholesterol esters, and released to the circulation.

VLDL

VLDL deliver hepatic triacylglycerols to extrahepatic tissues. They achieve this through lipoprotein lipase, which acts on VLDL as it does on chylomicrons. As the VLDL particles shrink, they become *intermediate-density lipoproteins* (IDL). Half of these are taken up by the liver (again by endocytosis), whereas the other half lose almost all of their triacylglycerols by the further action of lipoprotein lipase to become LDL.

LDL

LDL are the main carriers of cholesterol (both free and esterified) in the plasma. Cholesterol is delivered to cells through endocytosis of the entire LDL particle (figure 10.21), a process that has been characterized more adequately than the endocytosis of other lipoproteins. Apoprotein B-100, lying at the surface of LDL, binds to a specific receptor in the plasma membrane, causing it to cave in and enclose the complex of LDL with its receptor. Once in the cytoplasm, the LDL particles are attacked by degradative enzymes, which release cholesterol and the other components of LDL for use by the cell.

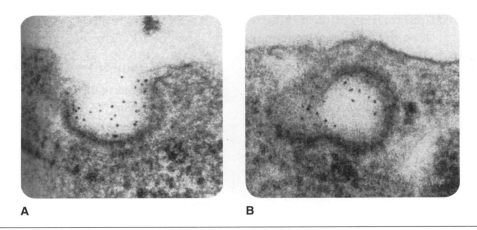

A　　　　　　　　　　　　　　　　**B**

▶ **Figure 10.21**　LDL endocytosis. LDL particles (artificially stained in order to be visible) appear as black dots in these electron micrographs showing the invagination of the plasma membrane *(a)* and the "swallowing" of the particles inside a vesicle *(b)*.

Reprinted from *Cell*, vol. 10(3), R.G. Anderson, M.S. Brown, and J.L. Goldstein, Role of the coated endocytic vesicle in the uptake of receptor-bound low density lipoprotein in human fibroblasts, pgs. 351-364, Copyright 1977, with permission from Elsevier.

HDL

HDL also carry cholesterol in the blood, but their mission differs from that of LDL. HDL set out as *nascent HDL* from the liver and small intestine to collect cholesterol that is released to the bloodstream from dead cells and restructured membranes. Cholesterol is first incorporated in the shell of nascent HDL; then it is esterified through the attachment of an acyl group derived from *lecithin,* a phospholipid.

Lecithin is the common name for phosphatidyl choline (figure 5.13).

cholesterol + lecithin ⇌ cholesterol ester + lysolecithin　　　　　　**(equation 10.15)**

Lysolecithin is what remains of lecithin after the removal of one of its two acyl groups. In reaction 10.15 (equation 10.15), the acyl group is removed from position 2 of glycerol.

The enzyme catalyzing the reaction is *lecithin-cholesterol acyl transferase* (LCAT). The cholesterol ester formed, being hydrophobic, is sequestered to the core of the lipoprotein particle. In this way, HDL become engorged with cholesterol esters, which they recycle in two ways.

- HDL deliver cholesterol esters to the liver by endocytosis.
- They pass cholesterol esters on to chylomicrons or VLDL with the aid of *cholesterol ester transfer protein* in exchange for triacylglycerols.

Thus, HDL accomplish what is known as *reverse cholesterol transport*.

Figure 10.22 summarizes the life cycle of the different lipoprotein classes, while table 10.2 summarizes their biological roles.

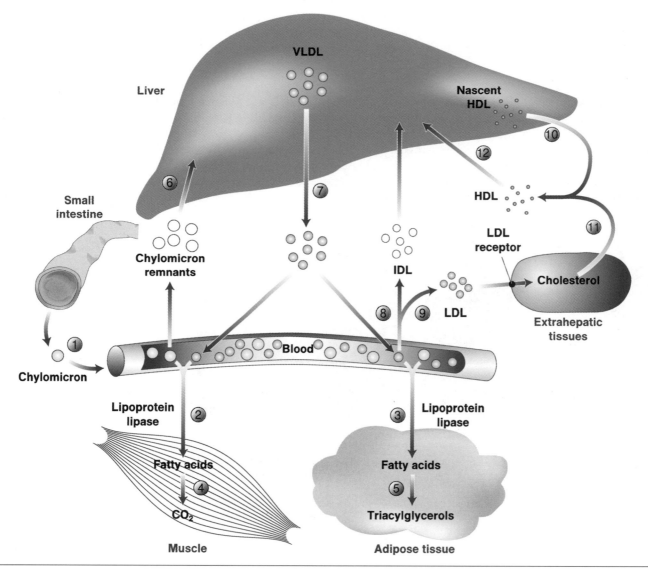

▶ **Figure 10.22** Lipoprotein circulation. Dietary lipids are packaged in chylomicrons inside the intestinal lining and are sent to the circulation (1). Lipoprotein lipase breaks down chylomicron triacylglycerols in the capillaries of muscles (2) and adipose tissue (3), letting the products in. Usually fatty acids are burned in muscle (4) and form triacylglycerols in adipose tissue (5). Chylomicron remnants are taken up by the liver (6), which synthesizes and secretes VLDL (7). VLDL follow the fate of chylomicrons, offering the components of their triacylglycerols to muscle and adipose tissue. Whatever escapes the action of lipoprotein lipase is converted to IDL (8), which return to the liver, or LDL (9), which bind to the LDL receptor and deliver their free and esterified cholesterol to extrahepatic tissues. Finally, nascent HDL from the liver (10) and the small intestine (omitted for clarity) are fed with surplus cholesterol from extrahepatic tissues (11) and become mature HDL, which deliver their cholesterol esters to the liver (12) or trade them for triacylglycerols with chylomicrons or VLDL (also omitted for clarity).

▶ **Table 10.2** Main Biological Roles of Lipoproteins

Chylomicrons	Transport of dietary triacylglycerols to extrahepatic tissues
VLDL	Transport of hepatic triacylglycerols to extrahepatic tissues
LDL	Transport of cholesterol to extrahepatic tissues
HDL	Transport of cholesterol from extrahepatic tissues to the liver, chylomicrons, and VLDL

10.12 Effects of Exercise on Plasma Triacylglycerols

The plasma triacylglycerol concentration is of interest with regard to health because it relates to a pathological condition called **atherosclerosis.** Atherosclerosis is the presence of *atherosclerotic plaques,* that is, deposits of lipids and other components, inside the blood vessels (figure 10.23). Atherosclerotic plaques narrow the vessels and diminish blood flow through them, resulting in inadequate supply of oxygen and nutrients to vital organs such as the brain and heart. This may cause irreparable damage in the form of stroke and myocardial infarction. The development of atherosclerotic plaques is favored by a high concentration of triacylglycerols—and cholesterol, as we will see in the next section—in the plasma. Thus, efforts are directed toward keeping it low.

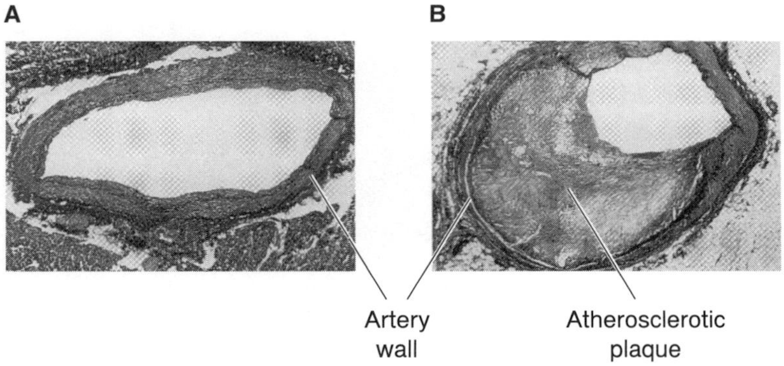

The plasma triacylglycerol concentration is markedly affected by heredity and nutrition, factors that lie outside the scope of this book. What about exercise? Examination of its effects on plasma triacylglycerols offers the opportunity to verify the last two basic principles of exercise metabolism outlined in the introduction to part III of this book (section III.1). It also offers the opportunity to apply the distinction between the acute and chronic effects of exercise.

▶ **Figure 10.23** Atherosclerosis. Comparison of a normal *(a)* and an obstructed artery *(b)* reveals what atherosclerosis does: About 80% of the artery interior in *b* is blocked by fatty deposits forming an atherosclerotic plaque.

Photo courtesy of In Touch Magazine.

Acute Effects

Let's begin with the effects of a single exercise bout on plasma triacylglycerols. A review of the literature reveals that most of the relevant studies have not shown a significant change. However, some studies showed a reduction. Generally, it seems that for plasma triacylglycerols to decrease, exercise must be prolonged (over 1 h), in which case the reduction relates to intensity. The reduction is probably due to an increase of lipoprotein lipase in the muscle capillaries. Epinephrine causes this increase, which lasts for many hours after the end of exercise. The products of plasma triacylglycerol hydrolysis can be used for ATP resynthesis in the exercising muscles. However, the contribution of plasma triacylglycerols to the energy expenditure of exercise is normally small (around 5%).

How exercise affects the plasma triacylglycerol concentration depends on the concentration value before exercise. Individuals with low values do not exhibit significant changes with exercise, whereas individuals with high values show considerable decreases. This may account for the smaller changes usually occurring in women, since women generally have lower plasma triacylglycerol concentrations than men (see section 15.6).

In those cases in which plasma triacylglycerols decrease at the end of exercise, an additional decrease is observed for several hours afterward, and plasma triacylglycerols remain below baseline for up to three days (figure 10.24). If one exercises again in the meantime, plasma triacylglycerols drop further. It is understood that all measurements are performed in the postabsorptive state so that dietary triacylglycerols do not interfere. Researchers accomplish this by asking the participants in a study to fast for around 12 h before providing a blood sample.

Exercise can lower not only the fasting but also the postprandial triacylglycerol concentration in the plasma. Studies have shown that *postprandial lipemia,* that is, the rise in the plasma triacylglycerol concentration after a meal, is diminished when one has performed moderate-intensity exercise for at least 1 h on the preceding day (figure 10.25). This may be due to prolonged activation of lipoprotein lipase in the muscles or to decreased secretion of triacylglycerols from the liver. The importance of lowering postprandial lipemia

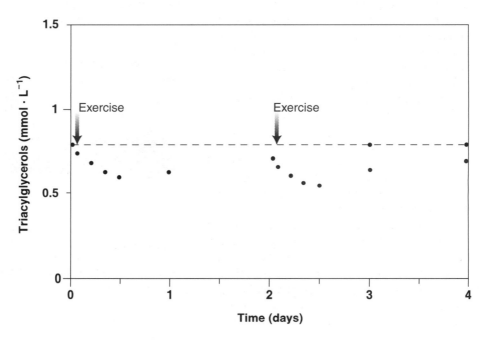

▶ **Figure 10.24** Exercise can decrease plasma triacylglycerols. An exercise bout of large energy expenditure can decrease the fasting plasma triacylglycerol concentration for several hours. The concentration returns to baseline (broken line) over a period of one to three days. If exercise is performed again in the meantime, the concentration drops even more (colored dots). The dots are not connected by lines because the concentration rises temporarily after each meal.

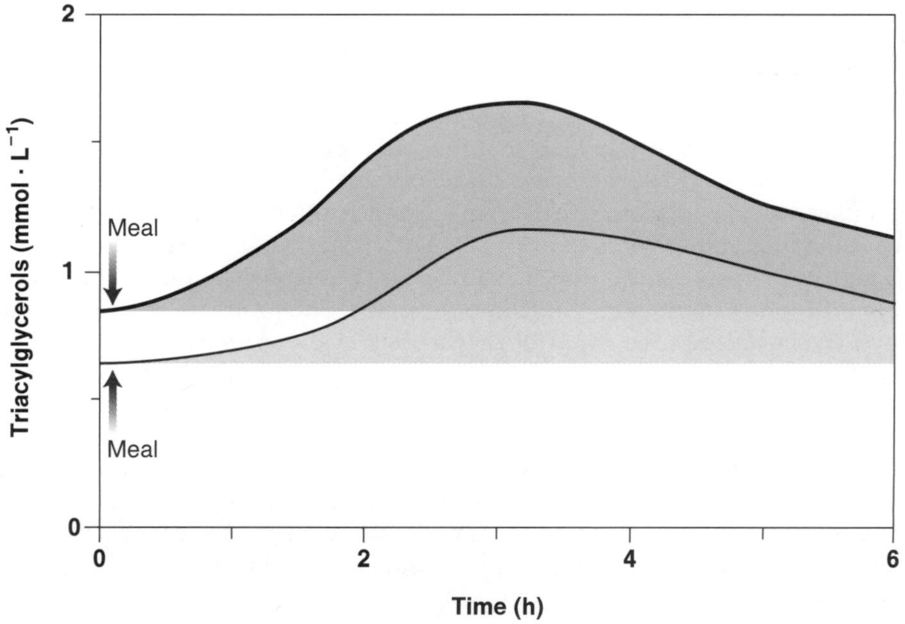

▶ **Figure 10.25** Effect of exercise on postprandial lipemia. The plasma triacylglycerol concentration rises after we consume a fatty meal and stays elevated for several hours, until the dietary lipids are delivered to the tissues. If the person being examined has exercised the day before (approximately 14 h before the meal), both the fasting concentration (at 0 h) and the postprandial concentration are lower (colored line) compared to values in the absence of previous exercise (black line). In addition, the so-called incremental area under the curve (shaded) is less after exercise.

lies in the reduction of the triacylglycerol load in the blood after a meal. Unfortunately, and similarly to the situation with lowering of the fasting concentration, the beneficial effect of exercise on postprandial lipemia is lost two days after exercise.

Some studies have shown a rise in the (fasting) plasma triacylglycerol concentration after moderate-intensity exercise. This is probably due to the increase in the plasma fatty acid concentration (figure 10.14) and a corresponding increase in the liver fatty acid concentration. This, in turn, could result in an acceleration of triacylglycerol synthesis, VLDL formation, and VLDL release to the bloodstream.

Chronic Effects

Now, what are the effects of regular exercise on the plasma triacylglycerol concentration? The answer to this question (as to any question regarding a chronic effect) can come from two kinds of studies, **cross-sectional** and **longitudinal.** In cross-sectional studies, researchers choose a point in time to examine and compare population groups differing in a certain characteristic. In longitudinal studies, on the other hand, researchers perform an experimental intervention on a population group for a period of time and compare the final to the initial state.

Cross-sectional studies have shown that endurance athletes (such as long-distance runners), soccer players, and, generally, individuals engaged in regular aerobic exercise have lower plasma triacylglycerol concentrations than untrained individuals. In contrast, the values of strength athletes (such as weightlifters), whose training is mainly anaerobic, do not differ from the values of untrained individuals.

In addition, several longitudinal studies have shown a decrease in the plasma triacylglycerol concentration of previously untrained individuals after aerobic training of usually more than a month. To eliminate the acute effect of the last exercise bout, investigators in such studies let at least two days go by after the end of training before taking the last blood sample. However, other studies have not shown a change in the plasma triacylglycerol concentration after aerobic training. In general, there appears to be a threshold of weekly energy expenditure over which significant reductions in the plasma triacylglycerol concentration can be detected. The threshold ranges from 1,200 to 2,200 kcal depending on the study. This translates to a few hours of moderate-intensity exercise scattered throughout the week.

> A cross-sectional study might compare the plasma triacylglycerol concentrations of a group of athletes and a group of nonathletes using blood samples taken today. A longitudinal study might compare the plasma triacylglycerol concentrations of a group at the beginning and end of a training program lasting three months.

Summary

Research on the acute and chronic effects of exercise on the plasma triacylglycerol concentration has produced conflicting results. It seems that a large energy expenditure is necessary for a significant acute decrease to take place. Finally, according to most studies, aerobic training for months or years reduces plasma triacylglycerols, provided there is considerable weekly energy expenditure.

10.13 Effects of Exercise on Plasma Cholesterol

Most studies measuring the plasma cholesterol concentration make no distinction between the free and esterified forms, as what seems to matter is the sum of the two. Thus, for the sake of brevity, from here on we will include cholesterol esters in the term cholesterol, unless otherwise stated. Cholesterol, then, relates to the risk for atherosclerosis in the same way triacylglycerols do: High concentrations in the plasma favor the formation of atherosclerotic plaques.

What is also important regarding atherosclerosis is the proportion of the two lipoprotein classes that carry the majority of cholesterol. LDL are the main *atherogenic* lipoproteins, whereas HDL inhibit the development of atherosclerotic plaques, thus protecting the blood vessels. It appears that HDL play their protective role by removing cholesterol from extrahepatic tissues, including the vessel walls.

> Because of their opposite effects on health, LDL and HDL are known as *bad* and *good cholesterol,* respectively.

Cholesterol does not serve as an energy source during exercise; thus one would not expect its plasma concentration to change with exercise. In reality, however, changes do occur. As in the case of triacylglycerols, they are neither spectacular nor unequivocal; nevertheless, they are interesting.

Acute Effects

Let's start again with the acute effects of exercise. A single exercise bout can lower the plasma cholesterol concentration provided that energy expenditure is large (in accordance with the lowering of the triacylglycerol concentration). A bout of this kind can, conversely, increase the cholesterol in HDL. Both changes are favorable for health.

The movement of the rise in HDL cholesterol after exercise is similar to the drop in triacylglycerols in terms of culmination (about one day after exercise) and disappearance (about three days after exercise, figure 10.26). The link between these opposite changes is probably the increased lipoprotein lipase activity, which augments the degradation of triacylglycerols in VLDL and causes the lipoprotein particles to shrink. This, in turn, creates a surplus of shell lipids (free cholesterol and phospholipids), which are transferred to HDL. Additionally, exercise activates LCAT, which, as we saw, "feeds" the HDL particles.

As with triacylglycerols, the acute response of plasma cholesterol to exercise seems to differ between sexes. Men usually exhibit an increase in HDL cholesterol, whereas women usually exhibit a decrease in total cholesterol (that is, cholesterol regardless of which lipoproteins it is in).

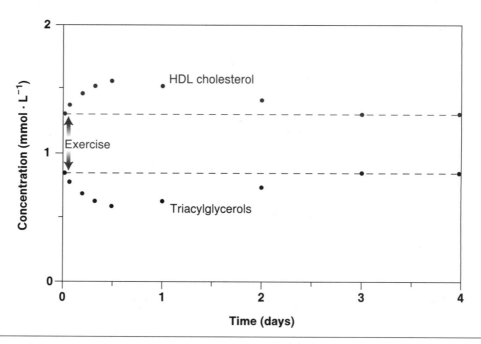

▶ **Figure 10.26** Opposite changes in blood lipids with exercise. Postexercise concentrations of HDL cholesterol and triacylglycerols in the plasma usually move in opposite directions. HDL cholesterol increases, triacylglycerols decrease, and both return to their baselines (broken lines) within a few days.

Chronic Effects

Chronic exercise can affect both the concentration of total cholesterol in the plasma and its distribution in LDL and HDL. Endurance athletes and other persons training aerobically have total cholesterol concentrations below those of untrained persons. However, this is apparently due to the lower body mass and fat mass usually seen in physically active people. The same holds for LDL cholesterol: It is lower in aerobically trained individuals, but this is apparently due to their lower body mass and fat mass. In addition, longitudinal studies in which the participants maintained their body mass or fat mass showed that total cholesterol and LDL cholesterol did not change with an aerobic exercise program.

In contrast to total and LDL cholesterol, HDL cholesterol is high in aerobically trained persons or persons who do physical work. The amount of physical activity per week and the duration of training affect HDL cholesterol in a positive manner. Longitudinal studies have confirmed the beneficial effect of aerobic training, although the differences measured are smaller than those in cross-sectional studies. As in the case of triacylglycerols, a minimal energy expenditure of 1,200 to 2,200 kcal per week seems to be necessary in order for favorable changes in HDL cholesterol to occur. It is worth mentioning that in contrast to the decrease in total and LDL cholesterol, the increase in HDL cholesterol and the decrease in triacylglycerols with aerobic training are independent of changes in body mass or fat mass.

All told, *the plasma lipid concentrations of aerobically trained persons are such that they run a lower risk than others for atherosclerosis.* For the sake of brevity, we say that they have a healthy **lipidemic profile** (see sections 15.6 to 15.10 for specific values). This is not the case with strength athletes. As with triacylglycerols, their total, HDL, and LDL cholesterol concentrations do not differ from those of untrained persons. There is, however, a note of consolation for those who love resistance training: HDL cholesterol increases when the training schedule includes moderate-resistance exercises with many repetitions, rather than high-resistance exercises with few repetitions.

The total and LDL cholesterol concentrations in the plasma relate to body mass and fat mass.

Summary

An exercise bout of high energy expenditure can decrease total cholesterol and increase HDL cholesterol in the plasma, while regular aerobic exercise involving considerable weekly energy expenditure increases HDL cholesterol.

10.14 Exercise Increases Ketone Body Formation

As we saw in section 10.4, the acetyl CoA produced from β oxidation enters the citric acid cycle, where the acetyl group is finally oxidized to CO_2. It is obvious that sufficient oxaloacetate must be present in order to bring the acetyl group into the cycle (refer to figure 9.20). Now remember that in the liver, oxaloacetate also participates in gluconeogenesis (figure 9.28), which is accelerated during exercise. This can cause a shortage of oxaloacetate. An inadequate dietary intake of carbohydrates will accentuate this shortage, as it lowers hepatic glycogen and blood glucose, resulting in increased secretion of glucagon and catecholamines, the very hormones that stimulate gluconeogenesis.

The lack of oxaloacetate in the liver drives acetyl CoA to a different metabolic pathway (figure 10.27), one that ends in the formation of **ketone bodies.** The first ketone body formed is *acetoacetate,* a four-carbon compound, the product of the net linking of two acetyl groups in the mitochondria (figure 10.28). Part of acetoacetate is then reduced to *3-hydroxybutyrate* (the other ketone body), with NADH serving as the reducing agent.

The two ketone bodies diffuse from the liver to the blood and are taken up by extrahepatic tissues (including muscle), where they can be used as fuel after being converted to acetyl CoA (two per ketone body, figure 10.29). Acetyl CoA enters the citric acid cycle readily in most extrahepatic tissues because, in contrast to the liver, these do not synthesize glucose and thus have adequate oxaloacetate.

The concentration of ketone bodies in the blood rises during prolonged exercise, although it does not become high enough to render them a worthwhile fuel for the exercising muscles. Even when their concentration is relatively high, the contribution of ketone bodies to energy expenditure in the exercising muscles does not exceed 7%.

▶ **Figure 10.27** Fats burn in the flame of carbohydrates. High and low carbohydrate states in the body promote different metabolic pathways in the liver. Boldface in the two panels signifies an abundant metabolite, while regular print signifies a scarce metabolite. Thick arrows denote a high rate, thin arrows a low rate. *(a)* When carbohydrates are adequate, there is sufficient oxaloacetate to support the entrance of acetyl CoA from the breakdown of fatty acids into the citric acid cycle. Thus, fatty acids are fully burned to CO_2. Ketone body formation is minimal and the rate of gluconeogenesis from pyruvate or oxaloacetate is low. *(b)* Exercise or a shortage of carbohydrates stimulates gluconeogenesis, diverting oxaloacetate to glucose synthesis. Then there is not enough oxaloacetate to join acetyl CoA in the formation of citrate, and the accumulation of acetyl CoA leads to increased ketone body formation.

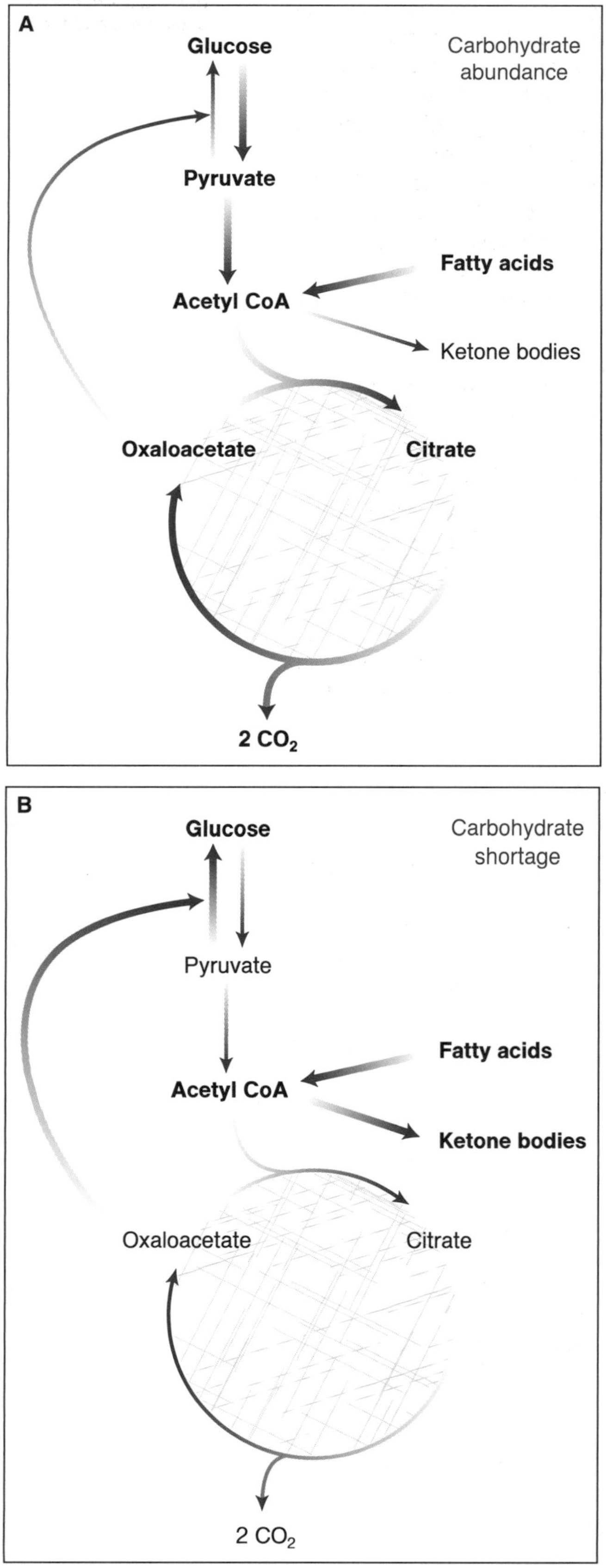

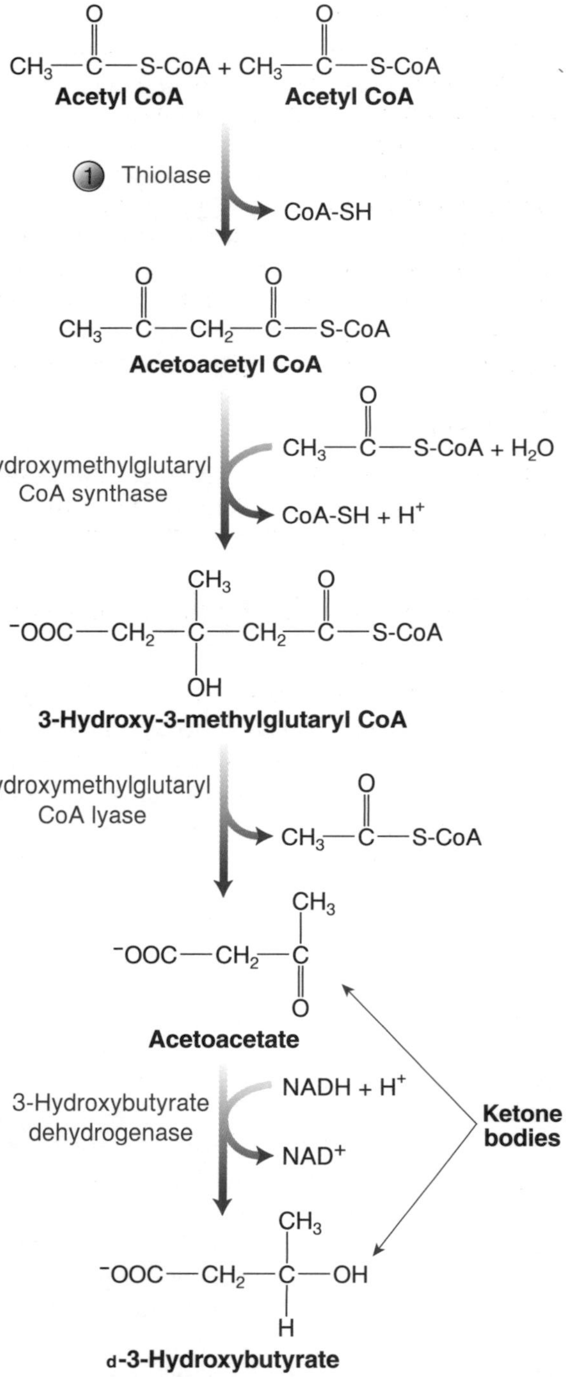

▶ **Figure 10.28** Ketone body synthesis. Ketone bodies are synthesized from two acetyl CoA in the liver.

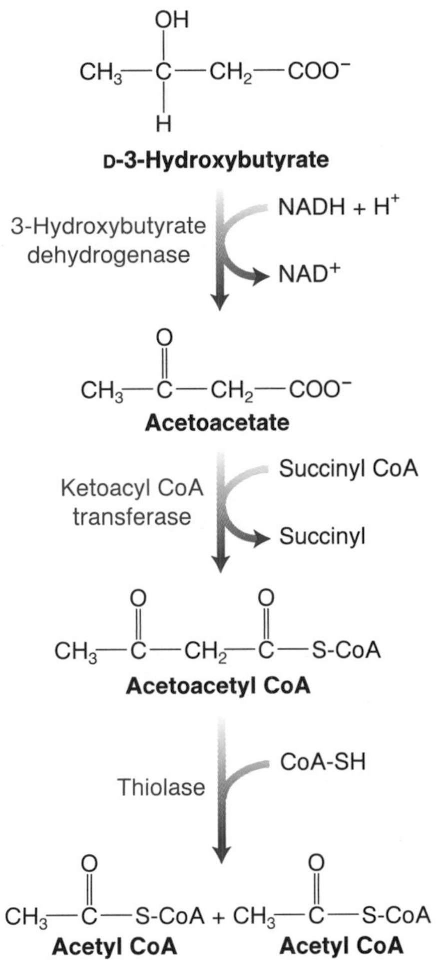

OH
|
CH_3—C—CH_2—COO^-
|
H
D-3-Hydroxybutyrate

3-Hydroxybutyrate
dehydrogenase

NADH + H$^+$

NAD$^+$

O
‖
CH_3—C—CH_2—COO^-
Acetoacetate

Ketoacyl CoA
transferase

Succinyl CoA

Succinyl

O   O
‖   ‖
CH_3—C—CH_2—C—S-CoA
Acetoacetyl CoA

Thiolase

CoA-SH

O   O
‖   ‖
CH_3—C—S-CoA + CH_3—C—S-CoA
Acetyl CoA   **Acetyl CoA**

▶ **Figure 10.29** Ketone body degradation.
Ketone bodies are broken down to two acetyl
CoA in extrahepatic tissues.

Problems and Critical Thinking Questions

1. What concentration changes speed up lipolysis in adipose tissue during exercise?

2. Calculate the ATP yield of the oxidation of each fatty acid in table 5.1.

3. Although arachidonate (table 5.1) has double bonds beyond C10, it is not classified as an essential fatty acid because our cells can synthesize it from linoleate. Suggest how.

4. Which triacylglycerol lipases have we met?

5. Which blood lipid does aerobic exercise training increase?

CHAPTER 11

Protein Metabolism in Exercise

In chapter 3 we explored proteins in terms of both structure and function. The chapters that followed were replete with them: We met a plethora of enzymes, motile proteins, transporters, receptors, and peptide hormones. Their participation in numerous biochemical processes confirms their pivotal role in the functions of life. Add to this the fact that in contrast to carbohydrates and lipids, we do not have protein depots in the body to compensate for a possible shortage, and you will understand why the survival, health, and performance of an organism depend heavily on the integrity of its proteins.

The present chapter deals with how exercise affects protein metabolism by examining the acute and chronic changes in the rates of protein and amino acid synthesis and breakdown.

11.1 Protein Metabolism

Proteins in a living organism take part in an ongoing cycle of formation and degradation (figure 11.1). These processes are called protein synthesis and proteolysis, respectively. Protein synthesis takes place in the ribosomes, which translate the mRNA that results from the transcription of a gene, as we discussed in chapter 4. Proteolysis is catalyzed by proteases, which hydrolyze a protein to its constituent amino acids. (We considered a case of proteolysis in section 7.4.)

The dominant mechanism of proteolysis inside eukaryotic cells appears to begin with the "marking" of a protein by the covalent attachment of ubiquitin chains to its

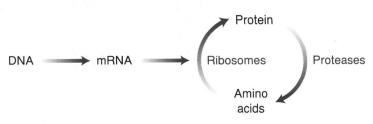

▶ **Figure 11.1** Life cycle of a protein. The synthesis and breakdown of a protein follow different metabolic pathways. Synthesis is more complex, as it requires the linking of amino acids in a precise sequence defined by DNA and transmitted to the ribosomes by mRNA.

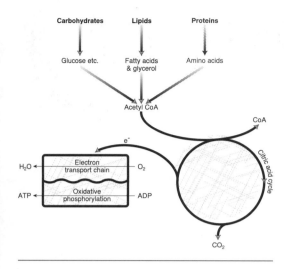

▶ You are here: proteolysis.

molecule. *Ubiquitin* is a small protein (76 amino acid residues, 8.5 kDa) that makes a protein recognizable by the *proteasome,* a bulky complex of proteins (over 60 subunits, over 2,000 kDa) hydrolyzing other proteins to small peptides. Both the attachment of ubiquitin and the operation of the proteasome are powered by ATP hydrolysis. Other proteases then hydrolyze the peptides released from the proteasome to amino acids. A healthy and adequately nourished adult has on average (in the whole body and over the day) approximately equal rates of protein synthesis and proteolysis. As a result, his or her protein mass remains constant.

Because amino acids are degraded in the body, we need a constant supply of them. Our primary source of amino acids is dietary proteins. Since our proteins differ from those of the animals and plants that we eat, and since protein molecules are too big to cross the intestinal wall, they have to be broken down along the digestive tract in order to be useful to the body.

The task of degrading dietary proteins is carried out by proteases secreted in the stomach and small intestine. *Pepsin* works in the stomach, while *aminopeptidase, carboxypeptidase, elastase, enteropeptidase, chymotrypsin,* and *trypsin* work in the small intestine. Although these proteases, like all proteases, catalyze the splitting of the same bond (the peptide bond), they are specific as to the amino acid residues that contribute the amino or the carboxyl side of the peptide bond.

As a result of the combined action of the proteases just listed, dietary proteins are converted to amino acids in the small intestine. These are then absorbed by the epithelial cells lining the small intestine and exit to the capillaries in the villi (see figure 9.4). Finally, the blood delivers the dietary amino acids to the rest of the body.

11.2   Effect of Exercise on Protein Metabolism

Exercise changes the rates of protein synthesis and proteolysis in muscle. The changes depend on the type of exercise. Let's begin with protein synthesis. If we perform hard resistance exercise, the rate of protein synthesis will increase (figure 11.2), provided there is an adequate amount of amino acids in the contracting muscle fibers to serve as substrates. This positive stimulus lasts one to two days. If resistance exercise is of moderate intensity, the rate of protein synthesis does not change considerably. Things are different with endurance exercise. If it is hard, the rate of protein synthesis decreases; if it is of moderate intensity or light, the rate of protein synthesis does not change significantly or may increase. These observations refer to muscle proteins as a whole. We do not know whether different proteins respond differently to a particular exercise.

The effect of exercise on the rate of proteolysis is less clear. Hard resistance exercise boosts it for one to two days but to a lower degree than the rate of protein synthesis. Thus, the net result is an increase in muscle proteins. Moderate-intensity or hard endurance exercise also seems to increase the rate of proteolysis. This, in combination with a decrease or no change in the rate of protein

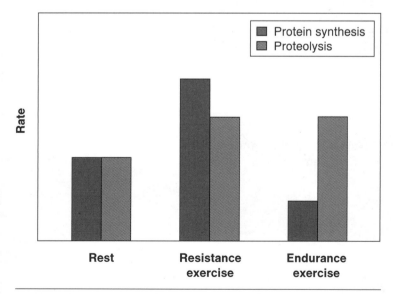

▶ **Figure 11.2**   How exercise affects the cycle of muscle proteins. The rates of protein synthesis and proteolysis are comparable at rest. Both increase during hard resistance exercise, but the rate of protein synthesis exceeds the rate of proteolysis, resulting in an increase of muscle proteins. During hard endurance exercise the rate of protein synthesis decreases, whereas that of proteolysis increases, resulting in an decrease of muscle proteins.

synthesis, leads to a net decrease in muscle proteins. However, the situation is reversed some hours after exercise, resulting in the restoration of protein mass.

Exercise affects protein metabolism in other organs besides muscle. Studies in humans have shown increased release of amino acids from the visceral area (intestine and liver) during prolonged exercise, suggesting a dominance of proteolysis over protein synthesis. Studies in dogs have demonstrated that the intestine is the main visceral source of the increased release of amino acids.

Connective tissue proteins too sense the exercise stimulus. Experiments showed that prolonged running decreased the rate of collagen synthesis and increased its rate of breakdown in the Achilles tendon. However, during the following four days the rate of synthesis was higher than baseline, whereas the rate of breakdown was similar to baseline, resulting in an increase of collagen. This change makes connective tissue more resistant to the mechanical stress of exercise.

11.3 Amino Acid Metabolism in Muscle During Exercise

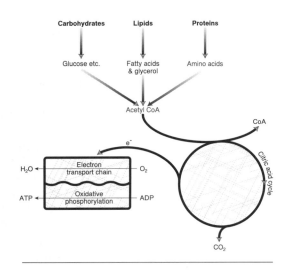

Although proteins make up a large part of muscle mass (about 20%, or 200 g · kg⁻¹), their constituent amino acids are present in free form (not incorporated in proteins) in very small amounts, approximately 3 g · kg⁻¹. Of the 20 amino acids, glutamine has the lion's share (around 60%), followed by glutamate (14%) and alanine (4%). The total amino acid concentration in the cytosol of muscle fibers increases during endurance exercise because of the prevalence of proteolysis over protein synthesis.

Amino acids can be broken down to produce energy. Each one follows a unique catabolic pathway, but all pathways start with the removal of the α-amino group. Two amino acids (serine and threonine) dispose of it as ammonia (figure 11.3) through deamination. The remaining amino acids except glutamate transfer their α-amino groups to α-ketoglutarate, an intermediate of the citric acid cycle (section 9.9). The equation of the transfer reaction is

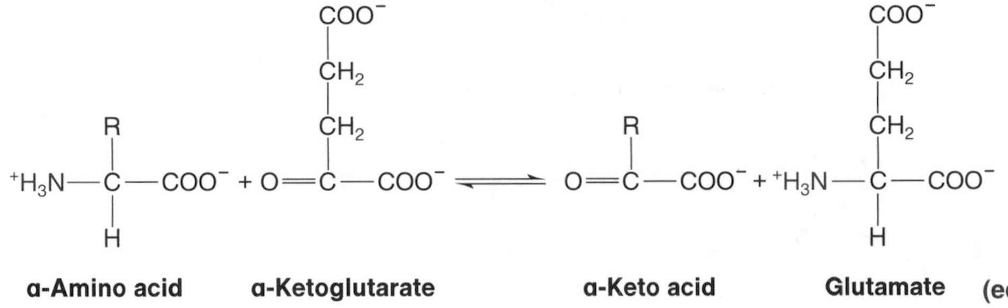

▶ You are here: amino acid degradation.

α-Amino acid **α-Ketoglutarate** **α-Keto acid** **Glutamate** (equation 11.1)

This reaction is a **transamination** catalyzed by an **aminotransferase**, or **transaminase**. The *α-keto acids* produced are (or can be) converted to intermediates of carbohydrate or lipid metabolism that we have considered in chapters 9 and 10. These intermediates are

pyruvate;

acetyl CoA;

acetoacetyl CoA, an intermediate of lipid metabolism (see figures 10.28 and 10.29); and

succinyl CoA, fumarate, and oxaloacetate, intermediates of the citric acid cycle (see figure 9.20).

What is the fate of the amino group that is loaded to glutamate in reaction 11.1 (equation 11.1)? Glutamate is one of the few amino acids that can discard the amino group as ammonia. This happens according to the equation

$$\text{glutamate} + NAD(P)^+ + H_2O \rightleftharpoons \alpha\text{-ketoglutarate} + NH_4^+ + NAD(P)H + H^+ \qquad \textbf{(equation 11.2)}$$

This is an *oxidative deamination,* as glutamate both loses the amino group and is oxidized by NAD$^+$ or NADP$^+$. α-Ketoglutarate, the product, can serve again as amino group acceptor in transamination reactions or can enter the citric acid cycle.

Figure 11.4 summarizes the fates of the carbon skeletons of the 20 amino acids. By merging with carbohydrate and lipid metabolism, amino acids can be catabolized aerobically to CO_2, yielding ATP. The yield varies from one amino acid to another and averages approximately 22 ATP.

Researchers have proposed another utility of the breakdown of muscle amino acids during exercise, that is, increase in the concentration of intermediates of the citric acid cycle. The supposition was that this increase promotes the aerobic catabolism of carbohydrates and lipids by offering more substrates for the processing of the acetyl groups entering the cycle. However, recent findings show that this is not the case. Thus, amino acid catabolism in muscle during exercise does not appear to facilitate ATP resynthesis through carbohydrate or lipid catabolism.

Glutamate dehydrogenase, an enzyme located in the mitochondrial matrix, catalyzes reaction 11.2 (equation 11.2). ATP and GTP inhibit, whereas ADP and GDP activate, glutamate dehydrogenase. The decrease in ATP and GTP, as well as the increase in ADP and GDP during exercise, activates the enzyme and speeds up glutamate deamination. Since the deamination regenerates α-ketoglutarate, it promotes transamination and the removal of amino groups from amino acids (see equation 11.1). This in turn opens the door to amino acid oxidation, although for most of the amino acids there is no evidence for increased oxidation during exercise.

However, the degradation of three amino acids does increase during exercise; these are leucine, isoleucine, and valine. All three have branched aliphatic side chains (refer to figure 3.1) and are collectively known as **branched-chain amino acids (BCAA).** BCAA are unique in that they are taken up directly by the muscles after a meal, whereas

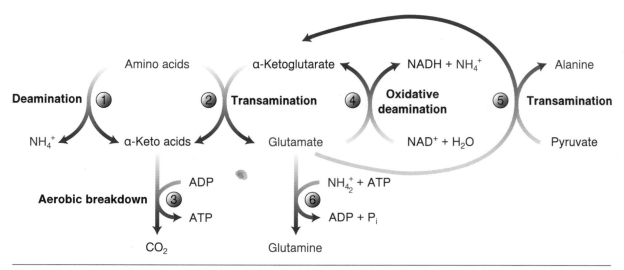

▶ **Figure 11.3** Overview of amino acid catabolism. Two amino acids discard their amino groups by deamination (1) and are converted to the corresponding α-keto acids. The other amino acids except glutamate transfer their amino groups to α-keto-glutarate by transamination (2), producing α-keto acids and glutamate. The α-keto acids can be degraded aerobically through the citric acid cycle to yield ATP (3). Glutamate can dispose of its amino group by oxidative deamination (4) or can transfer it to pyruvate by transamination (5). Both processes regenerate α-ketoglutarate. Alternatively, glutamate can be converted to glutamine through acquisition of another amino group (6). Compounds in color carry amino groups.

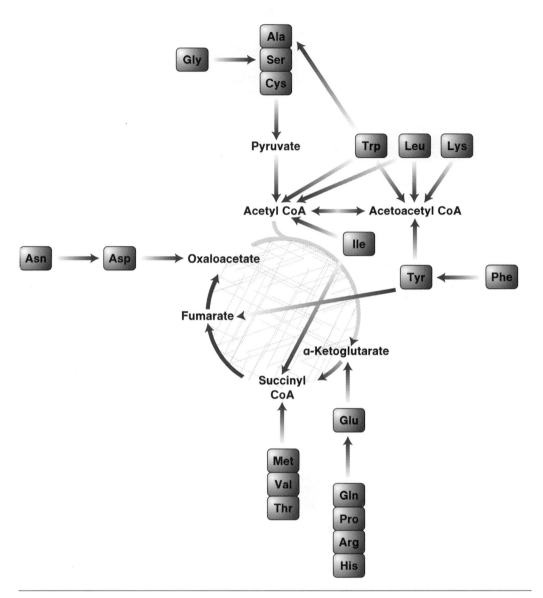

▶ **Figure 11.4** Fates of the carbon skeletons of amino acids. After ridding themselves of their α-amino groups, the 20 amino acids of proteins (enclosed in boxes) follow catabolic routes that lead to intermediates cf carbohydrate or lipid breakdown.

most amino acids are taken up by the liver before being distributed to extrahepatic tissues.

To be oxidized, BCAA are first transaminated to *branched-chain α-keto acids.* One of these is α-*ketoisocaproate,* derived from leucine.

$$\text{Leucine} + \alpha\text{-Ketoglutarate} \rightleftharpoons \alpha\text{-Ketoisocaproate} + \text{Glutamate}$$

Leucine　　**α-Ketoglutarate**　　　**α-Ketoisocaproate**　　**Glutamate**　　(equation 11.3)

Investigators usually measure α-ketoisocaproate as an index of BCAA catabolism. The concentration of branched-chain α-keto acids in muscle increases during moderate-intensity exercise.

Next, branched-chain α-keto acids are decarboxylated by the *branched-chain α-keto acid dehydrogenase complex*. In humans, this enzyme complex is located mainly in the skeletal muscles (which contain 60% of the total amount in the body) and is activated during moderate-intensity exercise. The branched-chain α-keto acid dehydrogenase complex resembles the pyruvate dehydrogenase complex (sections 9.7 and 9.8) in terms of the reaction catalyzed, location (the mitochondrial matrix), and control by reversible phosphorylation. Thus, an increase in its substrates (the branched-chain α-keto acids) and a decrease in ATP during exercise favor the active *a* form of branched-chain α-keto acid dehydrogenase (cf. figure 9.19).

The products of branched-chain α-keto acid decarboxylation participate in reactions leading to their complete oxidation to CO_2. In this way, they offer energy for exercise.

Reaction 11.2 (equation 11.2) relieves glutamate and, by extension, most amino acids of their amino groups permanently but creates a new problem: ammonia accumulation. Indeed, high concentrations of this compound are toxic to the brain. Therefore glutamate participates in two additional reactions.

- Rather than undergoing oxidative deamination, it transfers the amino group to pyruvate, thus converting it to alanine (notice how structurally similar pyruvate and alanine are).

| Glutamate | Pyruvate | α-Ketoglutarate | Alanine | (equation 11.4) |

Alanine aminotransferase, one of the most important aminotransferases, catalyzes the reaction.

- Glutamate reacts with ammonia to become glutamine at the expense of ATP.

$$\text{glutamate} + NH_4^+ + ATP \rightarrow \text{glutamine} + ADP + P_i + H^+ \qquad \text{(equation 11.5)}$$

Glutamine synthetase catalyzes the reaction.

Both reactions (included in figure 11.3) produce harmless amino acids (alanine and glutamine), which then exit to the bloodstream and travel to the liver for further processing.

You realize from all this that the predominance of glutamate, glutamine, and alanine among muscle amino acids is not accidental, since these three are the major recipients of amino groups from the other 17. Glutamate holds the central position in the intracellular network of amino group disposal, as it collects amino groups through one reaction (equation 11.1) and dispatches them through three others (equations 11.2, 11.4, and 11.5). This probably explains why glutamate decreases in exercising muscles, whereas, as we saw, the sum of the amino acids increases.

In addition to amino acids, ammonia too increases in muscle during exercise. We need to consider, however, that ammonia is also produced from AMP deamination, which is accelerated during hard exercise (section 8.5). It is believed that *AMP is the main source of ammonia during hard exercise, whereas amino acids are the main source of ammonia during moderate-intensity exercise.*

11.4 Amino Acid Metabolism in the Liver During Exercise

The liver has a free amino acid content similar to that of muscle. Part of this pool is used for glucose synthesis. How? As you can see in figure 11.4, several amino acids are converted to pyruvate or oxaloacetate, which are substrates for gluconeogenesis (figure 9.27). Additionally, many amino acids are converted to α-ketoglutarate, succinyl CoA, or fumarate, which yield oxaloacetate through the citric acid cycle. Thus, the carbons of most amino acids can find their way into glucose. These are the **glucogenic amino acids.** During exercise the increased secretion of glucagon and epinephrine, as well as the decreased secretion of insulin, speeds up gluconeogenesis (sections 9.20 and 9.23), with the result that more hepatic amino acids are used for glucose synthesis.

A few amino acids are degraded to acetyl CoA or acetoacetyl CoA (or both), which cannot produce glucose (section 10.9). These amino acids are termed **ketogenic,** since acetyl CoA and acetoacetyl CoA form ketone bodies (section 10.14). Fourteen of the 20 amino acids are glucogenic; two are ketogenic; and four are partly glucogenic and partly ketogenic (table 11.1).

Amino acids from muscle can also be precursors for glucose synthesis in the liver. As mentioned in the previous section, alanine and glutamine transport amino groups safely from the muscles to the liver. There, glutamine is converted back to glutamate (a glucogenic amino acid) by *glutaminase.*

▶ **Table 11.1** Classification of Amino Acids Based on Their Metabolic Fate

Glucogenic (14)	Ketogenic (2)	Glucogenic and ketogenic (4)
Alanine	Leucine	Isoleucine
Arginine	Lysine	Phenylalanine
Asparagine		Tryptophan
Aspartate		Tyrosine
Cysteine		
Glutamate		
Glutamine		
Glycine		
Histidine		
Methionine		
Proline		
Serine		
Threonine		
Valine		

Note: Several amino acids follow more than one metabolic route, which can change their character. For this reason there is no full agreement in the literature as to which amino acids are glucogenic and which are ketogenic. This table, and figure 11.4, are based on the routes that prevail in humans.

glutamine + H_2O → glutamate + NH_4^+ **(equation 11.6)**

The fate of alanine deserves a more extended discussion. This amino acid is converted to pyruvate by a reversal of equation 11.4 and by the catalytic action of the same enzyme, alanine aminotransferase.

alanine + α-ketoglutarate ⇋ pyruvate + glutamate **(equation 11.7)**

Hepatocytes then use the pyruvate produced to synthesize glucose, which they release to the circulation. This flow of glucose contributes to euglycemia in the postabsorptive state, during which some protein degradation takes place. The degradation of hepatic and muscle proteins in the postabsorptive state underlines the importance of euglycemia: Although proteins are valuable for survival, the body sacrifices a portion of them, since it cannot synthesize substantial amounts of glucose from the abundant and less vital triacylglycerols (section 10.9).

A common misconception is that the body begins to break down protein when it runs out of carbohydrate and fat. However, the adequacy of fat does not affect the protein status—let alone the fact that it is impossible to run out of fat. It is carbohydrate adequacy that affects the protein status. In particular, if liver glycogen is depleted because of inadequate carbohydrate intake, the body is deprived of its major source of blood glucose and has to resort to proteins.

The increase in muscle alanine production during exercise and the resulting increased liberation of alanine in the bloodstream result in its higher contribution to glucose synthesis in the liver. Part of the glucose produced is taken up by the exercising muscles, thus completing the **glucose-alanine cycle** (figure 11.5). The glucose-alanine cycle resembles the Cori cycle (figure 9.31); it differs only in that it has alanine in place of lactate.

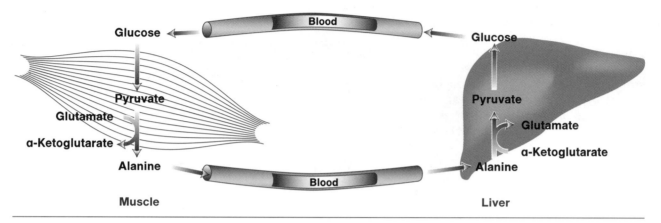

▶ **Figure 11.5**   The glucose-alanine cycle. The cycle supplies a working muscle with glucose synthesized in the liver from alanine of muscle origin.

The difference in ATP yield per cytosolic NADH between muscle and liver is due to the prevalence of different shuttles in the two organs (section 9.14).

The importance of the glucose-alanine cycle for muscle metabolism is dual. (1) It transports amino groups from muscle to the liver, and (2) every turn of the cycle yields two ATP for muscle (the yield of glycolysis) and charges six ATP to the liver (the cost of gluconeogenesis). Thus, as with the Cori cycle, muscle shifts part of its metabolic burden to the liver. In fact, the glucose-alanine cycle benefits muscle and burdens the liver more than the Cori cycle does, since the two NADH that are produced in the muscle cytosol during glycolysis can yield three ATP through the electron transport chain and oxidative phosphorylation (section 9.14), while gluconeogenesis in the liver *demands* two NADH, thus depriving the hepatocytes of five potential ATP. In contrast, there is no net production or consumption of NADH in the interconversion of glucose and lactate (the compounds participating in the Cori cycle).

Of course, on the other hand, by being converted to alanine, pyruvate misses the opportunity to move toward lactate and quickly regenerate the NAD^+ that is needed for the fast anaerobic production of ATP in glycolysis. At any rate, however, the glucose-alanine cycle requires several minutes to operate, which makes it useless in (necessarily brief) anaerobic efforts. But even in extended efforts, the importance of the cycle is limited. The amount of alanine entering the liver from muscle accounts for 4%, at the most, of the glucose produced during prolonged moderate-intensity exercise. An equally low amount is contributed by the sum of the other glucogenic amino acids that are taken up by the liver.

11.5   The Urea Cycle

Most (approximately 90%) of the nitrogen excreted from the human body is incorporated in urea, a simple nontoxic compound bearing two amino groups—a compound that we first met in section 1.7. Urea is synthesized in the liver through a series of four reactions called the **urea cycle.** The cycle was discovered in 1932 by Hans Krebs (the same researcher who later discovered the citric acid cycle) and Kurt Henseleit.

The urea cycle starts with *carbamoyl phosphate,* a compound of high phosphoryl transfer potential, derived from the linking of ammonia, bicarbonate, and a phosphoryl group at the expense of two ATP (figure 11.6). Urea is produced after four more reactions, which include the contribution of an amino group by aspartate. The four intermediates of the pathway are not consumed but recycled. The synthesis of carbamoyl phosphate and the first reaction of the cycle take place in the mitochondrial matrix, whereas the remaining three reactions take place in the cytosol of the hepatocytes.

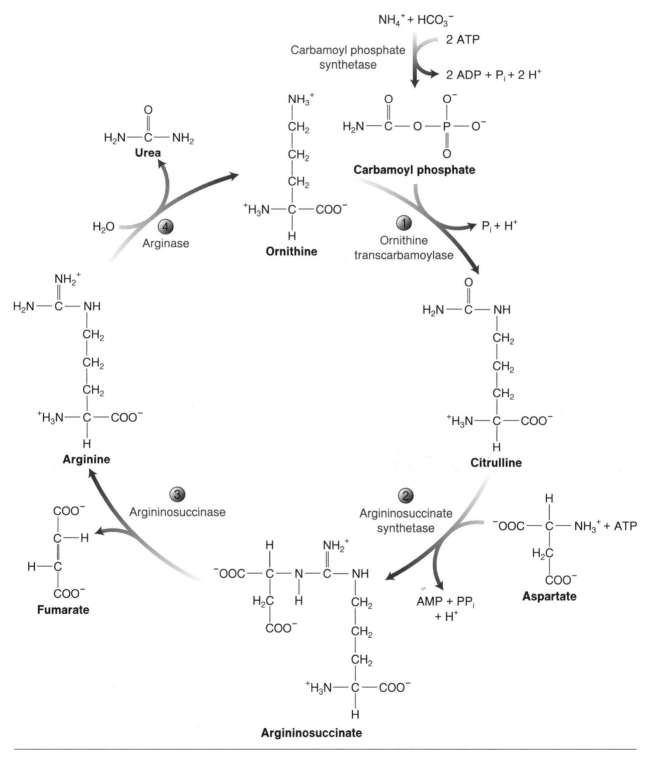

▶ **Figure 11.6**   The urea cycle. Two amino groups (one from ammonia at the top and the other from aspartate in the second reaction of the cycle) are incorporated in urea (the product of the fourth reaction) in order to be safely excreted from the body.

Although aspartate is the only amino acid that directly disposes of its amino group in the urea cycle, other amino acids can funnel their amino groups to urea through two consecutive transamination reactions (figure 11.7). The first is reaction 11.1 (equation 11.1), which transfers the amino group to glutamate, and the second is

glutamate + oxaloacetate $\rightleftharpoons$ α-ketoglutarate + aspartate   **(equation 11.8)**

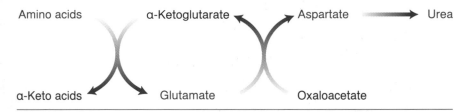

▶ **Figure 11.7** Aspartate as an amino group acceptor. Amino groups from various amino acids end up in aspartate through the intervention of glutamate. Aspartate then delivers its amino group to urea in the urea cycle.

which transfers the amino group to aspartate. The reaction is catalyzed by another important aminotransferase, *aspartate aminotransferase.*

The urea cycle is fueled by the hydrolysis of one ATP to AMP and PP_i (considered as expenditure of two ATP) in the second reaction. By adding the two ATP that are hydrolyzed in the synthesis of carbamoyl phosphate, we get a total cost of four ATP per urea.

After it is formed in the liver, urea diffuses to the blood. The kidneys then excrete it in the urine, after which it was named.

The rate of urea production does not change during short hard exercise. A rise is measured in prolonged efforts and appears to relate to exercise duration and intensity. Evidently, the increased urea production under such conditions is due to the accelerated removal of amino groups from alanine and other amino acids used in glucose synthesis in the liver. There are, however, studies that did not detect any significant rise in urea production during prolonged exercise. This may be due to the small magnitude of the changes in amino acid metabolism.

11.6 Amino Acid Synthesis

▶ **Table 11.2** Essential and Nonessential Amino Acids for Adult Humans

Essential (9)	Nonessential (11)
Histidine	Alanine
Isoleucine	Arginine
Leucine	Asparagine
Lysine	Aspartate
Methionine	Cysteine
Phenylalanine	Glutamate
Threonine	Glutamine
Tryptophan	Glycine
Valine	Proline
	Serine
	Tyrosine

Besides obtaining amino acids from dietary proteins, our bodies also synthesize amino acids. As is the case with degradation, the synthesis of each fatty acid follows a separate route. Glutamate plays a central role here too by serving as amino group donor to α-keto acids in transamination reactions. These reactions can thus be viewed as part of both catabolic and anabolic pathways. For example, reaction 11.4 (equation 11.4) produces alanine from pyruvate, and reaction 11.8 (equation 11.8) produces aspartate from oxaloacetate.

Mammals cannot synthesize all the amino acids needed to make their proteins. Humans lack the enzymes to synthesize nine of them (table 11.2), which are thus termed **essential amino acids** (as in essential fatty acids, section 10.6). Unless we obtain all essential amino acids in adequate amounts through the food we eat, protein synthesis will suffer, since almost every protein contains all 20 amino acids.

Unlike mammals, most bacteria and plants synthesize all amino acids. Herbivores obtain their essential amino acids by eating plants; carnivores, by eating herbivores; and humans, by eating foods of both plant and animal origin.

11.7 Plasma Amino Acid, Ammonia, and Urea Concentrations During Exercise

The total amino acid concentration in the plasma is 3 to 4 mmol $\cdot$ L^{-1}. The most abundant plasma amino acids are glutamine and alanine, as one would expect from the discussion of their role in section 11.3. Their concentrations are around 0.5 and 0.4 mmol $\cdot$ L^{-1}, respectively.

Exercise lasting less than 1 h does not affect the total amino acid concentration in the plasma to a great extent. At the level of individual amino acids, glutamate decreases and alanine increases. These changes are similar to the ones observed in the exercising muscles (section 11.3). When exercise duration exceeds 2 h, the plasma amino acid concentration drops by as much as 30%. This shows that, under such demanding conditions, there is an increase in their rate of disappearance toward the tissues relative to their rate of appearance in the plasma.

Moderate-intensity or hard exercise is accompanied by a rise in the plasma ammonia concentration, which relates to the plasma lactate concentration. Finally, the plasma urea concentration also increases in exercises lasting at least a half hour and having an intensity of at least 60% of $\dot{V}O_2$max. However, the plasma urea concentration may not change during hard exercise despite an increased production, as the blood flow through the liver decreases; thus, there is not enough blood to carry urea out of the hepatocytes. See sections 15.12 and 15.13 for specific values of plasma urea and ammonia concentrations.

11.8 Contribution of Proteins to the Energy Expenditure of Exercise

As discussed in previous sections, during prolonged exercise the rate of protein synthesis decreases, the rate of protein degradation increases, and the rate of BCAA degradation also increases in the contracting muscles. Additionally, the rate of gluconeogenesis from amino acids increases in the liver. All these changes point to an elevated utilization of proteins for energy supply to the exercising muscles. However, as mentioned in several instances, the changes are small. It seems that the body spares proteins because they are of paramount importance for all biological functions and because it lacks a protein reservoir from which it could draw in case of emergency.

All in all, the contribution of proteins to the energy expenditure of prolonged—even extremely prolonged—exercise is small. Most studies estimate it at 3% to 6% of the total energy expenditure. Finally, the input of proteins in ATP resynthesis during short efforts of any intensity is negligible.

The fact that proteins contribute little to energy expenditure during exercise suggests that athletes do not need the excessive protein intakes that nonexperts often advocate. In fact, most researchers agree that athletes are well advised to obtain about 10% to 15% of their energy from dietary proteins. (Carbohydrates, the main fuel for most sport activities, should contribute over 50% to the energy intake, and lipids should be confined to around 30%.) More accurate recommendations on protein consumption take into account the weight of an individual. Thus, while the recommended daily allowance for the general adult population is 0.8 g $\cdot$ kg^{-1}, athletes are advised to obtain 1.2 to 1.8 g $\cdot$ kg^{-1}. Most athletes usually meet this quota when they eat more food than the general population in order to meet the high energy demands of training and competition. Thus, they do not need to resort to protein supplements.

11.9 Effects of Training on Protein Metabolism

In section 11.2 we discussed the acute effects of exercise on protein metabolism. What about the chronic effects? They are more spectacular than the acute effects, as you can judge from two extreme examples: muscle hypertrophy caused by resistance training (also known as strength training) and muscle atrophy caused by immobilization because of injury or surgery. It is no exaggeration to say that regular physical activity is the most powerful factor controlling muscle protein metabolism.

As became evident from the discussion of how training influences the lipidemic profile (sections 10.12 and 10.13), and as will be reinforced in the following two chapters, *different kinds of training elicit different adaptations.* Most research data on the adaptations of protein metabolism refer to resistance training (figure 11.8). When combined with adequate protein intake, resistance training increases the amount of proteins, particularly the contractile proteins (myosin, actin, etc.), in the trained muscles. This causes a bulging of the myofibrils in the muscle fibers, resulting in the longitudinal fission of the myofibrils and an increase in their number. Along with the number of myofibrils, the volumes of the cytosol and sarcoplasmic reticulum also increase.

As a result of these changes, the cross-sectional area of the muscle fibers (assessed by light microscopy) increases, which is what we refer to as **hypertrophy** (meaning "overfeeding" in Greek). Then the cross-sectional area of the entire muscle and maximal strength increase. This adaptation is measurable a few weeks after the beginning of training and proceeds at a rate of 1% to 3% per week. Muscle hypertrophy reaches a plateau after about six months of hard resistance training. Sprint training also causes muscle hypertrophy, although hypertrophy in this case is smaller and requires about two months to appear.

Note that the number of muscle fibers does not change in hypertrophy. A rise in muscle fiber number, called **hyperplasia** (meaning excess formation), has been observed in animal models of resistance training, but there is little evidence that this occurs in humans. In the case of atrophy, the cross-sectional area of human muscle fibers decreases but, again, their number does not seem to change.

Endurance training is usually not accompanied by muscle hypertrophy unless it employs high intensities, in which case it may cause some increase in muscle proteins and a small degree of hypertrophy. The characteristic adaptation of muscle protein metabolism to endurance training is different: the increase in mitochondrial proteins, leading to an increase in mitochondrial number and size (figure 11.9). The generation of new mitochondria is termed **mitochondrial biogenesis.**

This adaptation allows the muscles to regenerate, through aerobic processes, a larger portion of the ATP they need for exercise. We will explore this important change in sections 13.12 and 13.13, in the wider context of how endurance training affects the selection of energy sources during exercise. For now, in the context of protein and amino acid metabolism, it is worth noting that the mitochondrial enzymes, which increase with endurance training, include aminotransferases and the branched-chain α-keto acid dehydrogenase complex. These adaptations enhance the capacity of the muscles to produce energy from BCAA degradation.

The biochemical mechanisms mediating the effects of training on protein metabolism are largely unknown. However, they appear to

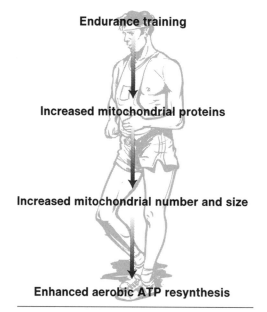

Resistance training

Increased muscle proteins

Increased cross-sectional area of muscle fibers (hypertrophy)

Increased cross-sectional area of muscle

Increased maximal strength

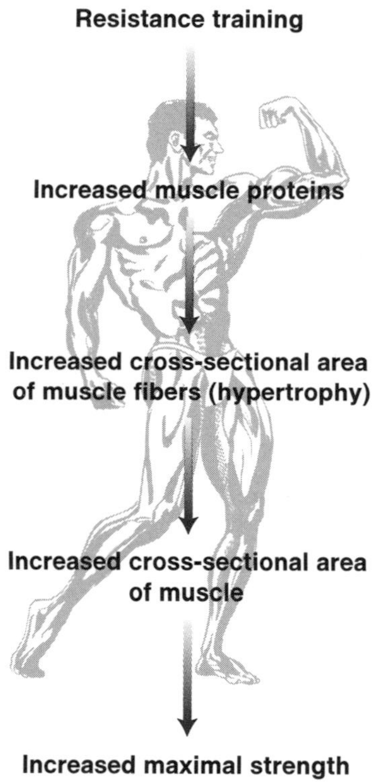

▶ **Figure 11.8** Effect of resistance training on muscle proteins.

Endurance training

Increased mitochondrial proteins

Increased mitochondrial number and size

Enhanced aerobic ATP resynthesis

▶ **Figure 11.9** Effect of endurance training on muscle proteins.

begin with mechanical, neural, hormonal, and metabolic stimuli that modify gene expression. Some of the things we know about these mechanisms are the subject of the next chapter.

Problems and Critical Thinking Questions

1. Which branch of protein metabolism in muscle do hard resistance and endurance exercises affect in a similar way?

2. Which branch of protein metabolism in muscle do hard resistance and endurance exercises affect in opposite ways?

3. Do the BCAA belong to the same or different categories of amino acids depending on whether the human body can synthesize them? What about categories depending on whether the human body can synthesize glucose from them?

CHAPTER 12

Effects of Exercise on Gene Expression

Skeletal muscle displays an astonishing ability to adapt to the ever-changing demands of the environment, particularly the demands of exercise. This ability is often expressed as *muscle plasticity* and is due to changes in gene expression. Remember that we have defined gene expression as the presence of a gene product (that is, RNA or protein) in a cell, organ, or tissue (section 4.11). Changes in gene expression because of exercise are the focus of this chapter.

Our knowledge about how exercise affects gene expression lags far behind our knowledge of how exercise affects energy metabolism. It is—one might say—similar to the knowledge of explorers in an uncharted territory (figure 12.1). This is a scientific field that is attracting many investigators and accumulating information at a rapid rate. In the ensuing sections, I will present as much of this information as is pertinent to a textbook, placing emphasis on skeletal muscle, where most of the relevant research has been

▶ **Figure 12.1** *Terra incognita.* What we know about the effects of exercise on gene expression, and about the control of gene expression in general, resembles—in terms of both quantity and accuracy—what cartographers of past centuries knew about the shape of the continents. Shown here is part of a world map drawn by Diego Ribero in 1529. Note the absence of most of the western coastline of North and South America. Also note, however, how much was known just 37 years after the discovery of the New World. The map is kept in the Vatican Library.

conducted. Note, however, that exercise affects gene expression in other tissues and organs as well.

12.1   Stages in the Control of Gene Expression

Exercise, like any natural stimulus, can affect the degree of expression of genes that are already expressed in a cell type; it cannot cause the expression of genes that are not expressed. A gene can be *induced* (its product increases) or *repressed* (its product decreases). Induction is the predominant way of regulating gene expression in eukaryotes.

Cells control gene expression by regulating the rate of the processes depicted in figure 12.2. Most are known to us from chapter 4 and are simply mentioned in the following list. Those that are introduced here are briefly described in the list.

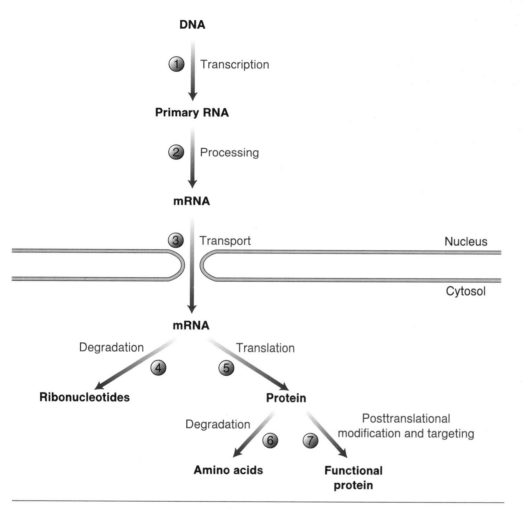

1. *Transcription of DNA to RNA.*

2. *Conversion of primary RNA to mRNA.*

3. *Messenger RNA transport from the nucleus to the cytosol for translation.*

4. *Messenger RNA stability.* Messenger RNA is rapidly hydrolyzed by *ribonucleases* like the one depicted in figure 3.10. The half-lives of the various mRNA range from 10 min to two days. Addition of a chain of AMP units to the 3' end of an mRNA delays its degradation and prolongs its use for translation.

5.   *Translation of mRNA to protein.*

▶ **Figure 12.2**   Control points in gene expression. The amounts of RNA and protein produced from the expression of a gene are affected by the rates of transcription (1), primary RNA processing (2), exit of mRNA to the cytosol through the pores of the nuclear envelope (3), mRNA degradation (4), translation (5), protein degradation (6), and posttranslational modification as well as targeting (7).

6. *Protein stability.* Although proteins are more stable than mRNA (their half-lives range from 10 min to decades!), they too are degraded.

7. *Posttranslational modification and protein targeting.* Many proteins are not ready to play their biological roles right after they are synthesized; instead, they require *posttranslational modification.* Two common kinds of such modification are the excision of one or more segments off the polypeptide chain and *glycosylation,* that

is, the covalent linking of one or more carbohydrate chains to the polypeptide chain. Additionally, many proteins have to be transported to specific intracellular compartments (such as the mitochondria), which requires *targeting*. Often a certain posttranslational modification of a protein is a prerequisite for its correct *targeting*.

Note that most of the processes listed consist of smaller steps, which may be individually controlled. For example, the rate of translation depends on the rates of initiation, elongation, and termination. Moreover, given that processes 1 through 7 are controlled by RNA, or proteins, or both (for example, translation is controlled by the ribosomes), changes in the expression of the genes encoding them can cause changes in the expression of the genes that these molecules control. The result is a complex meshwork of interactions, which explains part of the difficulty of studying gene expression.

12.2 Which Stages in the Control of Gene Expression Does Exercise Affect?

Which of the stages itemized in the previous section does exercise affect, thus modifying gene expression? There is evidence for intervention in most of them. Let's examine it.

- *Transcription.* This appears to be the most important control point of gene expression. Because of technical difficulties, few studies have measured transcription rates in muscle after exercise. These studies have revealed enhanced transcription rates of genes encoding structural proteins, transport proteins, and enzymes. This is mainly due to increased concentrations of transcription factors (section 4.10) or structural alterations in transcription factors (for example, by phosphorylation) that result in their increased binding to DNA.

- *RNA processing.* An impressive finding related to the effect of exercise on RNA processing is that hard resistance exercise, stretching, or electrical stimulation causes skeletal muscle to produce an unusual form of the peptide hormone **insulin-like growth factor 1, or IGF1,** termed **mechano growth factor, or MGF** (Goldspink 2005). MGF is derived from the IGF1 gene by alternative splicing of the primary RNA (section 4.12), synthesis of a different mRNA, and hence synthesis of a different protein. (Section 12.5 has more on IGF1 and MGF.)

- *Translation.* Translational control after modified contractile activity seems to be particularly important. The amount of many proteins increases in the beginning of such activity without corresponding increases in mRNA, which points to an increase in the rate of translation or a decrease in the rate of degradation for these proteins. As far as translation is concerned, initiation seems to be its most important control point. Initiation requires the presence of a group of proteins called *eukaryotic initiation factors* in addition to the components presented in section 4.16 (mRNA, ribosomes, charged tRNAs, ATP, and GTP). Increased concentrations of eukaryotic initiation factors have been found after both endurance and resistance exercise.

- *Protein stability.* As mentioned in section 11.2, little is known about the effect of exercise on the rate of muscle protein degradation. In general, the rate increases but the mechanism is not known.

- *Posttranslational modification and protein targeting.* Investigators have found increased amounts of proteins participating in the passage of other proteins from the cytosol to the mitochondria after chronic contractile activity. This change can contribute to the increase in mitochondrial biogenesis observed with endurance training.

12.3   Kinetics of a Gene Product After Exercise

The fact that there are changes in the rates of protein synthesis and degradation well after exercise implies that gene expression is modified not only during exercise but also in the recovery period. In fact, the changes that take place during this time account for a large part of the adaptations to training.

We can monitor the changes in gene expression during recovery from exercise by measuring the concentrations of mRNA and proteins in muscle or other tissues at different time points following an exercise bout, for example, during a 24 h period. In this way, we determine the *kinetics* of the gene products, that is, the progress of their concentrations over time. As you might expect, the kinetics differs from one mRNA to another and from one protein to another.

Figure 12.3 depicts two frequently appearing cases of mRNA kinetics during a week of training that comprised one exercise session per day. One mRNA is the product of a gene that is induced quickly, for example the gene encoding pyruvate dehydrogenase kinase, a key regulatory enzyme in carbohydrate metabolism (figure 9.19). Such mRNA increase spectacularly in the first postexercise hours (when their rate of synthesis exceeds their rate of degradation) and then usually decrease to baseline during the remainder of the day. This is repeated in the subsequent days.

The other mRNA is the product of a gene that is induced slowly. Such genes encode proteins directly involved in carbohydrate, lipid, amino acid, and nucleotide metabolism. An example is cytochrome *c* of the electron transport chain (section 9.11). Messenger RNA of this kind increase slowly after exercise, probably because the acceleration of transcription of their genes is rather complex and time-consuming. On the other hand, the subsequent decrease in these mRNA is also slow because they are rather stable (they have a long half-life). As a result, the mRNA concentration does not return to baseline within 24 h, and the next training session takes place in the presence of a slightly elevated concentration compared to the day before. This is repeated day after day, resulting in a considerable mRNA increase at the end of the week.

Now, what is the kinetics of the corresponding proteins? It is reasonable to think that a protein encoded by an mRNA having a long half-life (second case above) will increase, since the mRNA concentration is above baseline during the entire week, provided that the protein's rate of degradation does not increase at the same time. But what about a protein encoded by an mRNA having a short half-life as in the first case? The answer depends on protein stability. If the protein has a long half-life, it is possible that the transient increases in mRNA will lead to an accumulation of the protein. However, if the protein has a short half-life, then its concentration may not change in the long run.

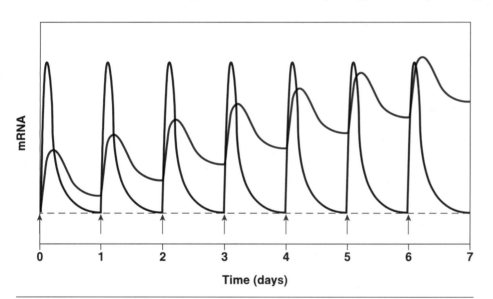

▶ **Figure 12.3**   Postexercise mRNA kinetics. Two mRNA display different kinetics during a week involving one training unit per day (indicated by the red arrows). One mRNA (black line) rises and falls so fast that it cannot accumulate. The other mRNA (red line) rises and falls more slowly. As a result, it is slightly elevated at the end of the day and accumulates gradually. Note that the graph assumes the response of each mRNA to the exercise stimulus to be identical day after day; in reality, this is not always the case.

Why are there mRNA and proteins of different stabilities? Compounds of short half-lives permit a fast response of the body to changing environmental conditions such as exercise. For example, proteins controlling pathways that produce energy have short half-lives so that their concentrations can change rapidly in response to changing energy demands. By contrast, structural proteins and proteins playing fundamental roles in metabolism have long half-lives. These proteins are economical, since the cells do not have to resynthesize them at rapid rates.

Tools in the Study of Gene Expression

A number of remarkable inventions in recent years have greatly facilitated our study of gene expression and of the ways in which factors such as diet, exercise, and disease affect it. To begin with, the base sequence of large DNA molecules is now easily and quickly determined in computerized instruments that rely on the pioneering work of Frederick Sanger back in 1977. Incidentally, Sanger was also the one to introduce the amino acid sequencing of proteins about a quarter-century earlier (in 1953) by deciphering the sequence of insulin. In 1983, the invention of the *polymerase chain reaction* (PCR) by Kary Mullis enabled the amplification and characterization of tiny amounts of DNA (consisting in even a single molecule), thus opening new avenues in genomics, diagnostics, forensics, archaeology, and paleontology.

Messenger RNA molecules synthesized in a tissue can be separated by electrophoresis and identified by base-pairing (*hybridization,* as we call it) with synthetic, radioactively labeled single-stranded DNA molecules of known sequence. This technique is known as *Northern blotting.* Minute amounts of mRNA (even from a single cell) can be analyzed by *reverse transcriptase PCR,* by which we first synthesize a DNA that has a base sequence complementary to the mRNA of interest (termed *cDNA*) and then amplify the cDNA by PCR. Transcription rates (rather than amounts of RNA, which reflect the balance between synthesis and degradation) can be assessed by the *nuclear run-on transcription assay,* which detects the amount of RNA molecules being transcribed at a particular time from a particular gene in the nuclei of a sample's cells.

Proteins, the end of gene expression, are analyzed by a great assortment of spectrophotometric, fluorometric, radioactive, mass spectrometric, chromatographic, electrophoretic, and other techniques (refer to the sidebar on page 23 for a description). All these offer tremendous possibilities for studying the immensely complex *proteomes* of living organisms and the ways in which gene expression at the protein level changes in response to a variety of conditions including exercise.

The ever-growing knowledge of the base sequence of genomes has enabled simultaneous examination of the expression of thousands of genes through *DNA microarray,* or *DNA chip,* or *gene chip* technology. Short deoxyribonucleotide sequences, characteristic of the genes of interest, are placed in minute amounts as dots on a plate measuring a couple of centimeters in each dimension. The mRNA molecules extracted from a tissue are then converted to cDNA with fluorescently labeled deoxyribonucleotides, and these cDNA molecules are permitted to hybridize to complementary sequences on the chip. The dots that fluoresce correspond to the genes that are expressed. It is even possible to treat the chip with two cDNA mixtures, each derived from the mRNA complement of a tissue under two conditions that we want to compare, for example, before and after exercise. In that case, each cDNA mixture is prepared with nucleotides that fluoresce in different colors. Dots on the chip emit one color (say, green) if expression of a gene increased with exercise, another (red) if expression decreased, and yet another (yellow) if expression did not change. Thus, DNA chips let us evaluate gene expression changes at the mRNA level on a global scale. Finally, to evaluate gene expression changes at the protein level, we may use *protein chips* containing antibodies to specific proteins rather than nucleotide sequences.

12.4 Exercise-Induced Changes That May Modify Gene Expression

The modification of gene expression in a cell by an external or internal stimulus is triggered by one or more changes taking place somewhere in the cell. Which of the many changes caused by increased contractile activity signal the modification of gene expression in a muscle fiber? The answer is not known with certainty, but research has pointed a finger at several suspects. The main ones are

- the increase in cytosolic $[Ca^{2+}]$ because of Ca^{2+} release from the sarcoplasmic reticulum upon excitation of the muscle fiber,
- the decrease in [ATP] and the concomitant increase in [ADP] and [AMP],
- the decrease in glycogen concentration,
- the decrease in oxygen concentration (**hypoxia**) because of increased consumption in the electron transport chain,
- the increase in radical concentration,
- the decrease in pH because of the anaerobic breakdown of carbohydrates,
- the application of tension (stretching) on the fiber, and
- the increased binding of proteins to cell receptors.

Some of these stimuli lie at the start of complex signal transduction pathways mediating the effect of resistance exercise and endurance exercise on gene expression. These pathways are being gradually deciphered and are the subject of the next two sections.

12.5 Mechanisms of Exercise-Induced Muscle Hypertrophy

As discussed in section 11.9, resistance training causes skeletal muscle hypertrophy resulting from enhanced protein synthesis. This increase in gene expression inside the muscle fibers is partly due to a higher DNA content thanks to the addition of nuclei derived from satellite cells.

Satellite cells are myoblasts that did not fuse during muscle development. They are closely associated with muscle fibers (figure 12.4), wrapped in the same *basal lamina,* a sheet of extracellular matrix surrounding

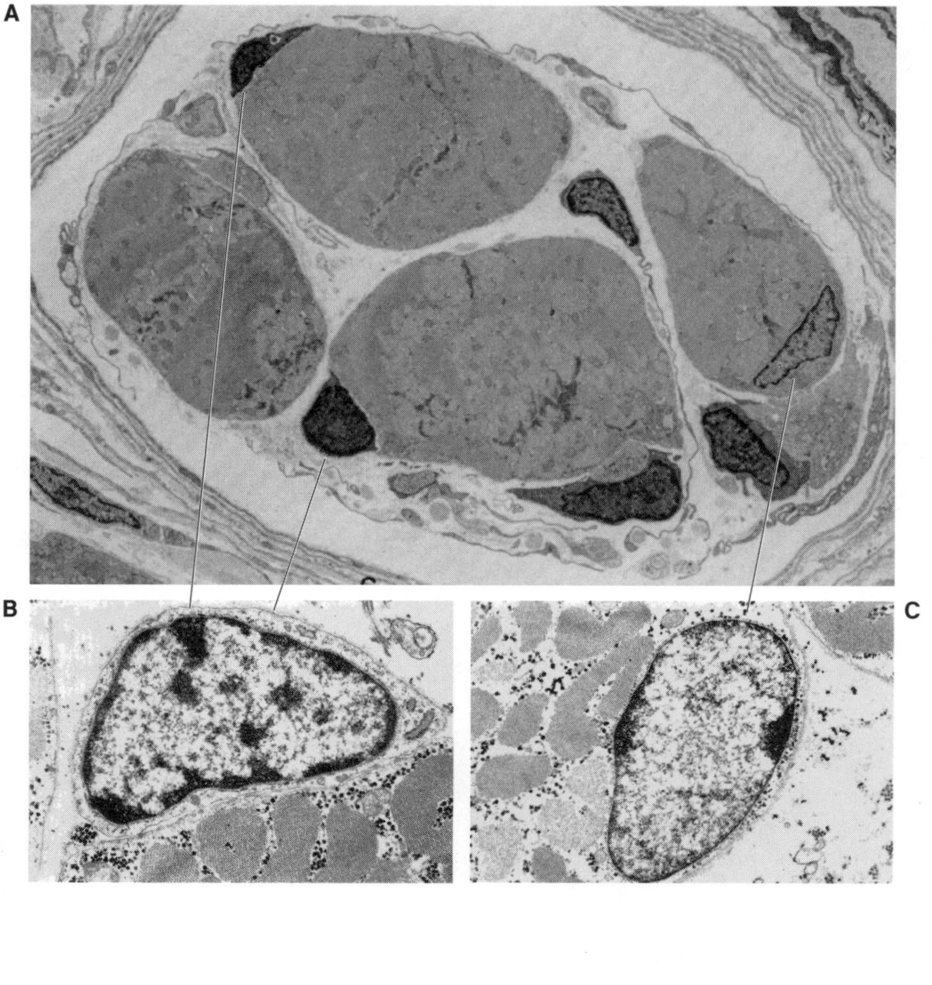

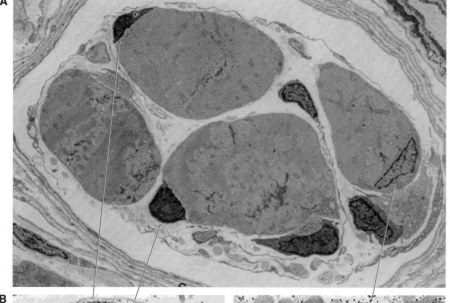

▶ **Figure 12.4** Satellite cells. *(a)* An electron micrograph of four muscle fibers shows two satellite cells stuck to two of the fibers. The satellite cells are stained dark owing to the presence of chromatin in their nuclei (cf. figure 1.5). *(b)* At a higher resolution, a satellite cell is seen to be almost entirely occupied by the nucleus. Note the distinct plasma membranes of the satellite cell and the muscle fiber underneath. *(c)* In contrast, a myonucleus lies inside a muscle fiber, close to the fiber's plasma membrane, but not surrounded by a plasma membrane of its own.

Figure 12.4a From I Berman, 2003, *Color atlas of basic histology* (New York: McGraw-Hill Companies). Reproduced with permission of The McGraw-Hill Companies.

Figure 12.4b-c Reprinted, by permission, from D.R. Campion et al., 1981, "Ultrastructural analysis of skeletal muscle development in the fetal pig," *Acta Anatomica* 110: 227-284. By permission of Karger AG, Basel.

each muscle fiber. Satellite cells play an important role in muscle growth and repair: When triggered by resistance exercise, mechanical loading, or damage, a satellite cell divides; one of the two daughter cells then fuses with the adjacent muscle fiber, thus providing an additional nucleus (figure 12.5). The other daughter cell remains a satellite cell, in reserve for future needs.

The biochemical events leading from resistance exercise to satellite cell proliferation and fusion with muscle fibers are not known with certainty. However, IGF1 (including its variant MGF introduced in section 12.2) seems to play a key role. IGF1 is synthesized in the liver and muscle. Besides serving the classical endocrine function of hormones (being transported by the blood to target organs throughout the body), muscle IGF1,

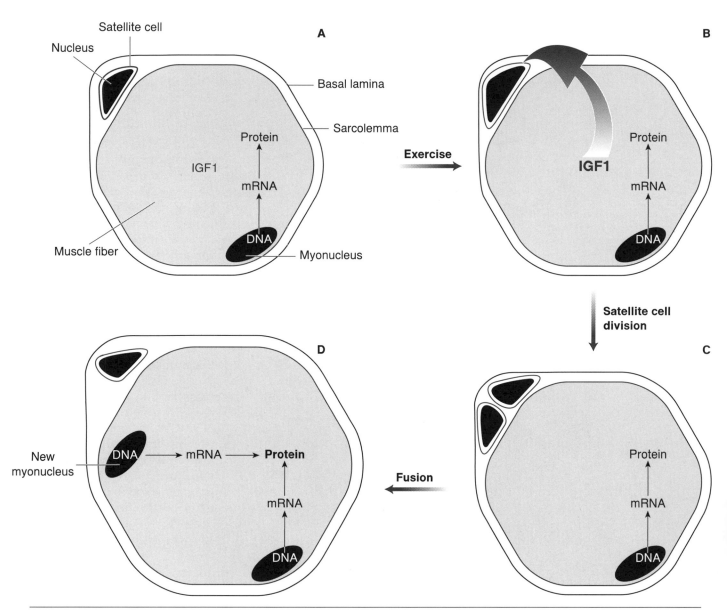

▶ **Figure 12.5** Satellite cell proliferation causes muscle hypertrophy. *(a)* Diagram of a cross section of a muscle fiber depicts a myonucleus lying close to the sarcolemma and offering the information contained in its DNA for mRNA and protein synthesis. A satellite cell lies close to the muscle fiber; both are enclosed in the basal lamina. The muscle fiber has a low resting IGF1 concentration. *(b)* After resistance or eccentric exercise, the IGF1 concentration rises and IGF1 binds to its receptor in the plasma membrane of the satellite cell, causing it to grow in preparation for mitosis. *(c)* The satellite cell divides. *(d)* One of the daughter cells fuses with the muscle fiber, thus providing a new nucleus and more DNA for mRNA and protein synthesis. This, in turn, increases the cross-sectional area of the muscle fiber.

particularly MGF, serves *autocrine* and *paracrine* functions: After its molecules are secreted, they act either on the same cells that synthesized them (autocrine function) or on contiguous cells (paracrine function).

The muscle IGF1 concentration increases with resistance or eccentric exercise, although how exercise does this is unclear. IGF1 is a potent *mitogen* (it induces mitosis), but again how it causes satellite cell proliferation is unclear. Several mechanisms are plausible, one of which involves the PI3K cascade utilized by insulin (figure 9.36). The IGF1 receptor is very similar to the insulin receptor (figure 9.35): Upon IGF1 binding, the receptor is activated by autophosphorylation and initiates a cascade of interactions leading to the phosphorylation and activation of PKB (figure 12.6). PKB in turn phosphorylates a transcription factor that controls the expression of a gene encoding a cell-cycle inhibitor. The latter is one of a number of proteins that keep cells from dividing. Phosphorylation of the transcription factor slows down transcription, thus decreasing the inhibitor concentration. This releases inhibition of cell division and permits satellite cell proliferation.

Activation of PKB in a muscle fiber by IGF1 binding to its plasma membrane contributes to hypertrophy by accelerating translation. PKB does so in three ways. First, it activates a eukaryotic initiation factor. Second, it facilitates binding of the small ribosomal subunit to certain mRNA, one of the first steps in translation (section 4.16). Third, it even increases the number of ribosomes in a muscle fiber by stimulating the transcription of genes encoding rRNA and the translation of mRNA encoding ribosomal proteins.

Because they induce muscle hypertrophy, IGF1 and MGF may be used by athletes to boost performance in strength or sprint events. The World Anti-Doping Agency (WADA), which is the body regulating doping matters worldwide, considers the use of IGF1 or MGF in sport unethical and prohibits it. Relevant to IGF1 is the alarming possibility of *gene doping*, that is, boosting IGF1 production in muscle fibers by

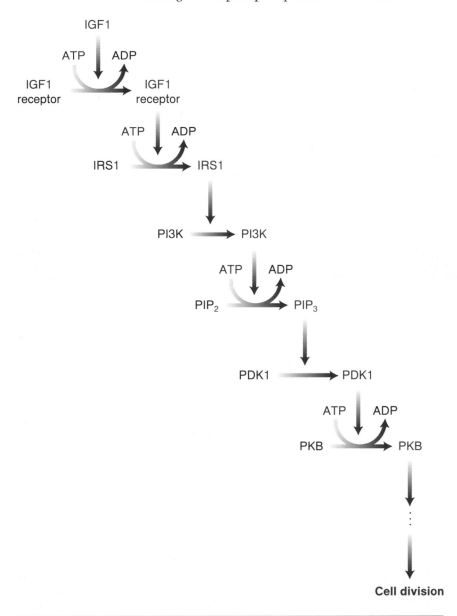

▶ **Figure 12.6** How IGF1 induces satellite cell proliferation. IGF1 binds to and activates its receptor, which phosphorylates IRS1, which activates PI3K, which phosphorylates PIP_2 to PIP_3, which activates PDK1, which phosphorylates and activates PKB. PKB, through a number of additional steps, lifts inhibition of cell division. See figure 9.36 for an explanation of abbreviations. Note that other pathways may also convey the signal of IGF1 for satellite cell proliferation.

either introducing additional copies of the IGF1 gene into the genome or inducing the IGF1 gene. Another related doping tool might be drugs designed to block *myostatin*, a protein that acts like a brake on muscle growth, possibly by inhibiting satellite cell proliferation. Such drugs would release the brake on muscle growth and permit muscle hypertrophy or even hyperplasia.

An additional mechanism of muscle hypertrophy in response to overloading (caused, for example, by ablation of synergic muscles) begins with a rise in cytosolic $[Ca^{2+}]$. Ca^{2+} binds to calmodulin, a calcium sensor presented as a subunit of phosphorylase kinase in section 9.2. Calmodulin also exists in free form. Upon binding Ca^{2+}, calmodulin can associate with several proteins and modify their biological activity. Two such proteins are *calmodulin-dependent protein kinase* and *protein phosphatase 2B*, the latter of which is commonly known as *calcineurin.* These enzymes phosphorylate and dephosphorylate, respectively, key transcription factors controlling certain genes involved in the hypertrophic response. Modification of the transcription factors results in their activation and induction of the genes in question.

Doping

WADA prohibits a substance if it meets any two of the following criteria:

- Scientific evidence or experience that the substance has the potential to enhance or enhances sport performance
- Scientific evidence or experience that the use of the substance represents an actual or potential health risk to the athlete
- WADA's determination that the use of the substance violates the spirit of sport

12.6 Mechanisms of Exercise-Induced Mitochondrial Biogenesis

Increased mitochondrial biogenesis is perhaps the most remarkable adaptation of skeletal muscle to endurance training. New mitochondria arise by growth and division of existing mitochondria (figure 12.7); they are not made from scratch. Their components are formed on the basis of genetic information contained not only in the nuclear DNA but also in the small circular mitochondrial DNA. This exists in multiple copies inside each mitochondrion and encodes 13 proteins of the electron transport chain. It also encodes the rRNA and tRNA needed for the synthesis of these proteins. How the two genomes cooperate in producing the right mix of mitochondrial components is being gradually unraveled.

Mitochondrial biogenesis is controlled by two classes of proteins, which regulate transcription of the genes encoding mitochondrial proteins. The first class consists of transcription factors that bind to specific DNA base sequences in the promoters of the genes in the nucleus (see section 4.10). These transcription factors include *nuclear respiratory factors 1* and *2* (NRF1 and NRF2), which bind to the promoters of genes encoding

- proteins responsible for mitochondrial protein import;
- proteins of the electron transport chain;
- proteins involved in heme biosynthesis (remember that several proteins of the electron transport chain contain heme; section 9.11); and

The existence of mitochondrial DNA attests to the origin of the mitochondrion as an aerobic bacterium that was engulfed by a larger anaerobic cell about a billion years ago. In this mutually beneficial event, the bacterium found nutrients to live on and proliferate, while the larger cell found energy-efficient machinery, since aerobic catabolism yields more ATP than anaerobic catabolism. The bacterium evolved into a mitochondrion and the larger anaerobic cell into an aerobic cell.

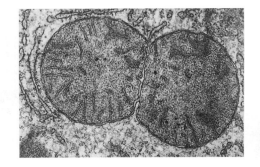

▶ **Figure 12.7** A mitochondrion divides. Almost split in half, a mitochondrion is still surrounded by a single outer membrane, but two distinct inner membranes have already formed.

Courtesy of Daniel S. Friend.

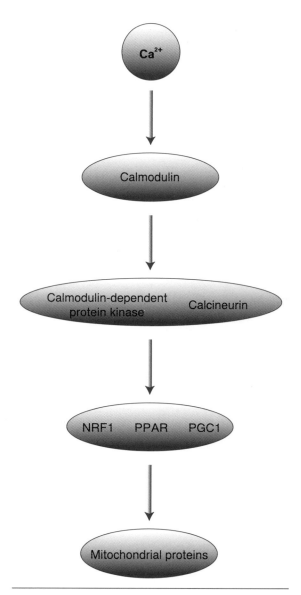

▶ Figure 12.8 Hierarchy of signal molecules in exercise-induced mitochondrial biogenesis. Although many details are missing, a sustained rise in cytosolic [Ca²⁺] in endurance exercise seems to activate calmodulin-dependent protein kinase and calcineurin, causing an increase in transcription factors (NRF1 and PPARα) and transcription coactivators (PGC1), which induce genes encoding mitochondrial proteins.

• proteins involved in the transcription and replication of mitochondrial DNA, as well as the processing of mitochondrial RNA.

Other transcription factors implicated in mitochondrial biogenesis are the *peroxisome proliferator-activated receptors* (PPAR), a family of transcription factors engaged in lipid metabolism. PPAR are activated when a ligand such as a fatty acid or fatty acid derivative binds to them. Two members of this family, PPARα and PPARδ, control the transcription of genes encoding enzymes of β oxidation.

The second class of proteins regulating mitochondrial biogenesis consists of *transcription coactivators*. These proteins bind to transcription factors rather than DNA and facilitate the initiation of transcription by RNA polymerase. A transcription coactivator involved in mitochondrial biogenesis is *PPARγ coactivator 1* (PGC1), which binds to NRF1 and PPARα. Endurance exercise or chronic electrical stimulation having a pattern characteristic of endurance training can increase NRF1, PPARα, PPARδ, and PGC1.

What triggers the change in gene expression that results in mitochondrial biogenesis with endurance exercise? Ca²⁺ seems to be the primary signal. Once again, Ca²⁺ acts by binding to calmodulin (figure 12.8), and calmodulin activates calmodulin-dependent protein kinase and calcineurin, the very enzymes implicated in the hypertrophic response with resistance training as discussed in the previous section. Calmodulin-dependent protein kinase and calcineurin are thought to mediate the increase in the transcription factors and transcription coactivators responsible for mitochondrial biogenesis, although we do not know how this happens.

If both endurance training and resistance training affect gene expression through Ca²⁺, calmodulin-dependent protein kinase, and calcineurin, how do they elicit such different responses (mitochondrial biogenesis vs. muscle hypertrophy)? The key seems to be the magnitude and duration of the cytosolic [Ca²⁺] increase. Endurance exercise causes a moderate, sustained rise, whereas resistance exercise causes large increases of short duration. How the same enzymes sense these differences is an intriguing question that awaits an answer.

Problems and Critical Thinking Questions

1. Which of the exercise-induced changes that may modify gene expression are most probably operating in endurance exercise?
2. Answer the same question for resistance exercise.
3. Answer the same question for sprint exercise.

CHAPTER 13

Integration of Exercise Metabolism

In previous chapters of part III we examined separately the effect of exercise on the metabolism of compounds of high phosphoryl transfer potential, carbohydrates, lipids, and proteins. Independent examination was dictated by the need to study each class of energy sources in detail. However, you will rarely encounter a kind of exercise that does not require a considerable contribution of energy from at least two classes. It is therefore necessary to integrate the information on the use of each energy source in a way that will provide a complete picture of exercise metabolism. This integration is the subject of the present chapter, which covers topics including the factors influencing the choice of energy sources during exercise, the metabolic specialization of different muscle fiber types, the adaptations to different kinds of training, the reversal of adaptations when training is interrupted, the coordination of metabolism in different tissues and organs by hormones, the causes of fatigue, and the restoration of the energy state in the body after exercise.

13.1 Interconnections of Metabolic Pathways

Integrating exercise metabolism will be facilitated by a panoramic view of the metabolic pathways described in previous chapters. Such a view (figure 13.1) will convince you of how interrelated the fates of carbohydrates, lipids, and proteins are.

Let's begin with carbohydrates. Glucose entering a cell is rapidly phosphorylated to glucose 6-phosphate. This compound can proceed through the glycolytic pathway and end up in pyruvate while ATP is produced. The same fate awaits the glucosyl units of glycogen, most of which are converted initially to glucose 1-phosphate and then to glucose 6-phosphate. It is also possible to synthesize glycogen from glucose through the two glucose phosphates.

Pyruvate holds a central position in metabolism in its ability to follow a variety of alternative routes. To begin with, it can be converted to lactate through a reversible

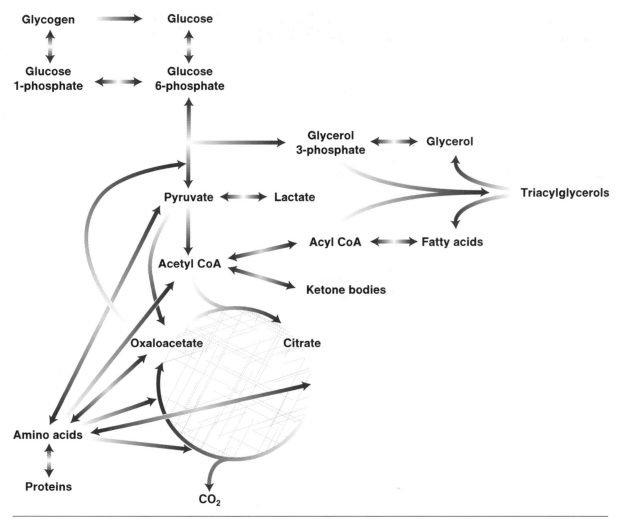

▶ **Figure 13.1** Metabolic interconnections. The metabolic pathways of carbohydrates, lipids, and proteins are connected by a multitude of intermediate compounds, the most important of which are presented.

reaction. In contrast, pyruvate's conversion to acetyl CoA is irreversible and paves the road to its complete oxidation to carbon dioxide through the citric acid cycle. Electrons produced in the process drive the production of many ATP molecules thanks to the coupling of the electron transport chain to oxidative phosphorylation. Pyruvate can also be converted to oxaloacetate, an intermediate of the citric acid cycle. This conversion is the first step in glucose synthesis through gluconeogenesis. Lastly, pyruvate can be easily converted to alanine and vice versa. This interconversion is but one of the bridges linking carbohydrate and protein metabolism.

Let's move now to lipids. Triacylglycerols, the most abundant lipid category, are hydrolyzed to glycerol and fatty acids. Glycerol communicates with glycolysis and gluconeogenesis through glycerol 3-phosphate bidirectionally; glycerol 3-phosphate is also required for triacylglycerol synthesis. Fatty acids, on the other hand, are activated by being converted to acyl CoA, which can be either channeled back to triacylglycerol synthesis or degraded to acetyl CoA through β oxidation. Apart from producing ATP by oxidation, acetyl CoA (also derived from carbohydrates) is utilized in fatty acid synthesis and ketone body formation.

We can fill in the picture with proteins. Proteins are hydrolyzed to amino acids, which follow various catabolic routes leading to the incorporation of their amino groups in urea (through the urea cycle) and the conversion of their carbon skeletons to pyruvate, acetyl CoA, or intermediates of the citric acid cycle. These compounds can be broken down to yield energy or they can be used in the synthesis of glucose and fatty acids. Conversely, pyruvate and intermediates of the citric acid cycle can produce amino acids, which are utilized in protein synthesis.

13.2 Energy Systems

We can classify the energy sources described in previous chapters into three energy systems. The systems include not only the sources themselves but also the enzymes that ensure ATP production from these sources. Each system is characterized by two parameters of pivotal importance for its usefulness in exercise:

The total energy it can yield

Its power, that is, the ratio of energy to the time within which it is produced (in simple terms, how fast the system yields energy)

The three energy systems are the following:

1. The **ATP-CP system,** which includes
   - in terms of energy sources, ATP, ADP, and CP; and
   - in terms of enzymes, creatine kinase and adenylate kinase, or myokinase.

The energy of the ATP-CP system is low, but its power is high. Thus, the system predominates in maximal bouts lasting up to about 7 s, such as weightlifting, jumps, throws, a 60 m dash, and explosive efforts in ball games or fight sports.

2. The **lactate system,** which includes
   - in terms of energy sources, carbohydrates;
   - in terms of enzymes, those of glycogenolysis (phosphorylase and the debranching enzyme), those of glycolysis, and lactate dehydrogenase.

The lactate system has intermediate energy and power. Thus, it predominates in maximal bouts lasting about 7 s to 1 min. Examples are the 100, 200, and 400 m runs; the 50 and 100 m swims; and the 200, 500, and 1,000 m cycling races.

3. The **oxygen system,** or **aerobic system,** which includes
   - in terms of energy sources, carbohydrates, lipids, and (to a small degree) proteins;
   - in terms of enzymes, pyruvate dehydrogenase; the enzymes of lipolysis, fatty acid degradation, the citric acid cycle, and the electron transport chain; and ATP synthase.

The energy of the oxygen system is high but its power low. Thus, the system predominates in exercise tasks exceeding 1 min in duration regardless of intensity. Examples are the 800 m run and upward, the 200 m swim and upward, rowing, cross-country skiing, and ball games as a whole.

13.3 Energy Sources in Exercise

The most appreciable energy sources of the three energy systems—in terms of quantity and rate of ATP resynthesis—are summarized in table 13.1. The table shows the quantity of each source at rest, the amount of ATP it yields, and the maximal rate of ATP

▶ **Table 13.1**   Energy Sources in the Human Body During Exercise

Source	Quantity (mmol)	ATP yield (mmol)		Maximal rate of ATP resynthesis (mmol · kg⁻¹ · s⁻¹)	
		Anaerobic	Aerobic	Anaerobic	Aerobic
Muscle ATP	168	168	–	–	–
Muscle CP	560	560	–	2.6	–
Muscle glycogen	2,160[a]	6,480	66,960	1.5	0.5
Liver glycogen	556[a]	1,112	16,680	0.2	0.1
Myocellular triacylglycerols	163	–	58,680	–	0.3
Adipose tissue triacylglycerols	11,628	–	4,186,080	–	0.2

[a]As glucosyl units.

resynthesis from the breakdown of the source itself or (in the case of liver glycogen and adipose tissue triacylglycerols) the source's products after the blood has transported them to the exercising muscles.

The data in table 13.1 are based on the following assumptions, most of which we have already presented in the appropriate sections of chapters 8, 9, and 10.

- They refer to a 70 kg (154 lb) individual having 28 kg (62 lb) of muscle tissue and a 1.8 kg (4 lb) liver.
- Muscle contains 6 mmol · kg⁻¹ ATP, 20 mmol · kg⁻¹ CP, 1.25% (12.5 g · kg⁻¹) glycogen, and 0.5% (5 g · kg⁻¹) triacylglycerols.
- The liver contains 5% (50 g · kg⁻¹) glycogen.
- Adipose tissue contains 10 kg (22 lb) triacylglycerols.
- Each glucosyl unit of muscle glycogen yields three ATP anaerobically and 31 ATP aerobically.
- Each glucosyl unit of liver glycogen yields two ATP anaerobically and 30 ATP aerobically, since it is delivered to muscle as glucose.
- Based on a usual fatty acid composition, the average molecular mass of a triacylglycerol is 860 Da, and the average yield of the complete oxidation of its components (three fatty acids and glycerol) is 360 ATP.

Obviously, the numbers in the table will differ with different tissue masses and energy source contents. Therefore, they should be used with knowledge of the assumptions they are based on and should not be treated as constants.

Not included in table 13.1 are the following energy sources:

- Blood glucose, fatty acids, and triacylglycerols, because the amounts of blood glucose and fatty acids are very low compared to their main sources (liver glycogen and adipose tissue triacylglycerols, respectively), which we have taken into consideration. Blood triacylglycerols, on the other hand, do not participate substantially in supplying energy for exercise.
- Muscle ADP, glucose, and fatty acids, because their quantities are minimal.
- Muscle and liver proteins, because their contribution to the energy expenditure of exercise is minuscule.

Table 13.1 shows that there is generally an inverse relationship between the quantity of ATP that a source can yield and the maximal rate at which it can do so. Thus, CP, the smallest source of ATP, is also the fastest, whereas adipose tissue triacylglycerols, the largest of all sources, are also the second slowest. This combination of an advantage and a disadvantage gives the various sources a *raison d' être,* since, if the largest source were also the fastest, the other sources would never be used. Therefore, thanks to these inverse gradients of quantity and rate, each energy source has a place in some kind of exercise.

13.4   Choice of Energy Sources During Exercise

What is the contribution of each of the available energy sources toward meeting the demands of exercise? The answer depends on several factors, the most important of which are the following:

Exercise intensity

Exercise duration

Exercise program

Heredity

Nutrition

Training state

Age

Two additional parameters, over which controversy exists, are sex and ambient temperature.

Knowing the factors that influence the body's choice of energy sources during exercise, and knowing how they alter the mixture of energy sources, are not only of theoretical but also of practical interest, since such knowledge allows one to modify the factors that can be modified, aiming at a desirable mixture. Let's examine all these factors.

13.5   Effect of Exercise Intensity on the Choice of Energy Sources

Fatty acids derived from the hydrolysis of adipose tissue triacylglycerols and, to a lower degree, of myocellular triacylglycerols are the main fuel of the body at rest and during light exercise (figure 13.2). The rate of ATP resynthesis by these two sources compensates for the low rate of ATP breakdown. As exercise intensity rises, changes in the concentrations of substances that we explored in chapters 9 and 10 (increase in P_i, ADP, AMP, Ca^{2+}, epinephrine, and norepinephrine; decrease in ATP) speed up carbohydrate and lipid degradation, thus counterbalancing the increased demand for ATP. The carbohydrates used are muscle glycogen and plasma glucose, the latter emanating mainly from the breakdown of liver glycogen.

The rise in the amount of energy from carbohydrates as intensity increases is larger than the rise in the amount of energy from lipids. Thus, the percentage contribution of carbohydrates to the total energy expenditure of exercise increases, whereas that of lipids decreases. As a result, the two percentages become equal somewhere between 45% and 65% of $\dot{V}O_2max$, while the contribution of carbohydrates exceeds that of lipids at higher intensities. In fact, not only the percentage but also the absolute contribution of lipids (in terms of grams oxidized per minute) declines at higher intensities.

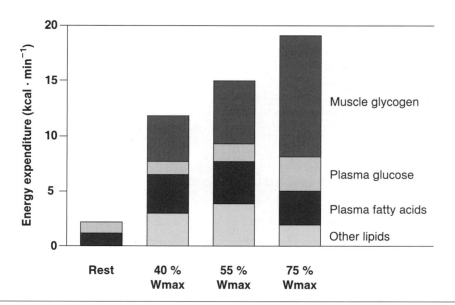

▶ **Figure 13.2** Energy sources at different exercise intensities. The percentage of total energy expenditure derived from carbohydrates and lipids relates to exercise intensity. At rest and during cycling at 40% of maximal power (Wmax) the majority of the energy spent by the body (about 55%) comes from the combustion of lipids. During exercise at 55% and 75% of Wmax, the share of lipids is limited to 49% and 24%, respectively. Muscle glycogen offers more energy than plasma glucose at all three exercise intensities, but muscle glycogen breakdown is minimal at rest. Plasma fatty acids offer energy equal to or greater than that contributed by myocellular and plasma triacylglycerols (reported together as "other lipids") at all four conditions. Note that not only the percentage but also the amount of lipids used at 75% of Wmax is less than that at 40% or 55% of Wmax. The data are derived from endurance cyclists who exercised for 30 min at each intensity. Percentages of Wmax are roughly equal to percentages of $\dot{V}O_2$max.

Reprinted, by permission, from L.J.C. van Loon et al., 2001, "The effects of increasing exercise intensity on muscle fuel utilization in humans," *The Journal of Physiology,* 536.1: 295-304.

The reason for the reduced contribution of lipids as exercise intensity increases is not known with certainty. Some authors attribute the reduction to the fact that carbohydrates yield more energy than lipids for a certain amount of oxygen. Indeed, if we write the equations describing the oxidation of glucose and a fatty acid (palmitic acid in particular), we get

$$C_6H_{12}O_6 + 6 O_2 \rightarrow 6 CO_2 + 6 H_2O \qquad \text{(equation 13.1)}$$

$$C_{16}H_{32}O_2 + 23 O_2 \rightarrow 16 CO_2 + 16 H_2O \qquad \text{(equation 13.2)}$$

Remember that the oxidation of glucose yields 30 ATP (section 9.14), while that of palmitate yields 106 ATP (section 10.5). Thus, glucose yields 30/6 = 5 ATP per O_2, while palmitate yields 106/23 = 4.6 ATP per O_2. Consequently, the supporters of the hypothesis maintain, the use of oxygen to burn carbohydrates is more energy efficient, and this becomes critical during hard exercise, when the availability of oxygen is limited.

There are at least three problems with this hypothesis. First, it assumes that the limiting factor of aerobic energy production is oxygen availability inside the mitochondria, which, as discussed in section 9.16, does not seem to hold true. Rather, the limiting factor seems to be the rate of acetyl CoA oxidation through the citric acid cycle, which is common to carbohydrate and lipid oxidation. Second, the difference between carbohydrates and lipids (5 vs. 4.6 ATP per O_2) is not large enough to justify the reversal of their proportion during the passage from light to hard exercise. Third, the hypothesis does not predict any biochemical mechanism for the selection of carbohydrates over lipids.

According to the existing evidence, the most likely reason for the fall in the contribution of lipids with increase in exercise intensity is the reduction in the concentration of carnitine that is available for the entry of fatty acids to the mitochondria. The reduction is due to the consumption of carnitine by acetyl CoA according to the reaction acetyl CoA + carnitine ⇌ acetyl carnitine + CoA   **(equation 13.3)**

This is a variation of reaction 10.10 (equation 10.10). The rise in acetyl CoA concentration during exercise (section 9.8) drives reaction 13.3 (equation 13.3) to the right, thus depleting carnitine and limiting its availability for the synthesis of acyl carnitine according to reaction 10.10. As a result, there is a decrease in the rate of fatty acid entry to the mitochondria and, consequently, the rate of fatty acid oxidation. There is also evidence that the decrease in cytosolic pH during hard exercise inhibits carnitine acyltransferase I (the enzyme catalyzing reaction 10.10). On the other hand, reaction 13.3 helps lower the [acetyl CoA]/[CoA] ratio and lift the inhibition of pyruvate dehydrogenase (section 9.8), thus speeding up carbohydrate oxidation.

Carbohydrates are basically broken down aerobically. However, as exercise intensity rises, so does the portion of carbohydrates that are broken down anaerobically to support the demand for high power output. Finally, at efforts of maximal intensity (which are necessarily very short) there is a substantial contribution of energy by CP, which exhibits the highest rate of ATP resynthesis.

13.6 Effect of Exercise Duration on the Choice of Energy Sources

The proportion of energy sources does not remain stable during exercise even if intensity is stable. This is due mainly to the limited quantity of most sources. Generally speaking, *during exercise of constant intensity the contribution of the smaller energy sources decreases, whereas the contribution of the larger sources increases.* Let's apply this principle to two different time frames.

First, we examine the proportion of energy sources during exercise of constant maximal intensity, which drives an athlete to exhaustion within a few minutes (figure 13.3). The rate of CP breakdown peaks almost instantaneously, to be followed by a rapid drop and zeroing at 25 s. The acceleration of glycogenolysis and glycolysis is a bit slower. Yet it is impressive, and rapid enough to permit the maintenance of a high power output as CP falls. At first, most of the glycogen that is broken down ends up in lactate. However, as seconds go by, more and more of the pyruvate produced through glycolysis follows the route of oxidation to CO_2. Upon completion of one-half minute of exercise,

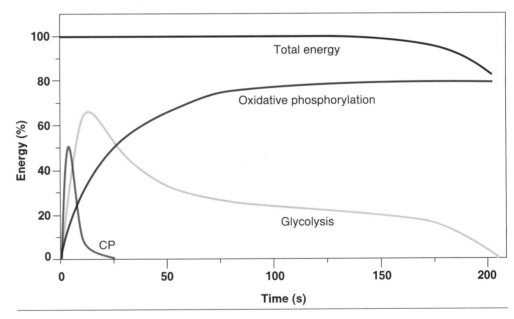

▶ **Figure 13.3** Energy sources during maximal exercise. CP is the main source for ATP resynthesis during the first seconds of maximal exercise (in a cycle ergometer). Glycolysis takes the lead at 7 s, only to give way to oxidative phosphorylation at 30 s. Energy supply is constant for 2.5 min and then decreases. Exhaustion of the athlete comes a minute later. Note how similar the curves of glycolysis and total energy are during the last half minute. This graph lets you find the percentage contribution of each source to total energy at any time point in the horizontal axis. Here's how. Draw a vertical line upward from the time point you are interested in and, from the points where it intersects the curves of the three energy sources, draw horizontal lines to the left. The point where each line meets the vertical axis is the percentage contribution of the corresponding energy source. The sum of the three percentages is 100% at any time, with two exceptions: the first few seconds, when it is short of 100% by the percentage contribution of ATP itself, and the last half minute, when total energy falls below 100%.

Adapted, by permission, from P.B. Gastin, 2001, "Energy system interaction and relative contribution during maximal exercise," *Sports Medicine* 31(10): 725-741.

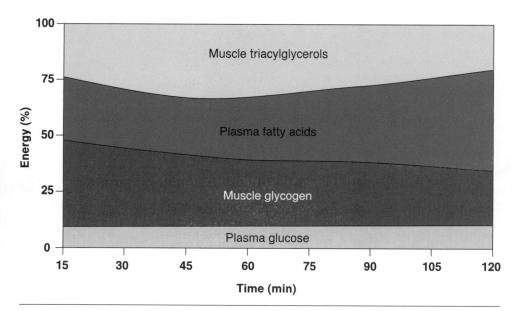

From J.A. Romijn et al., 1993, "Regulation of endogenous fat and carbohydrate metabolism in relation to exercise intensity and duration," *American Journal of Physiology* 265: E380-E391. Adapted by permission from The American Physiological Society.

▶ **Figure 13.4**   Energy sources during moderate-intensity exercise. The percentage contribution of carbohydrates to energy supply drops gradually during prolonged efforts. The data are derived from endurance cyclists who exercised at 65% of $\dot{V}O_2$max. The horizontal axis starts at 15 min because it was not possible to make reliable estimates of the contribution of each energy source during the initial minutes of exercise.

the curves of glycolysis and oxidative phosphorylation intersect. From that point onward, most of the ATP is produced by oxidative phosphorylation. The power of the oxygen system gradually increases, whereas that of the lactate system decreases. When the latter plunges to zero (after approximately 3 min), the athlete is unable to maintain the required power output and gives up.

Let's move now to the proportion of energy sources during exercise of constant moderate intensity, which can be maintained for hours, being fed almost exclusively by the aerobic breakdown of carbohydrates and lipids (figure 13.4). The proportion of these sources does not stay constant but shifts gradually to lipids. Thus, the rate of glycogen degradation falls, whereas that of fatty acid degradation rises.

13.7   Interaction of Duration and Intensity: Energy Sources in Running and Swimming

The events of running and swimming offer a first-rate opportunity to study the interaction of the two main factors affecting the proportion of energy sources during exercise, that is, duration and intensity. The reason is that running and swimming events comprise a wide variety of distances, which in turn impose a variety of speeds—since no matter how hard they try, athletes cannot run or swim a long distance at the speed they run or swim a short distance.

What then is the proportion of energy sources in running and swimming events? Our guide to answering the question will be figure 13.5, which presents the percentage contribution of each energy system to total energy supply depending on the duration of a maximal exercise. The graph was constructed by Gastin (2001) based on over 40 studies and differs from relevant graphs in other books (even new ones) in that the curves of the three energy systems are shifted to the left. This happens because, as mentioned in section 9.17, the experimental data of recent years have upgraded the capacity of the aerobic system and degraded the capacity of the two anaerobic systems.

The graph in figure 13.5 differs from the one in figure 13.3 in that the horizontal axis represents total exercise duration rather than time points. For example, whereas figure 13.3 shows that 50% of energy at the 30th second of maximal exercise is derived from the lactate system and another 50% from the oxygen system, figure 13.5 shows that 23% of the total energy for a maximal exercise bout lasting 30 s is derived from the ATP-CP system, 52% from the lactate system, and 25% from the oxygen system. This is reasonable, since, although the contribution of the ATP-CP system at the 30th second is null, the system has offered a considerable amount of energy at the beginning of exercise.

Having clarified how the two graphs differ, let's explore figure 13.5. Most of the energy in maximal exercise tasks lasting a few seconds comes from the ATP-CP system. However, its curve displays such a rapid decline that it intersects the rapidly rising curve of the lactate system at only 7 s. The lactate system reigns in efforts of longer duration but, after a fast peak in efforts lasting 10 to 15 s, it gives way to the oxygen system. The curves of the two systems cross at 60 s. If we consider the ATP-CP and lactate systems together as the anaerobic system, then this system overshadows the aerobic system in maximal exercise tasks lasting up to 75 s.

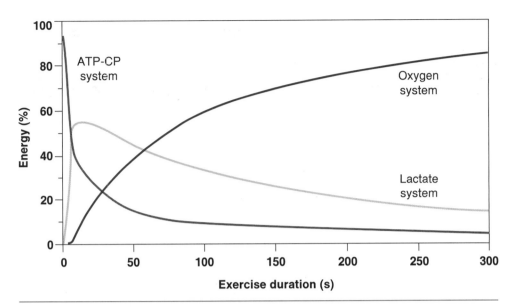

▶ **Figure 13.5** Percentage contribution of the energy systems to total energy supply depending on the duration of maximal exercise. The graph lets you find the contribution of each energy system for any duration of maximal exercise up to 5 min in the same way as described in the caption for figure 13.3. The ATP-CP system dominates in efforts lasting up to 7 s, while the lactate system dominates in efforts lasting 7 to 60 s. Together the two anaerobic systems contribute the majority of energy in exercises lasting up to 75 s, when the gradually rising (after an initial lag) curve of the aerobic system reaches 50%.

Adapted, by permission, from P.B. Gastin, 2001, "Energy system interaction and relative contribution during maximal exercise," *Sports Medicine* 31(10): 725-74.

Now is the time to examine the proportion of energy sources in running and swimming. Table 13.2 presents data on eight running events. As expected, speed (therefore, intensity) decreases as distance increases, with the exception of 100 and 200 m. The lactate system dominates in the 100, 200, and 400 m runs, followed by the ATP-CP system in the 100 and 200 m runs and the oxygen system in the 400 m run. The oxygen system prevails in the 800, 1,500, 5,000, and 10,000 m runs, as well as the marathon run, followed by the lactate system. The percentage contribution of each system may vary slightly from one athlete to another.

▶ **Table 13.2** Energy Sources in Running Events

Distance (m)	Time (s)[a]	Speed (m · s⁻¹)	Contribution of energy systems (%)[b]		
			ATP-CP	Lactate	Oxygen
100	9.77	10.24	39	56	5
200	19.32	10.35	30	55	15
400	43.18	9.26	17	48	35
800	101.11	7.91	9	33	58
1,500	206	7.28	4	20	76
5,000	759.36	6.58	1	6	93
10,000	1,582.75	6.32	1	3	96
42,195	7,538	5.60	0	1	99

[a]Men's world record or all-time record in the case of the marathon run (last line).
[b]Calculated on the basis of figure 13.5.

▶ **Table 13.3** Energy Sources in Freestyle Swimming Events

Distance (m)	Time (s)[a]	Speed (m · s⁻¹)	Contribution of energy systems (%)[b]		
			ATP-CP	Lactate	Oxygen
50	21.64	2.31	28	55	17
100	47.84	2.09	16	46	38
200	104.06	1.92	9	33	58
400	220.17	1.82	4	18	78
800	459.16	1.74	2	9	89
1,500	874.56	1.72	1	5	94

[a]Men's long-course world record.
[b]Calculated on the basis of figure 13.5.

Table 13.3 presents the corresponding data on six swimming events. The lactate system provides most of the energy in the 50 and 100 m swims, followed by the ATP-CP system in the 50 m swim and the oxygen system in the 100 m swim. The oxygen system provides most of the energy in the 200, 400, 800, and 1,500 m swims, followed by the lactate system.

13.8 Effect of the Exercise Program on the Choice of Energy Sources

So far, we have discussed the factors affecting the selection of energy sources during continuous exercise bouts of relatively constant intensity. Nevertheless, many sporting activities (like ball games) are characterized by periods of hard or maximal activity interspersed with intervals of rest, light exercise, or moderate-intensity exercise. This also happens during training sessions. In cases like these, the proportion of energy sources exhibits great fluctuations within very short periods.

In tennis, for example, the athlete exercises hard for intermittent periods of a few seconds followed by periods of rest lasting, on average, three times as long. CP and the anaerobic breakdown of carbohydrates are the major energy sources during the hard efforts. During the periods of rest, the aerobic breakdown of carbohydrates and lipids is the major energy source.

The proportion of energy sources during intermittent exercise (for example, repeated sprints) changes from one bout to the next. Because intervals are usually not sufficient to restore the energy state of the muscle (see section 13.22), there is a gradual decrease in the proportion of anaerobic to aerobic energy supply (figure 13.6).

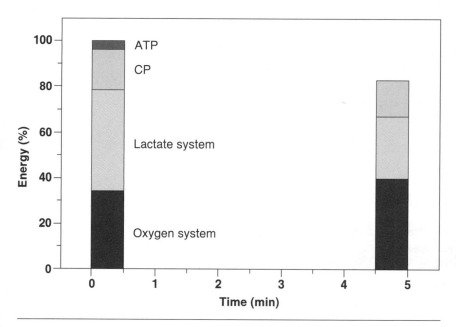

▶ **Figure 13.6** Energy sources change in successive exercise bouts. Repetitive efforts with short periods of rest in between, such as during interval training, rely on a mixture of energy sources that shifts gradually from anaerobic to aerobic. When maximal cycling for 30 s was performed twice with a 4 min interval, it drew more energy from ATP, CP, and the lactate system in the first compared to the second bout. The ratio of anaerobic to aerobic energy was 1.9 in the first bout and only 1 in the second bout. Power output dropped by 18% in the second bout. The graph was constructed based on data from Bogdanis and colleagues (1998).

13.9   Effect of Heredity on the Choice of Energy Sources in Exercise

Heredity affects the choice of energy sources in exercise mainly by determining the proportion of muscle fiber types. In section 7.8 we saw that human muscles contain three fiber types, I, IIA, and IIX. We also saw that, compared to type I fibers, type IIA and IIX fibers contain myosin isoforms of higher ATPase activity, have a higher maximal shortening velocity, are less economical, and are recruited at higher forces of contraction. It is now time to examine the metabolic differences of the three fiber types. As you will see, *metabolic differences serve the functional specification of fiber types.*

Type I Fibers

Type I fibers are designed to contract repeatedly for prolonged periods without getting tired. They produce ATP primarily through the oxygen system; thus, they are also known as *oxidative.* They have many mitochondria and thus a high concentration of mitochondrial enzymes. Biochemists measure the activities of these enzymes as markers of the oxidative capacity of a muscle and its fiber type composition. Some of the most frequently measured enzymes are citrate synthase and succinate dehydrogenase of the citric acid cycle (figure 9.20), cytochrome *c* oxidase of the electron transport chain (figure 9.22), carnitine palmitoyltransferase I (section 10.4), and hydroxyacyl CoA dehydrogenase of β oxidation (figure 10.9). Relying heavily on aerobic processes for ATP production, type I fibers use plenty of oxygen. They are nourished by many capillaries and have a high myoglobin concentration. Finally, they have a high triacylglycerol concentration.

Type IIX Fibers

Type IIX fibers are the opposite of type I fibers. They are designed to perform fast contractions, but they get tired easily. They regenerate ATP mainly through anaerobic processes. They have high concentrations of CP, glycogen, phosphorylase, glycolytic enzymes (phosphofructokinase being the one most frequently measured), and lactate dehydrogenase. In contrast, they are surrounded by few capillaries, display a low **mitochondrial density** (mitochondrial volume as a percentage of the total tissue volume), and have low myoglobin and triacylglycerol concentrations. Type IIX fibers produce a lot of lactate during contraction and are characterized as *glycolytic.*

Type IIA Fibers

Type IIA fibers lie between type I and type IIX fibers, although they are closer to the latter. They feature fast myosin and have high CP and glycogen concentrations. Nevertheless, they also display a substantial oxidative capacity because of substantial blood supply, mitochondrial density, and myoglobin as well as triacylglycerol concentrations. Thus, they are characterized as *oxidative-glycolytic.* Type IIA fibers can produce fast contractions and are fairly resistant to fatigue.

Table 13.4 summarizes the characteristics of the three muscle fiber types.

Human muscles contain a mixture of the three fiber types in proportions differing from one muscle to another. At rest and during light exercise, it is the type I fibers that contract primarily. As intensity increases, type IIA and IIX fibers are gradually recruited. This sequential recruitment contributes to the fine-tuning of a muscle's response to the demands of different kinds of exercise.

▶ **Table 13.4** Properties of the Main Muscle Fiber Types in Humans

Property	Muscle fiber type		
	I	IIA	IIX
ATPase activity	Low	High	High
Maximal shortening velocity	Low	Intermediate	High
Resistance to fatigue	High	Intermediate	Low
Metabolic character	Oxidative	Oxidative-glycolytic	Glycolytic
Blood supply	High	Intermediate	Low
Myoglobin concentration	High	Intermediate	Low
Mitochondrial density	High	Intermediate	Low
Oxidative enzyme activity	High	Intermediate	Low
Phosphorylase activity	Low	Intermediate	High
Glycolytic enzyme activity	Low	Intermediate	High
CP concentration	Low	High	High
Glycogen concentration	Low	High	High
Triacylglycerol concentration	High	Intermediate	Low

The proportion of muscle fiber types in a muscle varies widely from one individual to another. The proportion is primarily determined genetically and affects the selection of energy sources during exercise. The more type I fibers a muscle has, the more energy the oxygen system supplies in a given exercise. Conversely, the more type IIA and IIX fibers a muscle has, the more energy the ATP-CP and lactate systems supply.

The proportion of muscle fiber types determines, to a great extent, whether a person can excel as an athlete in certain sports. For example, type I fibers prevail in the leg muscles of endurance athletes, whereas type IIA and IIX fibers prevail in the leg muscles of sprinters. In particular, the vastus lateralis muscle has been found to consist of about 80% type I fibers in marathon runners but of only 30% type I fibers in sprinters.

13.10 Conversions of Muscle Fiber Types

Can a muscle fiber switch from one type to another? Yes, if the functional demands placed on the muscle change. This is part of muscle plasticity as mentioned in the beginning of chapter 12. Muscle fibers change types in a thorough manner: Along with the isoforms of myosin heavy chain (the trademark of fiber type), the other properties of a fiber type also change. Thus, there appear to be *programs of gene expression* specific for each fiber type.

The transition from one fiber type to another follows the sequence dictated by the gradient of their properties, that is, from type I to IIA to IIX, and from type IIX to IIA to I. Two isoforms of myosin heavy chain may coexist in a muscle fiber during the transition state; these will be either I and IIa or IIa and IIx (section 7.8).

What causes conversions of muscle fiber types? The transition from oxidative to glycolytic fibers is brought about by the decrease in neuromuscular activity as a result of, for example, muscle denervation or lowering of aerobic training. In contrast, the transition from glycolytic to oxidative fibers is brought about by the increase in neuromuscular activity as a result of, for example, chronic electrical stimulation or aerobic training. Figure 13.7 summarizes these conversions.

You may have noticed that anaerobic training is missing from these examples. It is not clear whether anaerobic training causes changes in muscle fiber type and if so, what these changes are. A frequently reported conversion with strength or sprint training is from types I and IIX to type IIA, that is, I → IIA ← IIX. Moreover, the muscle hypertrophy caused by anaerobic training is more manifest in type IIA and IIX fibers than in type I fibers. Thus, type IIA and IIX fibers together occupy more volume relative to type I fibers in a muscle experiencing hypertrophy.

Increased neuromuscular activity

Decreased neuromuscular activity

▶ **Figure 13.7** Conversions of muscle fiber types and their causes.

13.11 Effect of Nutrition on the Choice of Energy Sources During Exercise

As we have seen on several occasions, the amount of an energy source affects its contribution to meeting the energy demands of exercise. Nutrition can modify the amount of an energy source; the most convincing example of this is the case of carbohydrates. Moderate-intensity or hard exercise exceeding 1 h in duration can reduce muscle glycogen to such an extent that the rate of ATP regeneration drops substantially and sport performance is compromised. A carbohydrate-rich diet in the days preceding such exercise approximately doubles the muscle glycogen content and results in the use of more glycogen and fewer lipids. Athletes usually combine eating more carbohydrates with tapered or no training before the critical event in order to curb glycogen loss. This joint nutritional and training manipulation, termed **carbohydrate loading**, usually increases endurance performance in events lasting over 90 min.

Carbohydrates taken *during* prolonged moderate-intensity or hard exercise, at a rate of about 60 g · h^{-1}, also affect the proportion of fuels used. The continuous entry of glucose from the small intestine to the blood results in its increased uptake by the exercising muscles and a spectacular rise in its contribution to ATP resynthesis (figure 13.8). Equally spectacular is the boost to performance. Carbohydrate supplementation during exercise also maintains euglycemia. Athletes take carbohydrates during prolonged sport activities commonly in the form of *sport drinks* containing 6% to 8% carbohydrates, mainly as sugar. Sport drinks also provide optimal hydration to the body.

Several years after the discovery of the effects of carbohydrate loading, it was found that **fat loading** also influences exercise metabolism, although it does not increase performance. A lipid-rich diet for several days or weeks decreases glycogen stores and increases the uptake of plasma fatty acids by the exercising muscles, as well as the use of plasma triacylglycerols, during prolonged moderate-intensity exercise. Because of these changes, more lipids and fewer carbohydrates are used in exercise.

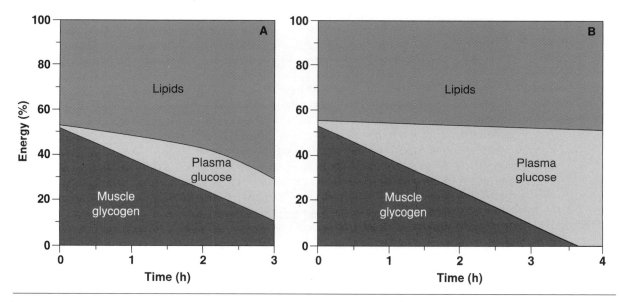

▶ **Figure 13.8** Carbohydrate intake during exercise. *(a)* Cycling at 70% to 75% of V̇O$_2$max drove experienced athletes to exhaustion after approximately 3 h, at which time the contribution of carbohydrates to total energy production had fallen below 30%. *(b)* When the athletes took carbohydrates during exercise, they were able to cycle for an additional hour, at the end of which carbohydrates still contributed half of the total energy. Note that plasma glucose, derived primarily from the ingested carbohydrates, was what made the difference. No muscle glycogen sparing (that is, slowing down of glycogen utilization) because of carbohydrate supplementation was evident in this or in similar experiments, although glycogen sparing has been found in other studies.

From E.F. Coyle, 1986, "Muscle glycogen utilization during prolonged strenuous exercise when fed carbohydrate," *Journal of Applied Physiology* 61: 165-172. By permission of The American Physiological Society.

The shift of exercise metabolism toward lipid utilization after fat loading persists even if we substitute carbohydrate loading in the final one to three days before an event, aiming at replenishing glycogen stores and maximizing performance. However, it is not clear whether fat loading followed by carbohydrate loading increases endurance performance any more than carbohydrate loading alone does.

What about fat intake *during* exercise? Efforts to mimic the positive effect of carbohydrate drinks have employed an unusual form of fat, *medium-chain triacylglycerols*, or MCT. These differ from the triacylglycerols in regular food in that their acyl chains are shorter, typically containing 8 to 12 carbons, as opposed to 16 or more carbons in most *long-chain triacylglycerols* (LCT). MCT are absorbed faster than LCT and have been hypothesized to enhance endurance performance by offering additional energy during exercise. However, research has shown that, whether taken alone or along with carbohydrates, MCT contribute little to the total energy of prolonged moderate-intensity exercise, partly because one cannot ingest a high amount of MCT without experiencing gastrointestinal distress. Moreover, MCT intake (alone or with carbohydrates) does not increase endurance performance compared to the intake of equal energy solely from carbohydrates.

13.12 Adaptations of the Proportion of Energy Sources During Exercise to Endurance Training

The training state affects the proportion of energy sources during exercise thanks to training-induced adaptations. The most spectacular adaptations of the proportion of energy sources accompany endurance training. In a nutshell, *endurance training increases the ratio of lipids to carbohydrates used during prolonged exercise.* Training programs that bring about these adaptations in humans have the following characteristics:

- Intensity: at least 40% of $\dot{V}O_2$max
- Duration: at least 30 min
- Frequency: at least three times a week
- Exercise mode: any mode involving continuous contractions of large muscle groups, such as running, cycling, swimming, and aerobics

Training load is a function of exercise intensity, duration, and frequency.

Changes in the proportion of energy sources can be detected as early as one week from the beginning of training, provided the training load is substantial. Adaptations culminate after several months of training.

How are adaptations of the proportion of energy sources detected? A person has to perform two exercise tests (one at the beginning and the other at the end of a training program) of the same mode (for example, treadmill running), duration, and intensity. In each test we measure the fuels used by employing a variety of laboratory techniques, such as those described in the sidebar on page 23 and later in this chapter (section 13.17). As far as duplicating the test parameters is concerned, duration presents no problem (all you need is an accurate timer). Duplicating intensity would also be straightforward (for example, set the same treadmill speed in the two tests) if it were not for endurance training to increase $\dot{V}O_2$max. This forces us to distinguish absolute from relative intensity. If, for example, $\dot{V}O_2$max is 45 mL · kg^{-1} · min^{-1} at the beginning of an endurance training program lasting three months, it can rise to 52 mL · kg^{-1} · min^{-1} in the end. The two tests can then be performed at either the same *absolute* intensity (say, 27 mL · kg^{-1} · min^{-1}, or 60% of baseline $\dot{V}O_2$max) or the same *relative* intensity (60% of current $\dot{V}O_2$max, which is 27 mL · kg^{-1} · min^{-1} before training and 31.2 mL · kg^{-1} · min^{-1} after training).

The increase in $\dot{V}O_2$max with endurance training is primarily due to the increase in maximal cardiac output, that is, the blood volume that the heart pumps per minute of maximal effort.

When the two tests are matched for absolute intensity, there is general agreement among researchers that the lipid-to-carbohydrate ratio is higher after training (figure 13.9). However, when relative intensities are equal, the lipid-to-carbohydrate ratio does not change significantly, according to well-designed longitudinal studies of recent years. Nevertheless, cross-sectional studies comparing untrained individuals and endurance athletes exercising at a fixed percentage of each individual's $\dot{V}O_2$max verify that endurance athletes utilize a higher lipid-to-carbohydrate ratio. This discrepancy between longitudinal and cross-sectional studies may be due to genetic differences between athletes and nonathletes as described in section 13.9. Alternatively, it may mean that the training loads usually employed in longitudinal studies are insufficient to raise the lipid-to-carbohydrate ratio during exercise at the same relative intensity. A third explanation might be that the athletes have trained over many years, whereas few longitudinal studies last more than one year.

The benefit from the increase in the ratio of lipids to carbohydrates used during exercise at the same absolute intensity after endurance training is the sparing of carbohydrates, the smaller and faster of the two energy sources. Carbohydrate sparing may prolong exercise. Looking at it another way, one could say that the benefit is the ability to exercise at a higher intensity while spending the same amount of carbohydrates. This will let an endurance athlete complete an exercise task earlier, thus increasing performance.

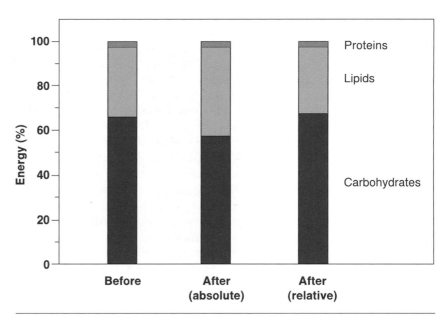

▶ **Figure 13.9** Effect of endurance training on the proportion of energy sources. The percentage contribution of lipids to the energy expenditure of prolonged moderate-intensity exercise (cycling for 90 min at 60% of $\dot{V}O_2$max) increased significantly (from 31% to 41%) after untrained men and women trained aerobically for seven weeks, when the posttraining test was conducted at the same absolute intensity as the pretraining test. However, when the posttraining test was conducted at the same relative intensity (60% of the new $\dot{V}O_2$max, which was higher than baseline by 20%), the percentages of lipids and carbohydrates did not change. Proteins offered 3% to 4% of total energy in all three tests.

From S.L. Carter, C. Rennie, and M.A. Tarnopolsky, 2001, "Substrate utilization during endurance exercise in men and women after endurance training" *American Journal of Physiology: Endocrinology and Metabolism* 280: E898-E907. By permission of The American Physiological Society.

13.13 How Does Endurance Training Modify the Proportion of Energy Sources During Exercise?

What makes a muscle shift toward lipid utilization and away from carbohydrate utilization with endurance training? In a nutshell, *endurance training increases the oxidative capacity of the muscles; it also increases their capacity to burn lipids more than it increases their capacity to burn carbohydrates.* More specifically, endurance training causes the following changes (some of which we have already considered) in the exercising muscles:

- *The number and, secondarily, the size of the mitochondria increase.* The number of mitochondria in a muscle fiber can more than double after a six-month program of endurance training. As a result, the enzymes of the citric acid cycle, carnitine acyltransferases I and II, the enzymes of β oxidation, and the components of the electron transport chain increase. These adaptations boost the capacity of the muscles to resynthesize ATP from the aerobic breakdown of carbohydrates and lipids.

- *The exercise-induced translocation of GLUT4 decreases.* Although endurance training increases the amount of GLUT4, it decreases the translocation from its intracellular reservoir to the plasma membrane during exercise at the same absolute moderate intensity. As a result, trained muscles take up less glucose from the plasma than untrained muscles during exercise.

Muscle insulin sensitiv-
ity is defined as the
insulin concentration
required to produce a
half-maximal increase
in glucose uptake,
although researchers
may use alternative
ways of expressing it.

- *The amount of muscle glycogen increases,* probably because exercise increases insulin sensitivity (see section 13.18). Insulin promotes not only glucose uptake for glycogen synthesis but also glycogen synthesis per se by activating glycogen synthase (section 9.23). Additionally, endurance training increases the amount of glycogen synthase. These adaptations result in increased glycogen availability. On the other hand, endurance training increases phosphorylase activity and thus the ability to catabolize glycogen. However, endurance training does not seem to affect the activity of the glycolytic enzymes. The fact that the muscles of endurance athletes usually have low glycolytic enzyme contents is apparently due to an abundance of type I fibers (section 13.9) rather than an effect of training.

- *The amount of myocellular triacylglycerols and the uptake of fatty acids from the plasma increase* because of increases in **capillary density** (the number of blood capillaries per muscle fiber or muscle cross-sectional area), lipoprotein lipase on the surface of the endothelial cells of the capillaries, and fatty acid transporters in the sarcolemma (section 10.3). In contrast, training does not seem to affect the lipolytic rate in adipose tissue either at rest or during exercise. These adaptations, combined with the increases in carnitine acyltransferases and the enzymes of β oxidation, increase the capacity of muscle to oxidize lipids.

The rise in the capacity of muscle to oxidize lipids with endurance training is larger than the rise in its capacity to oxidize muscle glycogen. This, along with the lower plasma glucose uptake by the trained muscles during moderate-intensity exercise, leads to a higher contribution of lipids to energy production during exercise.

Which lipids are broken down to a higher degree during moderate-intensity exercise after endurance training? We do not know for sure. Most research data point to an increased breakdown of myocellular triacylglycerols and not plasma fatty acids; however, there is respectable evidence for the opposite. It is possible that factors like exercise mode, exercising muscle mass, nutritional state, and the methodology used affect the experimental findings. Another lipid that could be broken down to a higher degree is plasma triacylglycerols, and this might contribute to their decrease with endurance training (section 10.12).

13.14 Adaptations of Energy Metabolism to Anaerobic Training

The effects of anaerobic training (both strength training and sprint training) on the proportion of energy sources during exercise are not as marked as those of endurance training. Nevertheless, anaerobic training does affect energy metabolism in general and each of the three energy systems in particular. Let's review these effects starting with strength training.

Strength training increases the muscle CP and glycogen concentrations at rest. It also increases the activity of the glycolytic enzymes and lactate dehydrogenase. These increases are apparently the result of muscle hypertrophy, which involves mainly type IIA and IIX fibers (section 13.10). Thus, the relative volume of type IIA and IIX fibers increases in a muscle biopsy sample. Since these fibers have higher concentrations of the previously listed components than type I fibers (table 13.4), the entire muscle appears to have a higher content of these components. In any case, their increased presence in a muscle boosts its anaerobic capacity. In contrast, strength training decreases the aerobic capacity of muscle because the resulting hypertrophy is not accompanied by an increase in mitochondria or capillaries. Thus, the aerobic system is "diluted."

The adaptations to sprint training are a bit different, probably because, although exercising hard, the muscles do not work against high resistance. Sprint training appears to raise the capacity of all three energy systems. Adenylate kinase and creatine kinase (the enzymes of the ATP-CP system) increase slightly. Phosphorylase, the glycolytic enzymes, and lactate dehydrogenase (the enzymes of the lactate system) increase more. Finally, the mitochondrial enzymes also increase. On the other hand, no changes in the concentrations of energy sources have been documented with sprint training.

These adaptations increase the total capacity to regenerate ATP in sprints. However, they depend greatly on parameters of the training program such as sprint duration, interval duration, and frequency of training. Thus, adaptations of the ATP-CP system have been observed with short-sprint (below 10 s) training, whereas adaptations of the aerobic system have been observed with long-sprint (over 10 s) training. An appropriate training program can also cause beneficial morphological changes such as an increase in type IIA fibers (section 13.10), muscle hypertrophy, and an increase in sarcoplasmic reticulum volume resulting in the liberation of more Ca^{2+} upon muscle excitation.

Attesting to the importance of the parameters of sprint training in achieving the desired adaptations is the fact that exceeding the optimal training load or the optimal frequency of training (or both) favors the development of slow contractile features. Unfortunately, as Ross and Leveritt (2001) remark in a relevant review, the complexity of the interaction between the variables of the training program and the adaptations, combined with individual differences, does not allow at present the transfer of knowledge and advice from the laboratory to the coach and the athlete.

Table 13.5 summarizes the adaptations of muscle to the types of training that we dealt with in the last three sections.

▶ **Table 13.5** Adaptations of Human Skeletal Muscle to Training

Endurance training
Increase in the proportion of lipids to carbohydrates used during exercise at a given absolute moderate intensity
Increase in mitochondrial number and size
Increase in the enzymes of the citric acid cycle
Increase in carnitine acyltransferases I and II
Increase in the enzymes of β oxidation
Increase in the components of the electron transport chain
Increase in GLUT4
Increase in glycogen
Increase in glycogen synthase and phosphorylase
Increase in myocellular triacylglycerols
Increase in capillary density
Increase in lipoprotein lipase
Increase in fatty acid transporters
IIX → IIA → I

Strength training
Hypertrophy (especially type IIA and IIX fibers)
Increase in CP
Increase in glycogen
Increase in the glycolytic enzymes and lactate dehydrogenase
I → IIA ← IIX (with reservation)

Sprint training
Hypertrophy (especially type IIA and IIX fibers)[a]
Increase in adenylate kinase and creatine kinase[a]
Increase in phosphorylase, the glycolytic enzymes, and lactate dehydrogenase
Increase in mitochondrial enzymes[a]
I → IIA ← IIX[a]

[a]Not all sprint training programs cause this adaptation. See text for details.

13.15 Effect of Age on the Choice of Energy Sources During Exercise

Age appears to affect the proportion of energy sources during exercise. Several studies have shown that children utilize proportionally more lipids than adults in prolonged exercise at a given absolute or relative moderate intensity. Researchers have proposed as reasons the lower muscle glycogen concentration of children (50-60% of the adult concentration) and the incomplete development of the lactate system and the sympathetic system. In contrast, children seem to have higher activities of oxidative enzymes in their muscles. There are, however, studies that do not show any differences between children and adults with regard to the proportion of lipids to carbohydrates burned during prolonged exercise at a given relative moderate intensity.

13.16 Do Sex and Ambient Temperature Affect the Choice of Energy Sources During Exercise?

In closing the examination of the factors affecting the selection of energy sources during exercise, it is worth mentioning that several investigators have proposed sex and ambient temperature as two additional factors. In particular, there is evidence that women, compared to men, derive a higher proportion of energy from lipids at a given relative intensity and that exercise in the cold enhances lipid oxidation. However, there is also evidence that sex and ambient temperature do not affect the choice of energy sources during exercise. In view of this controversy and the limited number of studies on either of these factors, one can make no safe statement at present.

13.17 The Proportion of Fuels Can Be Measured Bloodlessly

We can determine the contribution of carbohydrates and lipids to the energy expenditure of prolonged exercise in a bloodless and relatively easy way through the **respiratory exchange ratio (RER)**. The RER is the ratio of the volume of expired CO_2 to the volume of O_2 consumed, and it is determined through measurement of the two gases in the inspired and expired air. The RER is used as an index of the **respiratory quotient (RQ)**, which is the ratio of the volume of CO_2 released to the volume of O_2 taken up at the cellular level.

Determining the proportion of fuels by measuring the RER is based on their differing chemical composition. Specifically, carbohydrates have approximately one oxygen atom per carbon atom, whereas lipids are more reduced (hydrogenated) and contain only a few oxygens. As a result, carbohydrates need less O_2 than lipids per carbon atom and thus per molecule of CO_2 produced to be burned completely. You may convince yourself of this by examining equations 13.1 and 13.2 (section 13.5), which describe the oxidation of glucose and palmitic acid. According to these, the molar ratio—and volume ratio when it comes to gases—of CO_2 to O_2 is 6/6 = 1 for glucose and 16/23 = 0.7 for palmitic acid. The corresponding ratio for amino acids is 0.82 on average.

On the basis of these differences, one can calculate the share of carbohydrates, lipids, and proteins in the energy expenditure of exercise. An RER value close to 0.7 indicates a high share of lipids, whereas an RER value close to 1 indicates a high share of carbohydrates (figure 13.10). Protein oxidation can be either ignored without seriously affecting the accuracy of the calculations (as it is minimal) or determined by other techniques, thus permitting a more accurate calculation of the percentages of carbohydrates and lipids.

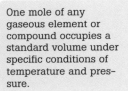

One mole of any gaseous element or compound occupies a standard volume under specific conditions of temperature and pressure.

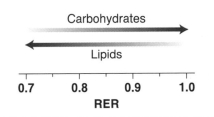

▶ **Figure 13.10**   A fuel breath test. By measuring the RER, that is, the ratio of the CO_2 that is added to the expired air, to the O_2 that is missing from the expired air, one can estimate the contribution of carbohydrates and lipids to energy expenditure. The higher the RER value, the higher the contribution of carbohydrates.

13.18 Hormonal Effects on Exercise Metabolism

From the very beginning of part III and through the previous chapter, we saw the important role of hormones in controlling exercise metabolism. This section integrates the hormonal effects on exercise metabolism and completes the picture by presenting the chronic effects of exercise on hormone concentrations.

Epinephrine

Epinephrine mediates many of the effects of exercise on metabolism. Secretion of this catecholamine by the adrenals increases in response to neural signals that prepare the body to face danger or apply muscular effort. Epinephrine has been aptly named a *fight-or-flight hormone,* as it helps animals get food or avoid becoming food. Its secretion also increases when the plasma glucose concentration drops. Figure 13.11 shows the metabolic effects of epinephrine on its target tissues and organs.

In muscle, epinephrine speeds up glycogenolysis (section 9.3), thus increasing ATP resynthesis from carbohydrates.

In the liver, it speeds up glycogenolysis and gluconeogenesis (sections 9.20 and 9.22), thus increasing glucose supply to the blood and, from there, to muscle.

In adipose tissue, epinephrine and norepinephrine stimulate lipolysis (section 10.2), thus increasing fatty acid supply to the blood and, from there, to muscle.

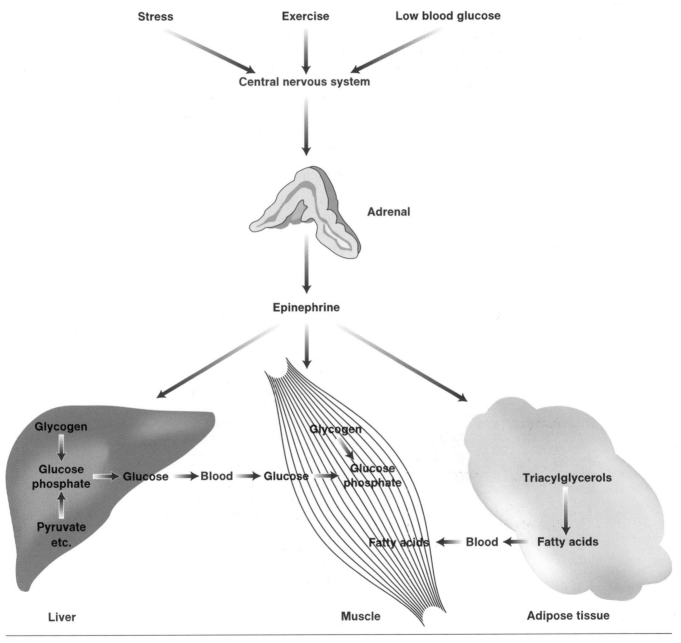

▶ **Figure 13.11**   A hormone for hard times. Stress, exercise, and low blood glucose act on the central nervous system to increase epinephrine secretion by the adrenals. Epinephrine stimulates glycogen breakdown to glucose phosphate (glycogenolysis) in muscle and the liver. Glucose phosphate is used as an energy source in muscle. In contrast, in the liver, where epinephrine also stimulates gluconeogenesis from pyruvate and other compounds, glucose phosphate is converted to free glucose, which is exported. In adipose tissue, epinephrine stimulates lipolysis, driving fatty acids to the circulation.

The action of glucagon to produce glucose is reflected in its name, which is derived from the Greek words *glykýs,* meaning "sweet," and *gennó,* meaning "give birth." Incidentally, the word glycogen has the same origin, as glycogen too produces glucose, although in a direct rather than indirect manner. Glycogen was discovered and named first; then the vowels were changed to create the term glucagon.

Through these effects epinephrine orchestrates the increased supply of substrates—both carbohydrates and lipids—to the exercising muscles. In addition, the augmented release of epinephrine and, mainly, norepinephrine from sympathetic nerve endings stimulates glucagon secretion and inhibits insulin secretion by the pancreas. This boosts glucose supply to the blood given the hyperglycemic action of glucagon and the hypoglycemic action of insulin.

Endurance training mitigates the rise in the plasma catecholamine concentration after exercise at a given absolute intensity. If exercise is at the same relative intensity (thus higher absolute intensity compared to pretraining), the rise in epinephrine is diminished but the rise in norepinephrine is not.

Glucagon

Glucagon acts like epinephrine but only in the liver (figure 13.12). This peptide hormone speeds up glycogenolysis and gluconeogenesis while slowing down glycogen synthesis and glycolysis (sections 9.20 and 9.22). All these effects raise the glucose concentration in the hepatocytes and then in the blood. The rise in plasma glucagon concentration when the plasma glucose concentration drops and during exercise contributes to glucose homeostasis. Endurance training mitigates the rise in glucagon caused by acute exercise.

Insulin

Insulin lies on the opposite side of epinephrine and glucagon. This other peptide hormone signals the fed state by speeding up carbohydrate and fat storage, as well as protein synthesis. How does it do all this? Starting with carbohydrates, insulin facilitates the entry of glucose into muscle (by enhancing the translocation of GLUT4 from intracellular vesicles to the plasma membrane, figure 9.15) and adipose tissue (in the same way). The hormone speeds up glycogen synthesis and slows down glycogenolysis in muscle and the liver. Additionally, it speeds up glycolysis and slows down gluconeogenesis in the liver. As pointed out in section 9.23, all these effects converge in the disappearance of (free) glucose.

Now here's how insulin increases fat storage. To begin with, it increases the amount of lipoprotein lipase in the capillaries of adipose tissue. Remember that this enzyme hydrolyzes the triacylglycerols in chylomicrons and VLDL to allow the lipolytic products (fatty acids and monoacylglycerols) to enter the cells (section 10.11). By also facilitating glucose entry to the adipocytes, insulin ensures the presence of the necessary precursors for triacylglycerol synthesis (section 10.1). At the same time insulin inhibits hormone-sensitive lipase, thus blocking lipolysis (section 10.2).

Finally, regarding proteins, insulin promotes the entry of BCAA to muscle, thus boosting the synthesis of muscle proteins. Apart from this, the hormone exerts a global stimulatory action on protein synthesis while inhibiting proteolysis. Because it promotes a positive protein balance in muscle, insulin is purportedly abused by athletes and is banned as a doping substance. Let me point out,

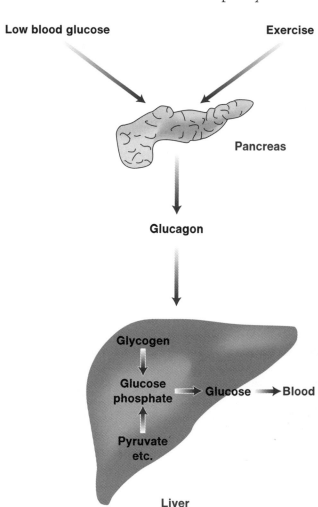

▶ **Figure 13.12**   A particularly picky hormone. Glucagon has the liver as its sole target organ. In response to low blood glucose and exercise, the pancreas secretes more glucagon, which, like epinephrine, stimulates glycogenolysis and gluconeogenesis. The two processes cooperate to increase glucose production and raise blood glucose.

however, that insulin administration to healthy athletes has no documented beneficial effects on muscle mass or strength. Moreover, one has to bear in mind the hormone's fat-storing capacity and the risk for developing severe acute hypoglycemia after insulin injection.

The importance of insulin for metabolism is stressed by the fact that insulin resistance is thought to lie at the basis of the **metabolic syndrome,** a condition characterized by visceral obesity, undesirable plasma lipid concentrations, high blood pressure, and hyperglycemia. Notably, insulin resistance is associated with an excessive concentration of myocellular triacylglycerols, although the molecular link between the two is not known.

As discussed in section 9.23, insulin secretion by the pancreas drops during exercise. This allows the deceleration of glycogen synthesis and acceleration of glycogenolysis in muscle and the liver, as well as the deceleration of glycolysis and acceleration of gluconeogenesis in the liver. The joint effect of these changes is an increase in glucose availability. Moreover, the decrease in insulin contributes to the deceleration of triacylglycerol synthesis and the acceleration of lipolysis in adipose tissue, resulting in increased fatty acid availability. However, the decrease in insulin does not seem to affect protein metabolism during exercise.

The only drawback of the decrease in insulin during exercise could be the mitigation of glucose uptake by muscle. However, as discussed in section 9.6, exercise increases glucose uptake by the exercising muscles through an insulin-independent mechanism. It is estimated that this mechanism accounts for 85% of the increase in glucose uptake by the exercising muscles.

Finally, there are clear chronic effects of aerobic exercise on insulin concentration and action. Trained persons experience a smaller decrease in plasma insulin concentration with exercise than untrained persons. In addition, endurance training increases insulin sensitivity by raising the number of GLUT4 molecules moving to the plasma membrane in response to a given moderate plasma insulin concentration. This is probably done through an increase in signal molecules of the PI3K cascade (figure 9.36). It is also because of this effect on insulin sensitivity (in addition to acutely increasing glucose uptake by muscle) that exercise helps persons with insulin resistance or overt type 2 diabetes mellitus control their plasma glucose concentration.

13.19 Fatigue

It is one of the most obvious corollaries of exercise. It has been studied extensively for over one and a half centuries. Yet **fatigue** remains a phenomenon of unclear etiology. Moreover, the etiology of fatigue in one exercise may be quite different from that in another. Thus, what we are going to examine in this and the following two sections are the *possible* causes of fatigue in various kinds of exercise.

We define fatigue as the *inability to maintain a prescribed exercise intensity*. Fatigue appears gradually, with repeated muscle contractions. It is usually divided into *central* and *peripheral*. Central fatigue emanates from the central nervous system, whereas peripheral fatigue results from inability of the muscle itself or the neuromuscular junction.

Fatigue is certainly a weakness of muscle function, but it may also be a mechanism of curbing muscle fiber damage caused by excessive contractile activity. Damage could result from at least two events:

- An excessive drop in [ATP], which endangers elementary cell functions such as biosynthesis and active transport
- An excessive rise in cytosolic [Ca^{2+}] (because of sustained excitation), which activates hydrolytic enzymes degrading cell components

Power production by a muscle or muscle group requires a series of biochemical processes including the neural transmission of the signal for contraction, muscle excitation, contraction itself, and energy production. A failure in any of these processes may act as the weak link that causes fatigue. Let's see what problems may appear both centrally and peripherally.

13.20 Central Fatigue

Central fatigue could arise from an inability of the central nervous system to maintain the ideal excitation frequency for motor neurons, that is, the frequency that elicits maximal contraction. This has indeed been observed during continuous contraction but is believed to be an effect, not the cause, of fatigue in the sense that the fatigued muscle sends negative feedback signals to reduce the excitation frequency.

A possible cause of central fatigue is the hypoglycemia that appears in moderate- or high-intensity endurance exercise tasks lasting over 1 h because of liver glycogen depletion and no or inadequate carbohydrate intake during exercise. Marathon runners often experience hypoglycemia. The condition is reversed quickly with intravenous glucose administration.

Two other possible causes of central fatigue are *dehydration* (loss of body water) and *hyperthermia* (high body temperature), which usually go hand in hand when one exercises for prolonged periods in the heat without ingesting adequate amounts of water or other fluids. Exercise at a high ambient temperature causes excessive sweating, which tends to remove water from the blood, the extracellular space, and the cytoplasm. If water intake during exercise does not compensate for the loss, body temperature rises, as there is not enough water to dissipate the heat produced. Hyperthermia compromises mental functions, including information processing, cognition, and memory, and this may lead to fatigue. Hyperthermia and dehydration may also cause peripheral fatigue by impairing muscle function. Both can be prevented via the consumption of ample amounts of water, sport drinks, or similar fluids during exercise.

Newsholme and coworkers have proposed that central fatigue may appear during prolonged exercise because of increased synthesis of the neurotransmitters norepinephrine, *dopamine* (another catecholamine), and *serotonin*, or *5-hydroxytryptamine,* in the brain (Blomstrand et al., 1988). It is believed that these compounds control mood and sleep, and that a drop in their concentration in certain areas of the brain causes depression. The catecholamines are synthesized from tyrosine, and serotonin is synthesized from tryptophan.

Tryptophan is carried in the blood bound to albumin, the protein that also carries fatty acids. In addition, both tyrosine and tryptophan enter the brain through a transport protein that is also used by the BCAA. According to Newsholme's group, prolonged exercise brings about two changes that increase tyrosine and tryptophan uptake by the brain.

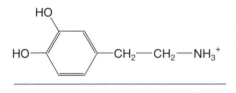
▶ Dopamine.

▶ Serotonin.

- Prolonged exercise increases the plasma fatty acid concentration. Fatty acids then compete with tryptophan for albumin binding, thus increasing the concentration of free tryptophan. Free tryptophan can enter the brain, whereas bound tryptophan cannot.

- It can decrease the plasma BCAA concentration by increasing BCAA uptake by the muscles, thus favoring the entry of tyrosine and tryptophan, which compete with the BCAA for entering the brain.

The rise in tyrosine and tryptophan concentration in brain neurons may augment the synthesis of the three neurotransmitters mentioned previously, leading to a sense of fatigue and drowsiness. However, there is no sufficient experimental evidence for this complex hypothesis.

Because BCAA are primarily handled by muscle, because their oxidation increases during exercise, and because their decrease in the plasma has been linked to central fatigue according to the hypothesis just presented, BCAA supplements have been tested as potential ergogenic aids in prolonged moderate-intensity exercise. However, research has shown that BCAA supplementation does not increase endurance performance any more than the intake of an equal amount of carbohydrates, the prime source of energy during most kinds of exercise. Thus, carbohydrate supplementation, as discussed in section 13.11, remains the most sound ergogenic practice during endurance exercise.

13.21 Peripheral Fatigue

The event that triggers contraction at the muscle level is the appearance of action potentials in the sarcolemma. In hardworking muscles, there is a large increase in the extracellular $[K^+]$ because of the multitude of action potentials generated and the inability of the muscle fibers to rapidly restore the $[K^+]$ gradient across the sarcolemma. This leads to a decrease in the excitability of the sarcolemma and may become a cause of fatigue. Indeed, when the frequency of nervous signals and power output are high, such as during resistance or sprint efforts of maximal intensity lasting up to about one-half minute, there is a decline in the amplitude and frequency of the muscle action potentials that coincides with the emergence of fatigue. One can expect this decline to compromise the coupling of excitation to contraction, since it is the action potentials that trigger the liberation of Ca^{2+} from the sarcoplasmic reticulum to the cytosol.

A recently proposed mechanism of fatigue during hard exercise involves the steep rise in the cytosolic $[P_i]$ because of rapid ATP breakdown. Researchers have proposed that P_i enters the sarcoplasmic reticulum, where it combines with Ca^{2+} to form calcium phosphate, an insoluble salt. Salt formation lowers the $[Ca^{2+}]$ in the sarcoplasmic reticulum and decreases the concentration gradient across its membrane. This, in turn, decreases the rate of Ca^{2+} influx to the cytosol when the Ca^{2+} channel opens, and results in a lower rise in the cytosolic $[Ca^{2+}]$ when the muscle is excited, thus contributing to the development of fatigue. In addition, P_i may cause fatigue by directly inhibiting the power stroke (section 7.7).

An obvious candidate cause of fatigue is the lack of sources for ATP resynthesis, specifically CP and glycogen. CP depletion may cause fatigue in brief maximal resistance or sprint efforts, while glycogen depletion may do so in endurance exercise tasks of moderate or high intensity lasting at least 1 h. Indeed, there is strong correlation between force reduction and CP reduction during hard exercise; and the smaller the glycogen stores in the body and muscle in particular, the earlier fatigue appears during moderate-intensity exercise lasting over 1 h.

The most publicized cause of fatigue is H^+ production and the subsequent drop in cytosolic pH that accompanies the anaerobic breakdown of glycogen to lactate. Lactate itself is not a cause of fatigue. However, the drop in pH can cause fatigue in several ways.

- It can inhibit myosin ATPase, which slows down ATP hydrolysis and decreases power output.
- It can attenuate the frequency and duration of opening of the Ca^{2+} channel in the sarcoplasmic reticulum membrane, which decreases the amount of Ca^{2+} liberated during muscle fiber excitation.
- It can block Ca^{2+} binding to troponin C, thus inhibiting the attachment of the myosin heads to actin.

Obviously, the drop in cytosolic pH is a cause of fatigue in efforts depending greatly on the lactate system for energy production, that is, hard or maximal exercise lasting about a half minute to a few minutes.

Efforts to mitigate fatigue caused by H⁺ production have focused on the use of sodium bicarbonate ($NaHCO_3$), the common baking soda, as a dietary supplement. Remember that bicarbonate accepts a proton (reverse equation 3.2) and buffers blood pH. Many (although not all) controlled studies have shown that performance is increased through ingestion of 0.3 g $NaHCO_3$ per kilogram body mass, a practice dubbed (**sodium**) **bicarbonate loading**, or **soda loading**, 1 to 3 h before maximal exercise lasting 1 to 3 min. Although such exercise tasks do not depend on the lactate system as their main energy source, they do nonetheless cause high H⁺ production. In fact, there are some reports suggesting that soda loading boosts performance in exercise tasks lasting even longer. A consistent finding is increased plasma pH after soda loading. It is also possible that soda loading mitigates the drop in muscle pH caused by the anaerobic breakdown of carbohydrates.

Table 13.6 summarizes the most probable causes of fatigue depending on the energy system that dominates in exercise.

▶ **Table 13.6** The Most Probable Causes of Fatigue During Exercise Depending on the Dominant Energy System

Energy system	Cause of fatigue
ATP-CP	CP depletion, P_i production, [K⁺] increase in the extracellular space of muscle fibers
Lactate	Drop in cytosolic pH
Oxygen	Hypoglycemia, dehydration, hyperthermia, muscle glycogen depletion

13.22 Restoration of the Energy State After Exercise

When exercise is over, it is desirable to bring back the body to the pre-exercise energy state as soon as possible in anticipation of a new exercise task. The time needed for the energy state to be restored is generally longer than the duration of exercise because the replenishment of energy sources is slower than their consumption. Depending on the kind of exercise, the sources that need replenishment may be

muscle ATP,

muscle CP,

muscle glycogen,

liver glycogen, and

myocellular triacylglycerols.

Adipose tissue triacylglycerols are of no concern, since they are barely affected by an exercise bout.

ATP Replenishment

Muscle ATP is resynthesized in the mitochondria from ADP through oxidative phosphorylation, which is fueled by the combustion of carbohydrates and lipids. The portion of ADP that may have been lost to AMP through the adenylate kinase reaction (equation 8.3) is replenished by reversal of the reaction, as now there is ample ATP.

It is also possible to resynthesize muscle AMP from IMP (the product of AMP deamination), but not by reversal of equation 8.7. Rather, two different reactions and the expense of one ~P are needed for AMP resynthesis.

IMP + aspartate + GTP → adenylsuccinate + GDP + P_i + 2 H⁺   **(equation 13.4)**

adenylsuccinate → AMP + fumarate   **(equation 13.5)**

Reactions 8.7, 13.4, and 13.5 (equations 8.7, 13.4, and 13.5) are often referred to as the *purine nucleotide cycle,* as they concern the recycling of AMP and IMP, which are nucleotides carrying purine bases (section 4.3). Note, however, that

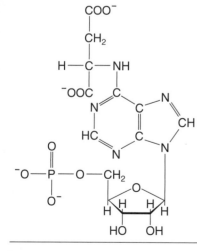

▶ Adenylsuccinate.

the three reactions are accelerated at different phases, or energy states, of a muscle fiber: the first one during exercise, the other two during recovery. In contrast, other metabolic cycles like the citric acid cycle are accelerated in a concerted manner.

CP Replenishment

CP is replenished through the phosphorylation of creatine by ATP in a reversal of reaction 8.4.

$$C + ATP \rightleftharpoons CP + ADP + H^+ \qquad \Delta G^{\circ\prime} = 3 \text{ kcal} \cdot \text{mol}^{-1} \qquad \textbf{(equation 13.6)}$$

Although the reaction has a positive $\Delta G^{\circ\prime}$, its ΔG becomes negative (and the reaction is shifted to the right) during recovery because of the ample supply of ATP. Remember that ΔG can be quite different from $\Delta G^{\circ\prime}$ depending on the actual concentrations of reactants and products (section 2.2).

Reaction 13.6 (equation 13.6) is catalyzed by an isoform of creatine kinase that is different from the ones we met in section 8.3. The isoform is called *mitochondrial* and is symbolized as mit-CK. Unlike CK1, CK2, and CK3, which are located in the cytosol, mit-CK lies on the outer surface of the inner mitochondrial membrane. Thus, it seems to be strategically located so as to ensure the immediate resynthesis of CP by the ATP that exits the mitochondria (figure 13.13).

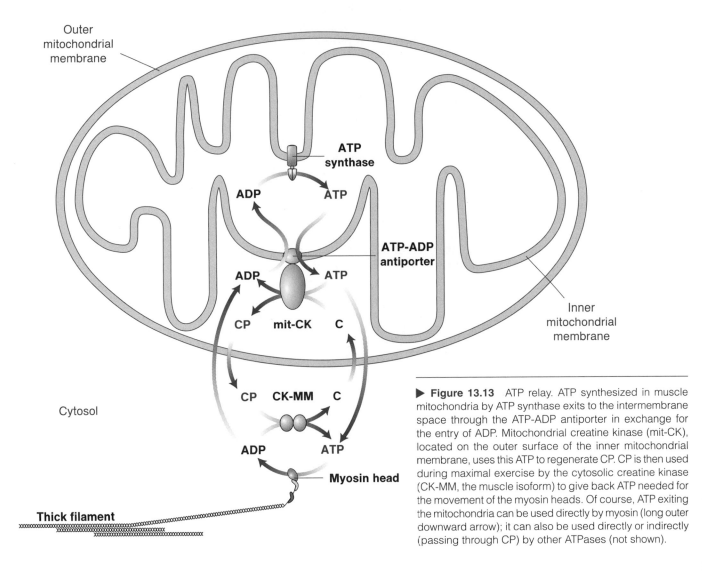

▶ **Figure 13.13** ATP relay. ATP synthesized in muscle mitochondria by ATP synthase exits to the intermembrane space through the ATP-ADP antiporter in exchange for the entry of ADP. Mitochondrial creatine kinase (mit-CK), located on the outer surface of the inner mitochondrial membrane, uses this ATP to regenerate CP. CP is then used during maximal exercise by the cytosolic creatine kinase (CK-MM, the muscle isoform) to give back ATP needed for the movement of the myosin heads. Of course, ATP exiting the mitochondria can be used directly by myosin (long outer downward arrow); it can also be used directly or indirectly (passing through CP) by other ATPases (not shown).

The time needed to resynthesize CP depends on the quantity that has been broken down. It also depends on the blood flow and oxygen delivery to the muscles that exercised, since the ATP used in equation 13.6 is derived from oxidative phosphorylation. Thus, CP resynthesis is slow when the blood flow is constricted, and fast when exercise is followed by active recovery, which maintains an increased blood flow to the muscles. There is also evidence that the rate of CP resynthesis relates to $\dot{V}O_2$max. This finding is in line with the dependence of the rate of CP resynthesis on aerobic processes. If most of CP has been depleted during exercise, the time for resynthesis is at best 3 min but may exceed 10 min in the worst case.

CP resynthesis and other aspects of CP action can be affected by creatine supplementation, a popular practice among athletes in our day. Many well-designed studies in recent years have shown that intake of usually 20 g of creatine per day for typically five to seven days elevates muscle creatine and CP concentrations and increases performance in maximal exercise tasks lasting up to 30 s. Remember that a considerable part of total energy (at least 23%) is drawn from the ATP-CP system during such tasks (section 13.7).

How creatine supplementation boosts performance is not clear. Reasons could be increased CP availability during exercise and increased creatine availability for CP resynthesis during recovery. Creatine supplementation also augments the increase in muscle mass observed with resistance training. Again, the mechanism for this effect is not known. Possibilities include water retention, the ability to train harder (and therefore to induce more muscle hypertrophy) because of increased strength, the stimulation of protein synthesis, and the inhibition of proteolysis.

Glycogen Replenishment

Replenishment of the muscle and liver glycogen stores after exercise requires the presence of sufficient raw material, that is, glucose. This can be achieved only through ample carbohydrate intake. Cessation of the exercise stimulus decreases epinephrine and glucagon secretion. As the plasma concentrations of the two hormones return to baseline, their molecules dissociate from the β-adrenergic and the glucagon receptors, and the receptors are deactivated. Then $G_{s\alpha}$ (figure 9.10) is deactivated, as it possesses a low intrinsic

As in the case of neurotransmitters (section 6.5), the hormone–receptor binding is reversible.

GTPase activity; that is, it slowly hydrolyzes GTP to GDP (figure 13.14). As a result, adenylate cyclase is also deactivated and cAMP synthesis slows down. The existing cAMP is hydrolyzed to AMP by the phosphodiesterase presented in section 10.2, and the entire cAMP cas-

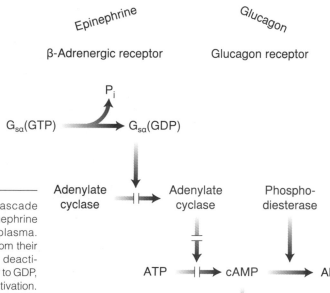

▶ **Figure 13.14** Cyclic-AMP cascade arrest. When exercise stops, epinephrine and glucagon decrease in the plasma. The two hormones are detached from their receptors, and the receptors are deactivated. $G_{s\alpha}$ hydrolyzes its bound GTP to GDP, resulting in adenylate cyclase deactivation. Cyclic-AMP synthesis slows down and cAMP degradation by phosphodiesterase prevails. This blocks the cAMP cascade.

cade is restrained. This lifts the activation of phosphorylase and the inhibition of glycogen synthase (figures 9.12 and 9.13). Thus, glyco-genolysis is decelerated and glycogen synthesis is accelerated. Glycogen synthesis is also boosted by the increase in insulin secretion because of exercise termination and carbohydrate consumption. If carbohydrate consumption is approximately 0.5 g per kilogram body mass per hour during the first 4 to 6 h postexercise, and approximately 10 g · kg^{-1} over the 24 h postexercise period, muscle and liver glycogen can be entirely restored within 24 h.

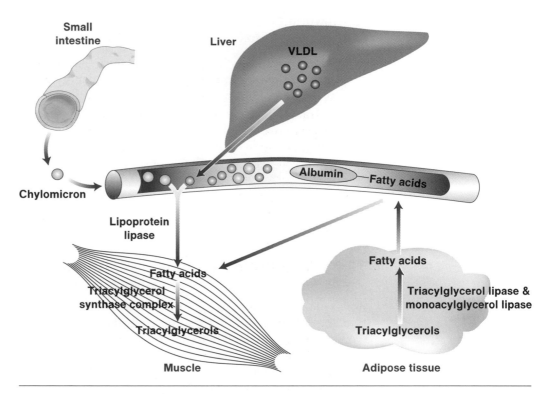

▶ **Figure 13.15** Replenishing myocellular triacylglycerols. Fatty acids from the small intestine, the liver, and adipose tissue enter the muscles to restore their triacylglycerol reservoirs after exercise.

Myocellular Triacylglycerol Replenishment

Myocellular triacylglycerols are replenished after exercise primarily from incoming fatty acids. Since there is no longer a need for increased fatty acid oxidation, most of the fatty acids are used to restore the intracellular triacylglycerol pool. Fatty acids enter the muscle fibers from three main sources, the intestine, the liver, and adipose tissue (figure 13.15). Dietary triacylglycerols packaged in chylomicrons and hepatic triacylglycerols packaged in VLDL are hydrolyzed by muscle lipoprotein lipase, whose activity increases after exercise (section 10.12). The fatty acids produced are then taken up by the muscle fibers. Finally, the muscle fibers receive fatty acids derived from lipolysis in the adipocytes.

EPOC

While recovering from exercise, one continues for some time to breathe at a rate higher than the regular resting rate (figure 13.16). The difference between the oxygen consumed during recovery and that normally consumed at rest is termed **excess postexercise oxygen**

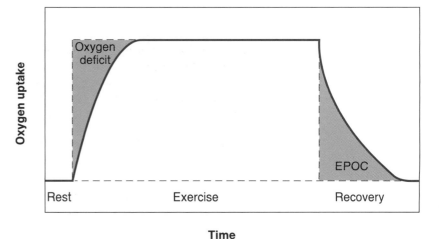

▶ **Figure 13.16** Unfinished business. Oxygen uptake does not return to baseline immediately after the end of exercise, although energy for movement is no longer needed. The excess postexercise oxygen consumption (EPOC) may serve to restore oxygen that was borrowed from hemoglobin and myoglobin during the initial lag phase of oxygen uptake (which resulted in an oxygen deficit), replenish energy sources, and support an elevated oxygen uptake as long as catecholamines and body temperature remain high after exercise.

consumption (EPOC), or **oxygen debt.** We do not know with certainty what leads to EPOC, but the following are some probable causes:

- Extra oxygen may be used to reload the hemoglobin and myoglobin molecules that offered their bound oxygen for aerobic energy production during the beginning of exercise, before oxygen uptake reached the steady-state exercise level.

- Extra oxygen may be used to oxidize fuels (mainly fatty acids) in the muscles in order to restore the ATP concentration and supply additional ATP for CP replenishment. Additional ATP would also be needed in both muscle and the liver to replenish glycogen either from ingested carbohydrates or from gluconeogenic precursors, notably lactate.

- Plasma catecholamine concentrations and body temperature rise during exercise and, naturally, do not return to resting levels immediately when exercise stops. Both have a positive effect on oxygen uptake.

13.23 Metabolic Changes in Detraining

Adaptations to training are not forever: Maintaining them requires the regular application of a substantial training stimulus. If the training load drops considerably or, even worse, ceases, the adaptations described in sections 13.12 to 13.14 will reverse. Athletes, and exercising people in general, often go through a voluntary or involuntary interruption of training (for example, during the summer period, during injury or illness, and, of course, when they quit participation in sport). It is therefore important to know how such changes in daily activity affect metabolism. Unfortunately, research on changes in the body with detraining lags behind research on changes with training.

The interruption of endurance training increases the proportion of carbohydrates to lipids that the body burns during exercise at a given absolute intensity. (Again, we have to distinguish the same absolute from the same relative intensity, since detraining is accompanied by a decrease in $\dot{V}O_2$max.) This is due to a decrease in mitochondrial number and size, resulting in lower concentrations of the enzymes of the citric acid cycle and β oxidation. It is also due to a decline in capillary density and lipoprotein lipase in the muscles. In contrast, lipoprotein lipase in adipose tissue increases, which favors fat deposition. Concomitant with all these biochemical and morphological changes, there is a drop in aerobic endurance. Changes are noticeable just a few weeks after training interruption.

Stopping endurance training also decreases muscle glycogen. This is probably due to a decline in insulin sensitivity (which in turn is probably due to a decline in the GLUT4 content of the muscles) and glycogen synthase activity. These changes are faster than those described in the previous paragraph: They take only a few days to appear. It is not clear whether phosphorylase activity is affected as well.

Finally, as we saw in section 13.10, stopping endurance training shifts the muscle fiber profile from type I to type IIX. This change is slower than those described in the previous paragraphs, taking months of inactivity to appear.

As for strength or sprint training, the most characteristic consequence of stopping is the reduction in muscle cross-sectional area and strength within a few weeks. The cross-sectional area of type IIA and IIX fibers is particularly affected. Muscle glycogen and the enzymes of the lactate system are also reduced, except for phosphorylase, whose response is again unclear. It is also unclear whether interrupting strength or sprint training causes any fiber type transitions.

Table 13.7 summarizes known changes in human muscle with detraining.

▶ **Table 13.7** Changes in Human Muscle with Detraining

Interruption of endurance training
Increase in the proportion of carbohydrates to lipids used during exercise at a given absolute moderate intensity
Decrease in mitochondrial number and size
Decrease in the enzymes of the citric acid cycle
Decrease in the enzymes of β oxidation
Decrease in GLUT4
Decrease in glycogen
Decrease in glycogen synthase
Decrease in capillary density
Decrease in lipoprotein lipase
I → IIA → IIX
Interruption of strength or sprint training
Decrease in cross-sectional area (especially type IIA and IIX fibers)
Decrease in glycogen
Decrease in the glycolytic enzymes and lactate dehydrogenase

Several of the adaptation losses with the interruption of aerobic training can be prevented or mitigated by maintaining a low training stimulus. These losses include the decrease in $\dot{V}O_2$max, the increase in the carbohydrate-to-lipid ratio during exercise, the decrease in oxidative enzymes, the decrease in aerobic endurance, the decrease in GLUT4, and the decrease in insulin sensitivity. Similarly consoling is the situation with anaerobic training: Maintenance of a low training stimulus can prevent or mitigate the decline in muscle cross-sectional area and strength.

By how much can one lower the training load and not experience losses of adaptations? According to different studies, the answer may be as much as 60% to 90%. One has to understand, however, that the larger the decrease in training load, the less the amount of time the adaptations are preserved. In particular, the frequency of training should not be curtailed by more than 30% for elite athletes (for example, 10 training sessions per week should not be decreased to below seven). It is therefore possible for an athlete to maintain his or her competitive level under conditions not favoring full training, provided he or she has the necessary willpower.

Problems and Critical Thinking Questions

1. On the basis of figure 13.3, deduce what percentage of the energy at the 50th second of a maximal exercise bout is derived from each energy system.

2. On the basis of figure 13.5, deduce what percentage of the total energy fueling a 50 s maximal exercise bout is derived from each energy system. Compare with the previous problem.

3. Does the RER go up or down
   a. during prolonged exercise at constant intensity?
   b. as exercise intensity increases?
   c. when exercise is performed after carbohydrate loading compared to a normal diet?
   d. when exercise is performed after fat loading compared to a normal diet?
   e. when exercise is performed after an aerobic training program, at the same absolute intensity as before training?

4. Name one probable cause of fatigue during each of the following exercises:
   a. A tennis match lasting close to 2 h
   b. A 200 m freestyle swim
   c. A set of eight weightlifting repetitions to exhaustion

Part III Summary

Exercise causes some of the most spectacular changes one can see in metabolism, especially within skeletal muscle, the liver, and adipose tissue. These changes are in the direction of supplying increased amounts of compounds for ATP resynthesis, since the muscles break down ATP during exercise at a much higher rate than at rest. Energy sources during exercise can be divided into four classes: compounds of high phosphoryl transfer potential, carbohydrates, lipids, and proteins.

Compounds of High Phosphoryl Transfer Potential

The first of these classes comprises ATP, ADP, and CP. The ATP concentration in muscle remains relatively stable both at rest and during exercise, as ATP is resynthesized from ADP and P_i thanks to the energy released from catabolic processes. ADP can offer some energy through the conversion of two of its molecules to ATP and AMP. This conversion is facilitated by the subsequent deamination of AMP to IMP and ammonia, although this hinders ATP resynthesis during recovery. IMP is further degraded to inosine, hypoxanthine, xanthine, and urate.

Compared to ATP and ADP, much more energy is stored in CP, which regenerates ATP instantaneously through the catalytic action of creatine kinase. CP is the main energy source in maximal efforts lasting up to 7 s, although it keeps offering energy up to 25 s. ATP returns the favor to CP by replenishing it during recovery from exercise. This requires some minutes of rest or active recovery, during which aerobic processes synthesize ample ATP. Creatine supplementation increases performance in maximal exercise tasks lasting up to 30 s, possibly by providing additional CP and creatine for CP replenishment.

Carbohydrates

The carbohydrates that serve as energy sources during exercise include muscle and liver glycogen, as well as muscle, liver, and plasma glucose. Glycogen is synthesized from dietary glucose entering the muscle fibers and hepatocytes after a meal. Exercise speeds up glycogen breakdown, or glycogenolysis, in muscle thanks to the increased cytosolic $[P_i]$, $[Ca^{2+}]$, and [AMP], as well as the increased epinephrine and decreased insulin secretion. These hormonal changes, along with the increased secretion of glucagon, also speed up glycogenolysis in the liver during exercise. In the liver, glucose phosphate produced from glycogenolysis is converted to glucose, which is exported to the blood and enters extrahepatic tissues, including muscle. In muscle, glucose phosphate from glycogen, and glucose from the blood, are broken down to pyruvate through the anaerobic pathway of glycolysis. Continuing in an aerobic route of degradation, pyruvate enters the mitochondria to become acetyl CoA and to end up in carbon dioxide through the citric acid cycle. This yields a sum of about 30 ATP per glucose and 31 ATP per glucosyl unit of glycogen. Of these, 26 are derived from oxidative phosphorylation, which is driven by the flow of electrons to oxygen in the electron transport chain.

Although it yields high amounts of ATP, the aerobic breakdown of carbohydrates does so at a rate that lags behind the rate of ATP consumption during hard exercise. Part of pyruvate is then converted to lactate anaerobically, thus regenerating NAD^+ needed for glycolysis. The harder the exercise, the higher the amount of lactate produced. The anaerobic breakdown of carbohydrates yields only two ATP per glucose and three ATP per glucosyl unit of glycogen. However, these are produced very fast, rendering the anaerobic route the major

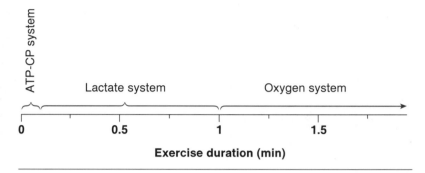

▶ **Figure III.5** Temporal dominance of energy systems. The ATP-CP system supplies most of the energy in maximal efforts up to 7 s. The lactate system prevails in maximal tasks lasting 7 s up to 1 min, and the oxygen system dominates in exercises of longer duration.

energy source in maximal efforts lasting from 7 s to 1 min (figure III.5). The aerobic route dominates in exercise tasks lasting over 1 min.

Lactate, produced in high amounts in a hard-working muscle, can be used for glycogen resynthesis in the early recovery period or can exit to the bloodstream and enter other tissues and organs. These either oxidize lactate to regenerate ATP (this is done in the heart and the resting muscles) or use it to synthesize glucose through the pathway of gluconeogenesis (done in the liver). Glucose synthesis requires 6 ~P and is accelerated during exercise because of the hormonal changes mentioned earlier. This, along with the acceleration of glycogenolysis in the liver and—if applied—carbohydrate intake during exercise, usually maintains euglycemia in spite of increased glucose uptake by the exercising muscles. Besides lactate, products of triacylglycerol and protein breakdown (specifically, glycerol and glucogenic amino acids) are used as substrates for gluconeogenesis.

Lipids

The lipids that serve as energy sources during exercise include the triacylglycerols of adipose tissue, muscle, and, to a small degree, plasma, as well as the fatty acids of muscle and plasma. Exercise speeds up triacylglycerol breakdown, or lipolysis, thanks to the increase in catecholamine (epinephrine and norepinephrine) secretion and the decrease in insulin secretion. Fatty acids produced from lipolysis in adipose tissue exit to the bloodstream and are taken up by the exercising muscles, whereas those produced from lipolysis in muscle are used on the spot. Fatty acids are degraded only aerobically in the mitochondria, where they enter with the help of carnitine. Fatty acids are first oxidized to acetyl CoA through the

pathway of β oxidation and then to CO_2 through the citric acid cycle. Fatty acids yield more energy than carbohydrates (for example, palmitate yields about 106 ATP), albeit at a lower rate. As a result, they offer considerable energy only in light or moderate-intensity efforts.

Triacylglycerols, cholesterol, and cholesterol esters are carried in the plasma by globular aggregates called lipoproteins, which are divided into chylomicrons, VLDL, LDL, and HDL. Chylomicrons transport triacylglycerols synthesized in the small intestine from dietary fat to extrahepatic tissues; VLDL transport triacylglycerols synthesized in the liver to extrahepatic tissues; LDL transport cholesterol to extrahepatic tissues; and HDL transport cholesterol from extrahepatic tissues to the liver, chylomicrons, and VLDL. The concentration of plasma lipids does not change much during or after exercise. However, regular aerobic exercise can decrease plasma triacylglycerols and increase HDL cholesterol. These changes lower the risk for atherosclerosis.

Two ketone bodies, acetoacetate and 3-hydroxybutyrate, are formed in the liver from fatty acids when there is a dearth of carbohydrates. The ketone bodies are transported to extrahepatic tissues, where they are burned for energy production. Ketone body formation increases during exercise but the ketone bodies' contribution to energy requirements is small.

Proteins

Protein metabolism is affected by acute exercise. During hard resistance exercise, the rate of protein synthesis in the exercising muscles increases more than the rate of proteolysis, resulting in an increase of muscle proteins. Conversely, during hard endurance exercise, the rate of protein synthesis decreases while the rate of proteolysis increases, resulting in a decrease of muscle proteins. However, this is reversed during recovery. The amino acids that result from proteolysis can contribute to the energy expenditure of exercise either by being oxidized in the exercising muscles or by being converted to glucose in the liver. Either way, they first dispose of their α-amino group, which ends up in urea through the urea cycle, a pathway operating in the liver, at the expense of four ATP. The carbon skeletons that remain are then converted to pyruvate, acetyl CoA, acetoacetyl CoA, or intermediates of the citric acid cycle. The contribution of proteins to the energy expenditure of exercise is up to 6%.

Chronic exercise has a more spectacular effect on protein metabolism than acute exercise. Adaptations depend on the kind of training. Strength training and, to a lesser degree, sprint training increase the amount of muscle proteins, in particular the contractile proteins. This is manifested as muscle hypertrophy. In contrast, endurance training increases the mitochondrial proteins, resulting in increased mitochondrial biogenesis.

Gene Expression

The adaptations to training just discussed are achieved through concerted changes in gene expression. Such changes can occur in transcription, the conversion of primary RNA to mRNA, the transport of mRNA from the nucleus to the cytosol, mRNA stability, translation, the stability of the synthesized proteins, the posttranslational modification of proteins, and protein targeting. There is evidence that exercise can modify most of these processes. The biochemical changes mediating the effects of exercise on gene expression are largely unknown but may be triggered by the increase in $[Ca^{2+}]$, the decrease in [ATP] and the subsequent increase in [ADP] and [AMP], the decrease in glycogen, hypoxia, radical formation, the drop in pH, the development of mechanical tension, and the increased binding of proteins to cell receptors. IGF1 and MGF play important roles in muscle hypertrophy, in part by stimulating satellite cell proliferation, while transcription factors such as NRF1, PPARα, and PPARδ, as well as transcription coactivators such as PGC1, mediate mitochondrial biogenesis.

Integration of Exercise Metabolism

The numerous acute and chronic effects of exercise on metabolism permit a better response of the body to the demands of increased contractile activity. The primary demand, a high energy supply, is satisfied by the cooperation of three energy systems, the ATP-CP, the lactate, and the oxygen systems. The systems rank inversely in terms of energy and power: The ATP-CP system has the lowest energy and highest power, whereas the oxygen system has the highest energy and lowest power. At least two of the energy systems participate in most exercise tasks, although one system usually stands out as the dominant one, that is, the system supplying more energy than the other(s).

The selection of energy sources during exercise is determined by intensity, duration, program, heredity, nutrition, training state, and age. The first two factors—and the most important—affect the proportion of energy sources in opposite ways: The higher the intensity, the higher the contribution of the faster sources, whereas the longer the duration, the higher the contribution of the larger sources (figure III.6).

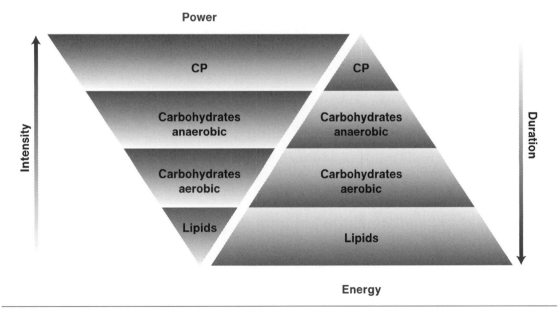

▶ **Figure III.6** Choice of energy sources. As exercise intensity increases, so does the contribution of the sources producing high power. As exercise duration increases, so does the contribution of the sources producing a lot of energy.

Heredity affects the selection of energy sources during exercise mainly by determining the proportion of muscle fiber types. A high proportion of type I, or oxidative, fibers favors aerobic function, whereas a high proportion of type IIA, or oxidative-glycolytic, and particularly type IIX, or glycolytic, fibers favors anaerobic function. The proportion of muscle fiber types is primarily determined genetically, but it can change over the long term if the contractile activity changes. An increase in contractile activity shifts the balance toward type I fibers, whereas a decrease in contractile activity shifts the balance toward type IIX fibers.

Nutrition affects the proportion of energy sources during exercise by increasing the contribution of carbohydrates or lipids, depending on which of the two is consumed in high quantities. Training state affects the proportion of energy sources thanks to training-induced adaptations. More spectacular are the adaptations accompanying endurance training, which increases the ratio of lipids to carbohydrates that are broken down during prolonged exercise at the same absolute intensity. This is due to an increase in the oxidative capacity of the muscles and to an increase in the capacity to burn lipids that is larger than the increase in the capacity to burn carbohydrates. Strength training increases the capacity for anaerobic energy production, whereas sprint training can increase the capacity of all three energy systems depending on the details of the training program. Finally, regarding age, children utilize relatively more lipids than adults in prolonged exercise at the same absolute or relative intensity, according to most of the relevant studies.

In addition to metabolism during exercise, of interest is the metabolism that relates to the end of exercise or training. The development of fatigue imposes a decrease in intensity or an end to exercise. The causes of fatigue are not known with certainty and seem to vary depending on the kind of exercise. The most probable causes of fatigue in maximal exercise tasks that are based mainly on the ATP-CP system are CP depletion and P_i production, whereas tasks based mainly on the lactate system are limited by the decrease in cytosolic pH. Finally, prolonged efforts based mainly on the oxygen system may be limited by hypoglycemia, dehydration, hyperthermia, or muscle glycogen depletion.

After exercise, the body restores its energy sources through the uptake of excess oxygen, the aerobic resynthesis of ATP, and the consumption of food. Cessation of training results in the slow loss of some adaptations and the rapid loss of others. However, several adaptations can be preserved through the maintenance of a moderate training stimulus.

Biochemical Assessment of Exercising Persons

"…Read and strive and fight" he said. "Each to his own weapons" he said…

—Odysseus Elytis (Nobel Prize in Literature 1979), *Worthy It Is*

The multitude of acute and chronic biochemical changes that are caused by exercise, and that were the subject of part III, permit the collection of precious information about an exercising person through measurement of substances in his or her tissues. I will use the term biochemical assessment of exercising persons to describe the measurement of substances performed not (or mainly not) for research purposes but rather to estimate the condition of the exercising individual. Such an assessment interests primarily athletes and the professionals who support them, such as trainers, physicians, and dieticians. Nevertheless, people exercising for recreational or therapeutic purposes also benefit from biochemical tests.

The most suitable tissue for the biochemical assessment of exercising persons is the blood, as it combines a moderate ease of sampling with a satisfactory amount of information (figure IV.1). Naturally, muscle tissue can provide more information, since muscle is the primary setting for energy metabolism. In addition, one can measure in muscle a number of substances that do not appear in the blood (for example, ATP, CP, and glycogen). However, muscle biopsy is painful and requires highly specialized person-

nel, as well as advanced analytical techniques for the measurement of substances in small samples. Urine lies at the other extreme: It is collected painlessly but contains few substances that are useful in evaluating an exercising individual.

Of the hundreds of biochemical parameters that can be measured in the blood and are treated by

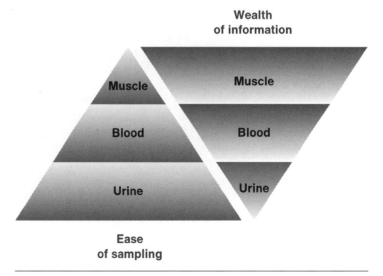

▶ **Figure IV.1** The golden section. The blood represents the golden section between ease of sampling and wealth of information regarding an exercising person. Muscle offers more information but is difficult to sample. Urine is obtained bloodlessly but does not offer much information.

273

clinical chemistry, I present those that I consider the most useful in assessing an exercising person. Measurement is relatively easy (these are, as we say, routine tests for a biochemistry laboratory), and the cost of the analysis is low to moderate. For each parameter, we will discuss the information it provides, its usual values, and the factors influencing these values.

IV.1 The Blood

Before I begin presenting the substances that are recruited in the biochemical assessment of exercising persons, I need to say a few words about the biological sample we use. Blood is not a homogeneous fluid. If we centrifuge it, it separates into roughly two parts of different densities: a dark-red sediment and a yellowish supernatant (figure IV.2). The sediment contains mainly erythrocytes. The supernatant is usually clear and is usually present in greater quantities than the sediment. The supernatant is called **serum** or **plasma** depending on the way the blood was collected.

> The serum or plasma collected after a fatty meal is turbid because of its high chylomicron content (see section 10.10).

What is serum and what is plasma? As the blood passes through the needle into the syringe or test tube used for collection, it begins to coagulate, forming a clot. (A modern method of blood sampling employs evacuated test tubes instead of syringes.) Coagulation is usually complete within one-half hour and, if the blood is subsequently centrifuged, the clot is squeezed to the bottom of the test tube. The supernatant is serum.

We can prevent coagulation by adding an *anticoagulant* to the blood right after its collection or, better yet, by collecting the blood directly in an evacuated test tube containing anticoagulant. Anticoagulants are substances that inhibit one or more steps in the complex coagulation process. The most frequently used anticoagulants are *heparin* and *ethylenedinitrilotetraacetate* (EDTA). Blood containing an anticoagulant can be centrifuged immediately, and the supernatant is plasma.

Plasma differs from serum with respect to certain components, but most substances have the same concentration in the two fluids. For most biochemical analyses, we prefer to let the blood coagulate and collect serum. When another choice is preferable, I note this in the appropriate section.

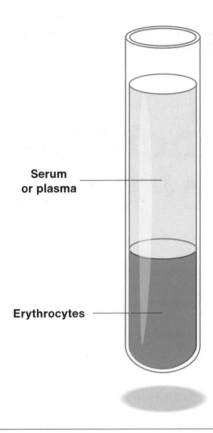

▶ **Figure IV.2** Separation of blood components. A test tube filled with blood presents two distinct layers after centrifugation.

IV.2 Aims and Scope of the Biochemical Assessment

The biochemical assessment of exercising persons aims at protecting or improving their health and increasing their performance. The first target covers all exercising individuals, while the second concerns primarily athletes. Good health is the sound foundation on which sport performance is built. Thus, good health ought to be the primary concern of a biochemist, as any scientist, who is monitoring an athlete.

As to the scope of the assessment, one can measure biochemical parameters in samples collected at different time points relative to exercise. We may discern three cases.

- *Samples collected at rest.* In this case we examine what the basal state of the body is or what adaptations previous exercise has caused.

- *Samples collected during exercise.* Such samples inform us about acute biochemical changes caused by physical activity.

- *Samples collected after exercise.* "After" can range from a few seconds to hours or even days from the end of exercise. Depending on the precise sampling time, these samples can show acute effects of exercise, the duration of the changes elicited by exercise, or changes of delayed onset. Of course, the farther we get from the end of exercise, the more we approach the first case.

IV.3   The Reference Interval

Suppose that the measurement of glucose in the serum of an athlete produced a value of 90 mg · dL^{-1}. In order to decide whether this value is normal or abnormal, we usually ask, "How much should it be?" or "What are the normal values?" If we are told that these range from 74 to 106 mg · dL^{-1}, we decide that 90 mg · dL^{-1} is a normal value. The range 74 to 106 mg · dL^{-1} is the **reference interval** for glucose. We therefore define (at first approach) the reference interval as *the range of values that a parameter usually gets.* The lower and upper ends of the reference interval are the lower and upper **reference limits.**

How are reference intervals determined? Investigators do this by analyzing biological samples taken from an adequate and representative part of the population. As you can expect, the value of a parameter differs from one individual to another. However, there is usually an accumulation of values around a central value; and the farther we get from it, the fewer values we encounter (figure IV.3). Within such a *frequency plot,* we can locate the lower 2.5% and the upper 2.5% of the values and reject them as being extreme. The range that includes the remaining 95% of the values is the reference interval.

You may have noticed that the reference interval is determined through a mathematical approach with no involvement of biochemistry (except, of course, for the measurement of the parameter). This is why we prefer the rather neutral and objective term reference interval instead of the more frequently used terms normal values and normal range. What I mean is that we are not certain whether the values exhibited by 95% of the population are normal (or, conversely, that the values exhibited by the remaining 5% are abnormal); they are simply values that we refer to in order to make comparisons. A case of reference intervals that are not normal (in the sense that they include values related to increased

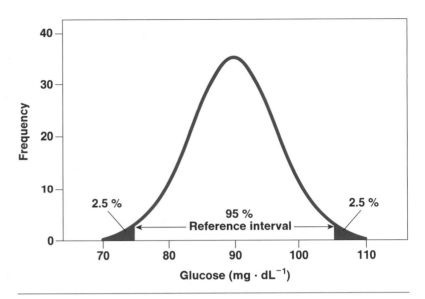

▶ **Figure IV.3**   The reference interval. By rejecting the lower and upper 2.5% of the values of a parameter such as the serum glucose concentration, which we have measured in a large number of individuals, we find the reference interval. The curve in this example depicts an ideal distribution called the normal distribution, which is not the case for all biochemical parameters.

health risks) is the reference intervals for serum lipids discussed in sections 15.6 to 15.9.

The reference interval depends on two factors.

- *The method of measurement.* Most parameters are measured by two or more methods, which may produce different results. A frequent cause of disagreement among methods is the occurrence of positive or negative errors because of substances coexisting in the sample with the substance of interest. Nevertheless, the differences among acceptable methods of measuring a parameter are generally small.

- *The population sample measured.* The values of many parameters depend on factors like sex and age. Such parameters have different reference intervals for different population groups. When this is the case, I cite reference intervals for both sexes and the entire age range of exercising persons (from children to older people). Additionally, one's way of life (nutrition and physical activity) may influence the value of a parameter. This is particularly true for athletes, as I will be noting frequently.

I close this section by clarifying two more things regarding reference intervals. First, in clinical chemistry, the concentration of a substance is usually expressed as mass per volume (for example, mg · dL^{-1}), although an effort is being made to establish units of chemical quantity per volume (for example, mmol · L^{-1}). Where the mass per volume unit is more customary, I cite reference intervals in both kinds of units, along with information on how to convert mass per volume to chemical quantity per volume.

Second, because the values of many biochemical parameters are influenced by food intake, reference intervals will refer to the fasted state (approximately 12 h after the last meal), the one we named postabsorptive state in section 9.23. Reference intervals are from Tietz (1995) unless otherwise indicated.

IV.4 Classes of Biochemical Parameters

In order to facilitate the examination of the parameters that are useful in the biochemical assessment of exercising persons, I have divided them into four classes.

- Parameters of the iron status
- Metabolites
- Enzymes
- Hormones

Table IV.1 summarizes the parameters that we will examine within the framework of these classes.

▶ **Table IV.1**   Parameters Useful in the Biochemical Assessment of Exercising Persons

Class	Parameters
Iron status	Hemoglobin Iron Total iron-binding capacity Transferrin saturation Soluble transferrin receptor Ferritin
Metabolites	Lactate Glucose Triacylglycerols (triglycerides) Total cholesterol HDL cholesterol LDL cholesterol Glycerol Urea Ammonia Creatinine
Enzymes	Creatine kinase Aminotransferases (transaminases)
Hormones	Cortisol Testosterone

CHAPTER 14

Iron Status

Iron holds a central place in oxygen transport, storage, and utilization for aerobic energy production, as it is part of heme, the prosthetic group of myoglobin and hemoglobin (sections 3.10 and 3.11). In fact, iron is precisely the element that binds O_2. In addition, iron is part of several proteins in the electron transport chain (section 9.11). The iron content of the tissues is one of the factors determining the aerobic capacity of an individual. As a result, the adequacy of iron in the body is closely linked to health, fitness, and sport performance, given the great importance of aerobic metabolism during exercise—even hard or maximal exercise—and recovery from exercise, as highlighted throughout part III. Thus, the interest of athletes in iron adequacy is justifiable.

Several parameters are used to assess the **iron status,** that is, the concentration of iron in the various compartments of the body. We will examine six of these parameters:

Hemoglobin

Iron

Total iron-binding capacity (TIBC)

Transferrin saturation

Soluble transferrin receptor

Ferritin

Why do we need so many parameters to assess the status of a single chemical element? The reason is that each parameter sheds light on a different facet of the iron status, specifically:

The hemoglobin concentration shows the capacity of the blood to absorb atmospheric oxygen in the lungs and carry it to the tissues.

The iron concentration (in the serum) informs us about the amount that is available for uptake by the tissues.

The TIBC represents the amount of iron that the plasma can carry.

The transferrin saturation shows how saturated the iron transport system is.

The soluble transferrin receptor and ferritin concentrations reflect the amount of iron stored in the tissues.

Let's examine these parameters one by one.

14.1 Hemoglobin

Hemoglobin is the oxygen-carrying protein in the blood. It is found in the erythrocytes and constitutes the most abundant blood component (about 14% of the blood) next to water. A minimal amount of hemoglobin is dissolved in the plasma because of the normal wear of the erythrocytes, termed *hemolysis*. Plasma hemoglobin does not take part in oxygen transport and is removed from the circulation by the liver.

We measure the hemoglobin concentration in whole blood after we have blocked coagulation by adding an anticoagulant. The reference interval depends on age and sex (table 14.1). Both the lower and upper reference limits rise as males enter adolescence owing to the anabolic action of testosterone (see section 16.6).

The blood hemoglobin concentration increases during living at high altitude because of the low oxygen concentration in the air. As the $[O_2]$ in the air drops, so does the $[O_2]$ in the blood, and this promotes **erythropoietin** production by the kidneys. Erythropoietin is a protein hormone causing erythrocyte formation (**hemopoiesis**) in the bone marrow.

The blood hemoglobin concentration (but not the amount of hemoglobin) also increases during exercise, as water exits the blood vessels to the sweat. This **hemoconcentration** lasts maximally 1 h after exercise and makes the measurement of hemoglobin useful in calculating blood volume changes. For example, if the hemoglobin concentration is 14.6 g · dL^{-1} at rest and 15.2 g · dL^{-1} after exercise, then, since concentration is inversely proportional to volume, the blood volume after exercise will be 14.6/15.2 = 0.96, that is, 96% of the blood volume at rest. We then declare that there is a hemoconcentration of 100 − 96 = 4%.

Sometimes, mainly for research purposes, we are interested not in the *blood* volume change but the *plasma* volume change with exercise. To calculate plasma volume changes we need to take into account changes in the volume of the blood cells. For this reason, and for the sake of the ensuing discussion on anemia, I will make a short detour to the hematologic parameters.

> Because hemoglobin possesses most of the body's iron (approximately 60%), it is often mistaken as an *iron* transporter. However, iron transport is not hemoglobin's biological function (instead, hemoglobin is an *oxygen* transporter). See sections 14.4 to 14.6 for discussion of the iron transporter.

▶ **Table 14.1**   Reference Intervals for Hemoglobin Concentration in Human Blood

| Age (years) | Reference interval (g · dL^{-1}) | | Reference interval (mmol globin · L^{-1})[a] | |
	Male	Female	Male	Female
5-9	11.5-14.5	11.5-14.5	7.2-9.1	7.2-9.1
9-12	12.0-15.0	12.0-15.0	7.5-9.4	7.5-9.4
12-14	12.0-16.0	11.5-15.0	7.5-10.0	7.2-9.4
15-17	12.3-16.6	11.7-15.3	7.7-10.4	7.3-9.6
18-44	13.2-17.3	11.7-15.5	8.3-10.8	7.3-9.7
45-64	13.1-17.2	11.7-16.0	8.2-10.8	7.3-10.0
65-74	12.6-17.4	11.7-16.1	7.9-10.9	7.3-10.1

[a]To convert g · dL^{-1} to mmol · L^{-1}, divide by 1.6.

Adapted from Tietz (1995).

14.2 Hematologic Parameters

Hematologic parameters are those that relate to the blood cells. They are measured in whole blood, and the most common ones are the **hematocrit, erythrocyte count, leukocyte count,** and **platelet count.**

The hematocrit (meaning "judge of the blood" in Greek) is the volume of the erythrocytes as a percentage of the blood volume. Table 14.2 presents the reference intervals. As expected, the hematocrit relates closely to the blood hemoglobin concentration and displays the same variation with age and sex, being highest in adult males. It too shows the oxygen-carrying capacity of the blood. Finally, exercise and altitude have the same effects on it as they have on the hemoglobin concentration.

The erythrocyte, leukocyte, and platelet counts express the number of the corresponding cells in 1 µL of blood. Erythrocytes are by far the most abundant blood cells and measure millions per microliter (M · μL^{-1}). In contrast, leukocytes and platelets measure thousands per microliter ($k \cdot \mu L^{-1}$). The erythrocyte count goes hand in hand with the hemoglobin concentration and hematocrit; it also has the same utility. The reference intervals are shown in table 14.2.

The erythrocyte count, along with the hemoglobin concentration and hematocrit, increases with **blood doping**, which is the transfusion of a large volume of blood to an athlete before an event. Injection of erythropoietin has the same effect. Both practices increase endurance performance and are prohibited in sport.

The leukocyte count is indicative of the state of the immune system, as leukocytes are part of this system. The leukocyte count increases in cases of infection; this condition is called *leukocytosis*. Conversely, it decreases when the immune system is suppressed; this is called *leukopenia* (meaning "lack of white cells" in Greek). The leukocyte count also increases for a few hours after hard or prolonged exercise. The reference intervals are

▶ **Table 14.2** Reference Intervals for Hematocrit and Erythrocyte Count in Human Blood

Age (years)	Hematocrit		Erythrocyte count ($M \cdot \mu L^{-1}$)[a]	
	Male	Female	Male	Female
6-8	33-41	33-41	3.8-4.9	3.8-4.9
9-11	34-43	34-43	3.9-5.1	3.9-5.1
12-14	35-45	34-44	4.1-5.2	3.8-5.0
15-17	37-48	34-44	4.2-5.6	3.9-5.1
18-44	39-49	35-45	4.3-5.7	3.8-5.1
45-64	39-50	35-47	4.2-5.6	3.8-5.3
65-74	37-51	35-47	3.8-5.8	3.8-5.2

[a]$M \cdot \mu L^{-1}$: millions per microliter.

Adapted from *Clinical Guide to Laboratory Tests*, N.W. Tietz, pg. 310, Copyright 1995, with permission from Elsevier.

4 to 13 $k \cdot \mu L^{-1}$ in children and adolescents and

4 to 11 $k \cdot \mu L^{-1}$ in adults.

Finally, the platelet count is indicative of the state of the blood clotting system, as platelets are part of this system. This parameter too may increase transiently after exercise. The reference interval is 150 to 400 $k \cdot \mu L^{-1}$.

Let's return now to the previous discussion regarding the plasma volume change with exercise. As mentioned, to calculate it we need to take into account not only the change in the blood volume but also the change in the volume of the blood cells. Taking the change in the volume of the erythrocytes as representative of the change in the volume of all blood cells, we can arrive through mathematical calculations at the following formula connecting postexercise to the pre-exercise values:

$$\frac{\text{plasma volume post}}{\text{plasma volume pre}} = \frac{\text{hemoglobin pre}}{\text{hemoglobin post}} \cdot \frac{(100 - \text{hematocrit post})}{(100 - \text{hematocrit pre})} \qquad \textbf{(equation 14.1)}$$

14.3 Does Sports Anemia Exist?

The condition in which the hematocrit and blood hemoglobin concentration of an individual are below the lower reference limits is termed **anemia.** Anemia can be inherited, as are *sickle cell anemia* and *thalassemia,* for example. In the absence of a genetic background the main cause of anemia is nutritional, namely, the inadequate intake of iron (see section 14.9), proteins, or certain vitamins (vitamin B_6 and folate) needed for hemopoiesis. It is particularly important for women losing a lot of blood during menstruation to be watchful about anemia.

Do athletes run a higher risk for anemia compared to the general population? The scientific literature of previous years has provided an affirmative answer leading to the introduction of the term *sports anemia*. Admittedly, there are reasons for reduction in an athlete's hemoglobin and hematocrit, such as erythrocyte destruction in the blood vessels of the feet when they hit the ground and erythrocyte loss in the stool as capillaries break in the digestive tract during exercise, particularly prolonged endurance exercise. Athletes may also display **hemodilution.** This is the increase in blood volume above normal levels following hemoconcentration; it is an adaptation to regular exercise, fading away three to five days after training interruption. Hemodilution does not compromise the oxygen-carrying capacity of the blood, since it does not affect the total amount of erythrocytes. Some have proposed the term *pseudoanemia* (meaning "false anemia" in Greek) to describe the "anemia" that is due to hemodilution.

In spite of all possible effects of exercise on the hematocrit and hemoglobin concentration, most of the studies that have compared athletes and properly matched nonathletes have shown similar incidences of anemia and similar hematocrit and hemoglobin values in the two groups. Thus, it seems that there is no considerable impact of exercise training on these parameters, and there is no sports anemia in reality.

14.4 Iron

Iron commutes among tissues as Fe^{3+} bound mainly to **transferrin,** or **siderophilin,** its transport protein in the plasma. The reference intervals for the serum iron concentration are

To convert $\mu g \cdot dL^{-1}$ to $\mu mol \cdot L^{-1}$, divide by 5.585.

50 to 120 $\mu g \cdot dL^{-1}$, or 9.0 to 21.5 $\mu mol \cdot L^{-1}$, in children and adolescents;

65 to 175 $\mu g \cdot dL^{-1}$, or 11.6 to 31.3 $\mu mol \cdot L^{-1}$, in men; and

50 to 170 $\mu g \cdot dL^{-1}$, or 9.0 to 30.4 $\mu mol \cdot L^{-1}$, in women.

The higher the iron concentration, the higher the amount of iron the tissues can absorb and incorporate in the proteins they synthesize. Thus, a high iron concentration (within the reference interval) is desirable. People can achieve this by eating foods rich in iron. The literature reports similar serum iron concentrations in athletes and nonathletes.

14.5 Total Iron-Binding Capacity

As mentioned in the previous section, iron is carried in the plasma primarily by transferrin. Transferrin is only partly saturated with iron (approximately by one-quarter to one-third). The serum iron concentration in the hypothetical case in which the iron-binding proteins in the plasma are fully saturated is the **total iron-binding capacity** **(TIBC).** The reference interval is 250 to 425 $\mu g \cdot dL^{-1}$, or 44.8 to 76.1 $\mu mol \cdot L^{-1}$.

To convert $\mu g \cdot dL^{-1}$ to $\mu mol \cdot L^{-1}$, divide by 5.585.

Interest in the TIBC stems from the fact that the body reacts to a shortage of iron by increasing the plasma transferrin concentration. Thus, TIBC values above the upper reference limit warn against iron depletion.

14.6 Transferrin Saturation

Transferrin saturation is the serum iron concentration as a percentage of the TIBC.

$$\text{transferrin saturation} = \frac{[Fe^{3+}]}{\text{TIBC}} \cdot 100 \qquad \text{(equation 14.2)}$$

For example, if $[Fe^{3+}]$ is 84 μg · dL^{-1} and the TIBC is 300 μg · dL^{-1}, then the transferrin saturation is 28%.

The reference intervals for the transferrin saturation are

20% to 50% in males and

15% to 50% in females.

The transferrin saturation relates positively to the adequacy of iron in the body, since an adequacy of iron will cause a high numerator and low denominator in equation 14.2. In fact, because it combines two parameters, the transferrin saturation changes relatively more than $[Fe^{3+}]$ or TIBC separately. As a result, it is considered a more sensitive index of the iron status and is used as a criterion of iron deficiency (see section 14.9).

14.7   Soluble Transferrin Receptor

The *transferrin receptor* is an integral protein of the plasma membrane of almost all cells (erythrocytes being a notable exception), consisting of two identical 95 kDa polypeptide chains. The receptor binds iron-laden transferrin and mediates iron uptake by endocytosis of the receptor–transferrin complex. Once in the cytoplasm, iron is released and used for cellular needs, whereas transferrin and the receptor are recycled to the plasma membrane.

Being a membrane protein, the transferrin receptor is poorly soluble in water. However, part of it is detected in the plasma as a water-soluble form produced by the removal of segments from the original polypeptide chains in the process of ridding the plasma membrane of the receptor. This process takes place mainly during erythrocyte maturation in the bone marrow: Once erythrocytes have synthesized their hemoglobin, they have no use for iron and thus need no transferrin receptor.

This <u>so</u>luble <u>t</u>rans<u>f</u>errin <u>r</u>eceptor (sTfR) is what we measure in the serum. Its concentration relates inversely to iron adequacy because, as with transferrin, the body reacts to iron depletion by increasing the amount of the transferrin receptor in the plasma membrane of its cells. By extension, the serum sTfR concentration also increases. Based on this, the sTfR concentration is also used as a criterion of iron deficiency, although, being a relatively new parameter, it has not been widely accepted yet. The reference intervals according to the manufacturer of a kit designed to measure the sTfR are

1.5 to 3.7 mg · L^{-1} in ages 4 to 10,

1.4 to 3.4 mg · L^{-1} in ages 10 to 16, and

1.3 to 3.3 mg · L^{-1} in ages over 16.

14.8   Ferritin

Ferritin is our iron-storing protein. It consists of 24 subunits having a total molecular mass of 445 kDa and forming a huge shell, which encloses as many as 4,000 Fe^{3+} ions in an internal cavity, 8 nm in diameter. Ferritin is found mainly in the spleen, bone marrow, and liver. Its quantity in the body relates positively to that of iron.

A minuscule portion of tissue ferritin, roughly proportional to its quantity, leaks into the plasma because of the natural wear of cells. Thus, the serum ferritin concentration serves as an index of the iron stores in the body and is used as a criterion of iron deficiency. Each microgram of ferritin per liter of serum is equivalent to 8 mg of stored iron in the body. The reference intervals for the serum ferritin concentration are

7 to 140 $\mu g \cdot L^{-1}$ in children and adolescents,

20 to 250 $\mu g \cdot L^{-1}$ in men, and

10 to 120 $\mu g \cdot L^{-1}$ in women.

14.9 Iron Deficiency

Iron deficiency is an alarming state, since it may lead to reduced synthesis of hemoglobin, myoglobin, and proteins of the electron transport chain. Eating iron-rich foods and taking iron supplements in moderate dosages usually rectifies iron deficiency.

Prolonged iron deficiency exhausts the iron stores in the body, leading to *iron deficiency anemia*. This condition (as with anemia from other causes) is characterized by paleness, fatigue, and reduced ability to keep the body warm in a cold environment. To reverse iron deficiency anemia one needs to take iron supplements at high dosages.

To declare a person iron deficient we usually examine two parameters of the iron status. (We do this in order to reduce the possibility of a wrong diagnosis because of an accidental drop in one parameter only.) As mentioned before, two such parameters are the transferrin saturation and ferritin concentration. Figure 14.1 clarifies the diagnostic criteria of the conditions discussed in the present chapter.

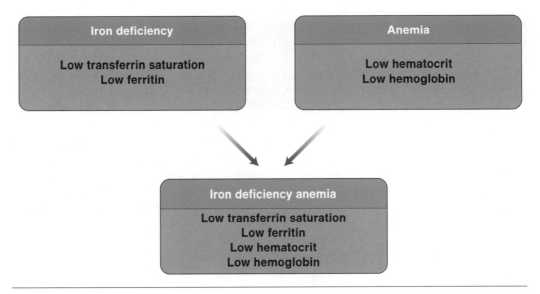

▶ **Figure 14.1** Iron deficiency and anemia. These are two different conditions having different indices. Nevertheless, iron deficiency can cause anemia, in which case we have iron deficiency anemia and a combination of indices. Note that there are also other biochemical indices of iron deficiency.

By analogy to the question posed in section 14.3 about anemia, we now ask, Do athletes run a higher risk for iron deficiency than the general population? The answer again seems to be negative. Although there is no agreement in the literature, the majority of the studies that used an appropriate control group showed no significant differences between athletes and nonathletes regarding the indices of the iron status or the prevalence of iron deficiency. The reason may be that exercise does not increase iron loss to any great extent. The only documented increased loss is through the stool, while the frequently mentioned loss through sweat is actually negligible. It is also possible that most athletes replenish any possible increased iron losses while eating more food to meet their increased energy demands.

Problems and Critical Thinking Questions

1. A runner had a blood hemoglobin concentration of 15.4 g · dL^{-1} and a hematocrit of 44.2 at rest. After exercise, these values changed to 15.8 g · dL^{-1} and 44.9. Calculate the change in plasma volume.

2. Which doping substance and which doping method affect the blood hemoglobin concentration, hematocrit, and erythrocyte count?

3. How does iron deficiency affect the serum TIBC, transferrin saturation, sTfR concentration, and ferritin concentration?

4. The serum iron concentration and TIBC of a person were 90 and 300 μg · dL^{-1}, respectively, six months ago. Because of poor nutrition, iron has now decreased by 10% and the TIBC has increased by 10%. Calculate the change in transferrin saturation.

5. An athlete has a high serum ferritin concentration but low transferrin saturation. What do you conclude and what would you advise him or her to do?

6. In contrast, another athlete has low ferritin and high transferrin saturation. What would you conclude?

CHAPTER 15

Metabolites

This chapter examines the utility of measuring biochemical parameters relevant to the four classes of substances that we explored in part III, in a slightly different order: carbohydrates, lipids, proteins, and compounds of high phosphoryl transfer potential. Table 15.1 itemizes the parameters that we will consider by class of energy sources in exercise.

15.1 Lactate

Lactate is the end product of the anaerobic carbohydrate breakdown. It is the metabolite displaying the most spectacular concentration changes in muscle and the blood with exercise. As a result, its measurement offers a wealth of information regarding the effect of exercise on metabolism. We usually determine lactate in whole blood rather than plasma or serum. In addition to the conventional laboratory equipment, there are practical portable devices that measure lactate within a minute or so of applying a drop of blood from a fingertip or earlobe.

The blood lactate concentration at rest is about 1 mmol $\cdot$ L^{-1}, whereas after maximal exercise lasting at least one-half minute, it can go over 20 mmol $\cdot$ L^{-1}. Efforts of lower intensity result in lower lactate concentrations, as carbohydrates are broken down primarily aerobically. The lactate concentration is also lower than 20 mmol $\cdot$ L^{-1} after maximal efforts lasting less than one-half minute, since the lactate produced in the exercising muscles is not sufficient to elicit such a high concentration in the blood.

When measuring lactate in the blood after short hard or maximal exercise it is necessary to remember that it takes some minutes to peak (figure 9.38). Thus, a blood sample taken right after the end of exercise will not produce the peak value. To trace it we need to perform repeated samplings spaced maximally 2 min apart, for about 10 min or until we see that the lactate concentration begins to drop (if

▶ **Table 15.1** Metabolites Useful in the Biochemical Assessment of Exercising Persons

Parameter	Relevant to the metabolism of
Lactate Glucose	Carbohydrates
Triacylglycerols Total cholesterol HDL cholesterol LDL cholesterol Glycerol	Lipids
Urea Ammonia	Amino acids
Ammonia Creatinine	Compounds of high phosphoryl transfer potential

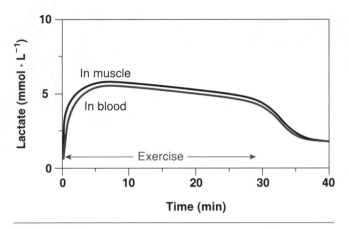

▶ **Figure 15.1** Lactate concentration in moderate-intensity exercise. Muscle and blood lactate concentrations peak a few minutes after the start of exercise at a constant moderate intensity and drop slightly as exercise continues. After the end of exercise, both return to baseline gradually.

we measure it on the spot with an automatic device). If we can take but one blood sample, the preferred time would be 5 min after the end of exercise.

During prolonged exercise at a constant moderate intensity, there is a gradual decline in blood lactate (figure 15.1). If intensity fluctuates, as in ball games, the lactate concentration corresponds roughly to the average intensity.

Apart from the exercise parameters (intensity, duration, and program), the blood lactate response depends on heredity, nutrition, training state, and age (the very factors affecting the choice of energy sources during exercise, as introduced in section 13.4). Here's how.

- *Heredity.* The higher the percentage of type IIA and IIX fibers in the exercising muscles, the higher the blood lactate concentration during exercise at a given absolute intensity.

- *Nutrition.* A carbohydrate-rich diet before or during exercise (or both) at a given intensity raises the blood lactate concentration, since relatively more

carbohydrates are utilized (section 13.11). Therefore a fair comparison of exercising persons in this respect requires the same diet for at least two days before the exercise test. One should take the same precaution when testing an individual at different times during a training period in order to evaluate adaptations to training (see section 15.4).

- *Training state.* Aerobic training lowers the blood lactate concentration during exercise at a given absolute moderate intensity.

- *Age.* Compared to adults, children have lower blood lactate concentrations during exercise at a given relative moderate intensity. Thus, while adults achieve a blood lactate concentration of 4 mmol · L^{-1} at about 70% of $\dot{V}O_2$max, children do so at about 80% of their $\dot{V}O_2$max. Children also have a lower maximal lactate concentration. These differences fade as children reach adolescence and then adulthood.

Because of the plethora of factors affecting the blood lactate concentration and the wide range of values lactate can show, it is a very sensitive and useful marker of the effect of exercise on metabolism. Let's be more specific. We can locate the utility of measuring lactate in three areas:

Estimating the anaerobic lactic capacity

Programming training

Estimating aerobic endurance

15.2 Estimating the Anaerobic Lactic Capacity

A high anaerobic lactic capacity interests athletes in events depending mainly or greatly on the lactate system, such as sprint or middle-distance running and swimming. Because sprint training increases the activity of the enzymes of the lactate system (table 13.5), it has the potential to increase the power of the system and thus the maximal rate of lactate production. However, this potential may not materialize if it "stumbles" upon inhibition of muscle function by a decrease in cytosolic pH. Thus, if a maximal effort elicits a blood lactate concentration that is higher than before, this has to be attributed to at least one of the following factors:

- Increased lactate production
- Increased buffering capacity (consult section 1.9) of the muscles, the blood, or both, which mitigates the drop in muscle pH and delays the development of fatigue
- Faster exit of lactate, along with H^+ (section 9.18), from the contracting muscle fibers

Any of these adaptations may increase performance. In fact, several researchers have found a positive relationship between the peak blood lactate concentration and performance in events such as 400 and 800 m running.

15.3 Programming Training

Success in sport usually depends on a combination of capacities, which the athletes develop through different kinds of training. Often the main factor determining the kind of training is intensity. In such cases, measuring blood lactate helps define the desired intensity.

Programming training based on intensities dictated by blood lactate concentrations is superior to programming training based on heart frequencies because lactate relates directly to muscle metabolism and muscle adaptations. Of course, this approach is less practical and carries a certain cost, as it requires blood sampling and technical equipment for measuring lactate. Thus, one could use lactate to determine training intensities at the beginning of a training program, monitor training through heart frequencies on a daily basis, and resort to lactate periodically (every few weeks) to fine-tune intensities.

How do we establish a relationship between training intensity and blood lactate concentration for an athlete? By constructing a lactate–intensity plot like the one in figure 9.39. How do we do that? By having the athlete perform rather brief (1 to 4 min long) exercise bouts of gradually increasing intensity (for example, from 60% to 100% of maximal intensity in increments of 5% to 10%) and taking blood samples after each bout to measure lactate. The amount of blood required is minimal (some microliters) and may be taken from a fingertip or earlobe. Remember that it is better to take more than one sample after each bout in order not to miss the peak lactate concentration (section 15.1).

Each bout must be performed at a steady intensity throughout, so that the lactate measured in the end reflects this intensity. The time between bouts should ideally be at least one-half hour to permit complete lactate removal from the blood and recovery of the athlete's strength. However, because this may result in an unacceptably prolonged trial, the interval may be cut down to 15 min without seriously affecting the results.

For maximal economy of time and money, tests of only two bouts have been developed. Obviously, one cannot construct an entire lactate–intensity plot with such a test. Nevertheless, one can determine the most useful intensity, which is the one corresponding to a concentration of 4 mmol · L^{-1}, since training around this intensity is considered the most efficient means of developing the aerobic capacity. A two-bout test requires the expertise of the coach in selecting two intensities, the first of which will elicit a lactate concentration slightly below 4 mmol · L^{-1} and the second a lactate concentration slightly above 4 mmol · L^{-1}.

Measuring lactate relative to exercise intensity may also prove useful in defining a training program for a person in the general population who wishes to obtain the most favorable health adaptations possible. Most investigators agree that intensities that hold the blood lactate concentration below 4 mmol · L^{-1} are the most effective in improving aerobic endurance, cardiac function, and the lipidemic profile.

15.4   Estimating Aerobic Endurance

It may at first seem odd how a product of anaerobic metabolism can be useful in estimating aerobic endurance. But if you think about how interlaced the energy systems are (as analyzed in chapter 13), you will realize that there is no oddity here. However, the connection between lactate and aerobic endurance is the opposite of that between lactate and anaerobic capacity: The higher the aerobic endurance, the lower the lactate concentration.

This line of reasoning has been confirmed by many studies, which show a strong relationship between performance in endurance events and the moderate exercise intensity (expressed in absolute terms such as speed) corresponding to a given blood lactate concentration (for example, $4 \text{ mmol} \cdot \text{L}^{-1}$). In simple terms, *a high endurance performance goes with being fast while keeping lactate low.* This connection is explained by the adaptations that accompany endurance training (section 13.12), in particular the increase in the proportion of lipids to carbohydrates that are broken down during moderate-intensity exercise. Because of the decreased carbohydrate breakdown, less lactate is produced in the muscles and less lactate appears in the blood. It is also possible that endurance training increases the rate of disappearance of lactate from the blood. Possible mechanisms include an increase in MCT1 in the plasma membrane of type I muscle fibers (section 9.18) and an increase in the rate of gluconeogenesis in the liver and kidneys.

A lower lactate concentration during exercise at a given intensity means a higher intensity for a given lactate concentration. It is exactly this higher intensity that characterizes the adaptations to aerobic training and the increase in aerobic endurance. In fact, the relevant studies find the intensity that corresponds to a given lactate concentration to be a better predictor of aerobic endurance than $\dot{V}O_2$max. This may be so because the intensity in question increases more rapidly than $\dot{V}O_2$max with endurance training. In addition, performance in an endurance event depends less on $\dot{V}O_2$max and more on how high an intensity an athlete can maintain for a long time.

The rise in the exercise intensity corresponding to a given blood lactate concentration can be used to monitor the adaptations to aerobic training. Indeed, if we construct a lactate–intensity plot before and several weeks after the beginning of a training program, the latter plot will be to the right of the former (figure 15.2). If we do not intend to use the plots to define training intensities but only to assess the improvement in aerobic endurance, it is not necessary to have long intervals between bouts in each test. In fact, there could even be no intervals at all, but instead continuous stages of gradually increasing intensity with a single blood sampling at the end of each stage. In any case, for the plots to be comparable, the test protocol (including the hour of the day) and the examinee's diet during the two days preceding each test must be the same. It is also advisable to avoid hard exercise during the preceding day.

The value of monitoring the adaptations to aerobic training is twofold: It lets us find out whether the training program works, and it lets us upgrade the training intensities in order to push the adaptations forward. Otherwise, there will be no further adaptations. This is the *progression principle* in training.

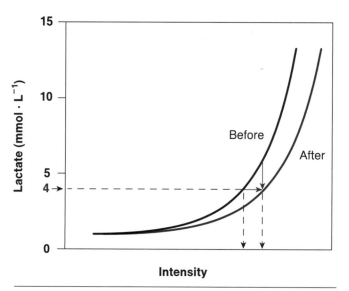

▶ **Figure 15.2**   Effect of endurance training on the lactate–intensity plot. Endurance training shifts the plot to the right. After training, a given blood lactate concentration is achieved at a higher exercise intensity (horizontal colored arrow), while a given intensity corresponds to a lower blood lactate concentration (vertical colored arrow).

15.5   Glucose

Glucose (often termed sugar in biochemistry lab reports) is kept at relatively constant concentrations in the plasma thanks to the opposing actions of glucagon and epinephrine on one hand, and insulin and glucose itself on the other. As analyzed in chapter 9, the former two tend to increase plasma glucose, whereas the latter two tend to decrease it. The importance of measuring glucose lies in the fact that it shows the balance between the mechanisms that raise and lower it.

The reference intervals for the serum glucose concentration are

60 to 100 mg · dL^{-1}, or 3.3 to 5.6 mmol · L^{-1}, in children and adolescents;

74 to 106 mg · dL^{-1}, or 4.1 to 5.9 mmol · L^{-1}, in ages 18 to 60; and

82 to 115 mg · dL^{-1}, or 4.6 to 6.4 mmol · L^{-1}, in ages 60 to 90.

> To convert mg · dL^{-1} to mmol · L^{-1}, divide by 18.

Values above the upper reference limits are characteristic of diabetes mellitus, which is the most common disturbance to glucose homeostasis. Exercise may also upset glucose homeostasis if an imbalance develops between glucose uptake by the exercising muscles and glucose supply by the liver. In case the uptake factor wins out, there is danger of hypoglycemia. Measuring blood glucose during exercise lets us spot such a danger and deter it by modifying the exercise parameters, nutrition, or both.

15.6   Triacylglycerols

Triacylglycerols are present in the plasma as part of lipoproteins (section 10.10). VLDL are the major carriers of triacylglycerols in the postabsorptive state. The reference intervals depend on age and sex (table 15.2). The reference limits, especially the upper ones, tend to increase with age and are generally higher in males than females.

The serum triacylglycerol concentration is of interest mainly because of its relationship with the risk for atherosclerosis (section 10.12). In 2001, the U.S. National Cholesterol Education Program (NCEP) proposed the following limits (in milligrams per deciliter) for the adult population:

< 150, normal

150 to 199, borderline high

200 to 499, high

≥500, very high

▶ **Table 15.2**   Reference Intervals for Triacylglycerol Concentrations in Human Serum

Age (years)	Reference interval (mg · dL^{-1})		Reference interval (mmol · L^{-1})[a]	
	Male	Female	Male	Female
0-9	30-100	35-110	0.34-1.13	0.40-1.24
10-14	32-125	37-131	0.36-1.41	0.42-1.48
15-19	37-148	39-124	0.42-1.67	0.44-1.40
20-24	44-201	36-131	0.50-2.27	0.41-1.48
25-29	46-249	37-144	0.52-2.81	0.42-1.63
30-34	50-266	39-150	0.56-3.01	0.44-1.69
35-39	54-321	40-176	0.61-3.63	0.45-1.99
40-44	55-320	45-191	0.62-3.62	0.51-2.16
45-49	58-327	46-214	0.66-3.69	0.52-2.42
50-54	58-320	52-233	0.66-3.62	0.59-2.63
55-59	58-286	55-262	0.66-3.23	0.62-2.96
60-64	58-291	56-239	0.66-3.29	0.63-2.70
≥65	55-260	60-240	0.62-2.94	0.68-2.71

Note: The reference intervals in this table were obtained after rejection of the lower and upper 5% rather than 2.5% of the values.
[a]To convert mg · dL^{-1} to mmol · L^{-1}, divide by 88.5.

Adapted from *Clinical Guide to Laboratory Tests*, N.W. Tietz, pg. 610, Copyright 1995, with permission from Elsevier.

The serum triacylglycerol concentration is a case in which the usual values differ from the desirable ones. Indeed, as you can see in table 15.2, a part of the reference intervals overlaps with the borderline high and high values. In simple terms, a considerable portion of the population has values that are not desirable. Thus, using the term normal values to describe the reference intervals would be misleading.

Measuring triacylglycerols in the serum of an exercising person is interesting after both acute and, particularly, chronic exercise. As mentioned in section 10.12, regular aerobic exercise can decrease the serum triacylglycerol concentration. Thus, measuring it from time to time contributes to the monitoring of exercising persons' health and helps to assess the effectiveness of training programs in modifying the lipidemic profile.

15.7 Cholesterol

Like triacylglycerols, cholesterol is present in the plasma as part of lipoproteins. LDL and HDL are the main carriers of cholesterol, while VLDL also contain a small amount of it. Table 15.3 shows the reference intervals in relation to age and sex. The upper and lower reference limits tend to increase with age.

The serum cholesterol concentration relates strongly to the risk for atherosclerosis. NCEP has proposed the following limits (mg · dL^{-1}):

< 200, desirable

200 to 239, borderline high

≥240, high

Note that there is a distinction between the reference interval and the desirable values in cholesterol as in triacylglycerols. Again, a large portion of the reference intervals lies within the borderline high and high zones. This is largely due to the prevalent lifestyle in developed countries, which is characterized by overeating, consumption of large quantities of animal fat (rich in cholesterol and saturated fatty acids, both of which raise plasma cholesterol), and physical inactivity.

▶ **Table 15.3** Reference Intervals for Cholesterol Concentrations in Human Serum

Age (years)	Reference interval (mg · dL^{-1})		Reference interval (mmol · L^{-1})[a]	
	Male	Female	Male	Female
5-9	121-203	126-205	3.13-5.26	3.26-5.31
10-14	119-202	124-201	3.08-5.23	3.21-5.21
15-19	113-197	119-200	2.93-5.10	3.08-5.18
20-24	124-218	122-216	3.21-5.65	3.16-5.60
25-29	133-244	128-222	3.45-6.32	3.32-5.75
30-34	138-254	130-230	3.58-6.58	3.37-5.96
35-39	146-270	140-242	3.78-6.99	3.63-6.27
40-44	151-268	147-252	3.91-6.94	3.81-6.53
45-49	158-276	152-265	4.09-7.15	3.94-6.87
50-54	158-277	162-285	4.09-7.18	4.20-7.38
55-59	156-276	172-300	4.04-7.15	4.46-7.77
60-64	159-276	172-297	4.12-7.15	4.46-7.69
65-69	158-274	171-303	4.09-7.10	4.43-7.85
≥70	144-265	173-280	3.73-6.87	4.48-7.25

Note: The reference intervals were obtained after rejection of the lower and upper 5% of the values.

[a]To convert mg · dL^{-1} to mmol · L^{-1}, divide by 38.6.

Adapted from *Clinical Guide to Laboratory Tests*, N.W. Tietz, pg. 130, Copyright 1995, with permission from Elsevier.

15.8 HDL Cholesterol

As mentioned in section 10.13, apart from total cholesterol, of interest is its distribution between LDL and HDL, the "bad" and "good" cholesterol, respectively. Of the two, HDL cholesterol is easier to measure. The reference intervals are shown in table 15.4. Adult females have higher upper and lower reference limits than males.

NCEP considers HDL cholesterol concentrations under 40 mg · dL^{-1} to be low and concentrations at or over 60 mg · dL^{-1} to be high. All lower reference limits in table

15.4 are below 40 mg · dL⁻¹, which means that a considerable portion of the population has undesirably low HDL cholesterol values. This is mainly due to the physical inactivity of people in modern Western societies.

15.9 LDL Cholesterol

The measurement of LDL cholesterol requires equipment that is not available in most laboratories performing routine tests. Thus, we usually calculate this parameter by subtracting HDL cholesterol and VLDL cholesterol from total cholesterol. VLDL cholesterol is calculated by dividing the triacylglycerol concentration by 5 if both are measured in mg · dL⁻¹, or 2.18 if both are measured in mmol · L⁻¹. For this approximation to be acceptable, the triacylglycerol concentration must not exceed 400 mg · dL⁻¹, or 4.52 mmol · L⁻¹. Otherwise, LDL cholesterol has to be measured directly.

Table 15.5 contains the reference intervals for LDL cholesterol. The upper and lower reference limits tend to increase with age.

NCEP has proposed the following limits (mg · dL⁻¹) with regard to LDL cholesterol:

< 100, optimal

100 to 129, near optimal/above optimal

130 to 159, borderline high

160 to 189, high

≥190, very high

As with the lipids discussed in the previous sections, a large part of the reference intervals for LDL cholesterol lies in the undesirable zones, the reasons being the same as those regarding total cholesterol.

▶ **Table 15.4** Reference Intervals for HDL Cholesterol Concentrations in Human Serum

Age (years)	Reference interval (mg · dL⁻¹) Male	Female	Reference interval (mmol · L⁻¹)ª Male	Female
5-9	38-75	36-73	0.98-1.94	0.93-1.89
10-14	37-74	37-70	0.96-1.92	0.96-1.81
15-19	30-63	35-74	0.78-1.63	0.91-1.92
20-24	30-63	33-79	0.78-1.63	0.85-2.05
25-29	31-63	37-83	0.80-1.63	0.96-2.15
30-34	28-63	36-77	0.73-1.63	0.93-1.99
35-39	29-62	34-82	0.75-1.61	0.88-2.12
40-44	27-67	34-88	0.70-1.74	0.88-2.28
45-49	30-64	34-87	0.78-1.66	0.88-2.25
50-54	28-63	37-92	0.73-1.63	0.96-2.38
55-59	28-71	37-91	0.73-1.84	0.96-2.36
60-64	30-74	38-92	0.78-1.92	0.98-2.38
65-69	30-75	35-96	0.78-1.94	0.91-2.49
≥70	31-75	33-92	0.80-1.94	0.85-2.38

Note: The reference intervals were obtained after rejection of the lower and upper 5% of the values.
ªTo convert mg · dL⁻¹ to mmol · L⁻¹, divide by 38.6.
Adapted from *Clinical Guide to Laboratory Tests*, N.W. Tietz, pgs. 334-335, Copyright 1995, with permission from Elsevier.

▶ **Table 15.5** Reference Intervals for LDL Cholesterol Concentrations in Human Serum

Age (years)	Reference interval (mg · dL⁻¹) Male	Female	Reference interval (mmol · L⁻¹)ª Male	Female
5-9	63-129	68-140	1.63-3.34	1.76-3.63
10-14	64-133	68-136	1.66-3.45	1.76-3.52
15-19	62-130	59-137	1.61-3.37	1.53-3.55
20-24	66-147	57-159	1.71-3.81	1.48-4.12
25-29	70-165	71-164	1.81-4.27	1.84-4.25
30-34	78-185	70-156	2.02-4.79	1.81-4.04
35-39	81-189	75-172	2.10-4.90	1.94-4.46
40-44	87-186	74-174	2.25-4.82	1.92-4.51
45-49	97-202	79-186	2.51-5.23	2.05-4.82
50-54	89-197	88-201	2.31-5.10	2.28-5.21
55-59	88-203	89-210	2.28-5.26	2.31-5.44
60-64	83-210	100-224	2.15-5.44	2.59-5.80
65-69	98-210	92-221	2.54-5.44	2.38-5.73
≥70	88-186	96-206	2.28-4.82	2.49-5.34

Note: The reference intervals were obtained after rejection of the lower and upper 5% of the values.
ªTo convert mg · dL⁻¹ to mmol · L⁻¹, divide by 38.6.
Adapted from *Clinical Guide to Laboratory Tests*, N.W. Tietz, pgs. 404-405, Copyright 1995, with permission from Elsevier.

15.10 Recapping Cholesterol

The purpose of measuring cholesterol and its distribution among lipoproteins in an athlete, or an exercising person in general, is to protect his or her health. Although one would expect the nutrition of athletes to be more healthful than that of the general population, many athletes consume large quantities of fat-rich animal foods such as red meat, dairy products, and eggs in order to benefit from their high protein, iron, and calcium contents. This practice, however, may raise serum cholesterol. Additionally, the use of anabolic androgenic steroids (a regrettably frequent practice among athletes, discussed in section 16.6) decreases HDL cholesterol. Both of these changes in the lipidemic profile are detrimental to health and justify our increased interest in the cholesterol parameters.

Table 15.6 summarizes the information presented in the previous four sections on the desirable serum cholesterol and triacylglycerol concentrations.

Apart from concentration values, the ratio of total cholesterol to HDL cholesterol and the ratio of LDL cholesterol to HDL cholesterol are often used to estimate the risk for atherosclerosis. These are called **atherogenic indices,** and we want them to be as low as possible. The desirable values are as follows.

$$\frac{\text{total cholesterol}}{\text{HDL cholesterol}} < 5$$

$$\frac{\text{LDL cholesterol}}{\text{HDL cholesterol}} < 4$$

▶ **Table 15.6** Desirable, Borderline, and Undesirable Serum Lipid Concentrations in Adults (in mg · dL^{-1})

	Total cholesterol	LDL cholesterol	HDL cholesterol	Triacyl-glycerols
Desirable	<200	<130	≥60	<150
Borderline	200-239	130-159	40-59	150-199
Undesirable	≥240	≥160	<40	≥200

As with triacylglycerols, chronic changes in cholesterol because of exercise are more interesting than acute ones. Remember that regular aerobic exercise increases HDL cholesterol and may decrease total and LDL cholesterol if it reduces body mass or fat mass (section 10.13). Thus, it is useful to measure these parameters frequently during a training program.

15.11 Glycerol

Plasma glycerol originates primarily from the complete hydrolysis of triacylglycerols in the cells. The reference intervals for the plasma glycerol concentration are 0.06 to 0.23 mmol · L^{-1} in ages 3 to 10 and 0.03 to 0.19 mmol · L^{-1} in ages 11 to 80. Glycerol is not measured in the serum. The acceleration of lipolysis during exercise raises the rate of appearance of glycerol in the plasma and the plasma glycerol concentration, which can reach 0.6 mmol · L^{-1}.

Although produced along with fatty acids upon triacylglycerol hydrolysis, glycerol exhibits a simpler kinetics in the plasma during exercise, that is, a relatively steady increase for as long as exercise lasts. Because of this, it is considered a better index of the lipolytic rate than fatty acids, which do not respond in a steady manner (figure 10.14). One can thus estimate how strong the lipolytic stimulus is by measuring plasma glycerol before, during, and after exercise.

15.12 Urea

Urea is the main product of nitrogen metabolism in humans (section 11.5). The reference intervals for the serum urea concentration are

11 to 39 mg · dL^{-1}, or 1.8 to 6.5 mmol · L^{-1}, in children and adolescents;

13 to 43 mg · dL^{-1}, or 2.2 to 7.2 mmol · L^{-1}, in ages 18 to 60; and

17 to 49 mg · dL^{-1}, or 2.8 to 8.2 mmol · L^{-1}, in ages 60 to 90.

To convert mg · dL^{-1} to mmol · L^{-1}, divide by 6.

Certain laboratories may report *urea nitrogen* instead of urea. Urea nitrogen has different reference intervals in terms of mg · dL^{-1}. These are

5 to 18 mg · dL^{-1} in children and adolescents,

6 to 20 mg · dL^{-1} in ages 18 to 60, and

8 to 23 mg · dL^{-1} in ages 60 to 90.

The serum urea concentration is affected by nutrition, in particular the amount of proteins consumed. If protein intake is low, the urea concentration is also low. If protein intake is high through ingestion of either the normal food or supplements, as is the habit of many athletes, the urea concentration rises because a large part of the proteins consumed are degraded. Thus, the measurement of serum urea serves to protect athletes against nutritional excesses.

The serum urea concentration also rises when the kidneys do not work properly because this hinders the removal of urea from the blood. Hence, the serum urea concentration serves as an index of kidney function.

15.13 Ammonia

The main sources of ammonia in the human body are AMP deamination (section 8.5) and amino acid deamination (section 11.3). This dual origin justifies the double presence of ammonia in table 15.1, that is, as relevant to the metabolism of both amino acids and compounds of high phosphoryl transfer potential (ATP to ADP to AMP). Ammonia is not measured in serum but in plasma, the reference interval being 19 to 60 μg · dL^{-1}, or 11 to 35 μmol · L^{-1}.

To convert μg · dL^{-1} to μmol · L^{-1}, divide by 1.7.

The plasma ammonia concentration rises during exercise and relates to the lactate concentration. However, ammonia does not increase as much as lactate does, in either absolute or relative terms: It reaches 130 μmol · L^{-1} at the most. Its measurement helps to estimate the degree of AMP and amino acid deamination. If exercise is hard, ammonia comes primarily from AMP; if exercise is of moderate intensity and prolonged, ammonia comes primarily from amino acids.

15.14 Creatinine

Creatinine is the product of creatine dehydration.

(equation 15.1)

Meat is the main dietary source of creatine.

The arrow indicates how the creatine molecule closes to form a ring. Creatinine is derived from tissue (mainly muscle) creatine and dietary creatine. The conversion of creatine to creatinine prepares it for removal through the kidneys to the urine. Table 15.7 contains the reference intervals for the serum creatinine concentration.

▶ **Table 15.7** Reference Intervals for Creatinine Concentrations in Human Serum

	Reference interval (mg · dL⁻¹)		Reference interval (μmol · L⁻¹)ᵃ	
	Male	**Female**	**Male**	**Female**
Children	0.3-0.7	0.3-0.7	27-62	27-62
Adolescents	0.5-1.0	0.5-1.0	44-88	44-88
18-60 years	0.9-1.3	0.6-1.1	80-115	53-97
60-90 years	0.8-1.3	0.6-1.2	71-115	53-106

ᵃTo convert mg · dL⁻¹ to μmol · L⁻¹, multiply by 88.4.

Adapted from *Clinical Guide to Laboratory Tests*, N.W. Tietz, pg. 186, Copyright 1995, with permission from Elsevier.

Like urea, creatinine serves as an index of renal function, as its plasma concentration rises if the kidneys do not remove it from the blood. However, a healthy athlete may have elevated creatinine for two other reasons.

- Because athletes usually have a larger muscle mass than the general population, they have more creatine and CP, and may thus form more creatinine. In addition, hard training causing muscle fiber damage may augment creatinine release in the plasma.

- Many athletes consume high amounts of meat or take creatine supplements. Both practices raise plasma creatinine.

Because of these particularities, the serum creatinine concentrations of many athletes approach the upper reference limit or may even exceed it without any other indication of a kidney problem.

Problems and Critical Thinking Questions

1. Draw a lactate–intensity plot for a sprinter and an endurance athlete. How do the two plots differ?

2. What is the serum LDL cholesterol concentration of an athlete whose total cholesterol is 220 mg · dL⁻¹, HDL cholesterol is 50 mg · dL⁻¹, and triacylglycerols are 100 mg · dL⁻¹?

3. Does the athlete in the previous problem have a healthy lipidemic profile?

4. What would you measure in the plasma to estimate the lipolytic rate during exercise?

5. Which substances would you measure in the serum to estimate renal function? Do these substances rise or fall when there is a problem?

6. Which common dietary practices of athletes could produce alarming values of the substances mentioned in the previous problem?

CHAPTER 16

Enzymes and Hormones

In this final chapter of part IV, and the book, we will examine how we can assess exercising persons by measuring enzymes (the catalysts of our chemical reactions) and hormones (our chemical messengers). The measurement of these substances lets one estimate the integrity of vital organs and the balance between catabolism and anabolism. At the end, I explain why I chose not to discuss certain substances in part IV.

16.1 Enzymes

We will examine three enzymes of interest in terms of assessing an exercising person: creatine kinase and two aminotransferases. These enzymes abound in organs such as the muscles and liver, and they are not destined to be secreted in the blood. Nevertheless, small amounts of them are detected in the plasma because of leakage from the cells that contain them. By measuring these amounts we can estimate the integrity of the organs the enzymes come from, in the sense that the more intact an organ is, the less of its characteristic enzyme will be in the plasma. Conversely, if part of the organ is damaged, the enzyme will have a high plasma concentration.

We measure the amount of an enzyme in a sample not by mass (say, in milligrams or millimoles), as we measure other biochemical parameters, but by enzyme activity (section 3.15), which is easier to determine. The oldest and most frequently used units of enzyme activity are called simply *units* and symbolized as U. Because different enzymes catalyze different reactions, the definition of U is not standard. Nevertheless, for many enzymes—including the three that we will examine—1 U is the amount that catalyzes the conversion of 1 μmol of substrate to product within 1 min. The unit of enzyme activity in the Système Internationale is the *katal;* it is symbolized as Kat and defined for all enzymes as the amount that catalyzes the conversion of 1 mol of substrate to product within 1 s.

A final introductory note is in order. Because the rate of enzyme reactions (and chemical reactions, in general) depends on temperature, a certain amount of an enzyme has different activities at different temperatures. It is therefore necessary, when reporting values of enzyme activity, to also report the temperature at which the measurement was performed. This is usually 37° C, the average temperature of the human body.

16.2   Creatine Kinase

Creatine kinase catalyzes the interconversion of CP and ATP according to equations 8.4 and 13.6. The enzyme is present in almost all tissues but is highest in the three kinds of muscle (skeletal, heart, and smooth) and in the brain. Different CK isoforms predominate in these tissues (section 8.3); this lets us detect the origin of CK in the plasma. The reference intervals for the serum CK concentration at 37° C are

> To convert $U \cdot L^{-1}$ to $\mu Kat \cdot L^{-1}$, divide by 60.

38 to 174 $U \cdot L^{-1}$, or 0.63 to 2.90 $\mu Kat \cdot L^{-1}$, in males and

26 to 140 $U \cdot L^{-1}$, or 0.43 to 2.33 $\mu Kat \cdot L^{-1}$, in females.

Over 94% of serum CK is of the CK3 (CK-MM) isoform, which predominates in skeletal muscle.

The serum CK concentration rises when an organ that contains the enzyme is damaged. Two typical cases are the following:

Acute myocardial infarction, in which CK2 (CK-MB), the heart isoform, rises

Myopathies, in which CK3 rises

Closer to our interests, serum CK increases in healthy persons after exercise, particularly eccentric exercise, because of increased muscle fiber damage. The serum CK concentration is probably the best biochemical marker of muscle fiber damage. It rises slowly after exercise and usually peaks after one or two days; then it declines even more slowly toward baseline (figure 16.1).

A frequent mistake is to associate the *serum* CK concentration with the power of the ATP-CP system. However, the two parameters are not related in any way. To estimate the power of the ATP-CP system one has to measure the *muscle* CK concentration.

Athletes, as a rule, have higher serum CK concentrations than nonathletes because of the regular strain imposed by training on their muscles. In fact, CK is one of the parameters affected the most by training, and values well in excess of the upper reference limits listed earlier (which apply to the general population) are very common among athletes. Such values should not be taken as pathologic without any other evidence of a health problem.

On the other hand, the increase in the serum CK concentration after a set exercise is lower in athletes than nonathletes thanks to what has been termed the **repeated-bout effect.** That is, repetition of an exercise, particularly eccentric exercise, after several days or even weeks causes less muscle fiber damage than the previous exercise. There is no agreement among investigators as to the explanation for this adaptation. The following are the most probable mechanisms:

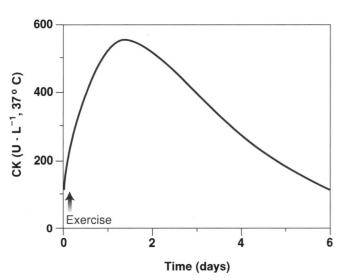

▶ **Figure 16.1**   Serum creatine kinase after eccentric exercise. Eccentric exercise causes muscle fiber damage, which results in increased serum CK concentrations for several days afterward.

- *The neural hypothesis.* Repetition of exercise increases the number of muscle fibers that are activated; as a result, more muscle fibers share a certain load, and damage is less.

- *The connective tissue hypothesis.* The rupture of connective tissue elements during exercise leads to their reconstruction, their increase during recovery, or both, leading to increased tolerance to damage.

- *The cellular hypothesis.* The number of sarcomeres along the myofibrils increases, which mitigates sarcomere strain and damage as the muscle is stretched.

Regardless of the actual cause of the repeated-bout effect, its existence offers the opportunity to estimate muscle tolerance to repeated bouts of exercise through regular measurement of the serum CK concentration.

16.3 Aminotransferases

Aminotransferases catalyze the transfer of an amino group from an α-amino acid to an α-keto acid. This transfer is necessary for amino acid metabolism, as we saw in chapter 11. Two are the most widely known aminotransferases.

- *Alanine aminotransferase* (ALT) catalyzes reactions 11.4 and 11.7. It is also called *glutamate-pyruvate transaminase* (GPT). The enzyme abounds in the liver and kidneys, while smaller amounts are present in the skeletal muscles and heart.

- *Aspartate aminotransferase* (AST) catalyzes reaction 11.8. It is also called *glutamate-oxaloacetate transaminase* (GOT). The enzyme abounds in the liver, heart, skeletal muscles, and erythrocytes.

The two aminotransferases are sometimes abbreviated as SGPT and SGOT to stress that they are measured in the serum.

Table 16.1 contains the reference intervals for the serum concentrations of the two aminotransferases.

The serum ALT and AST concentrations are low unless the organs that contain them are harmed, in which case the contents of the damaged cells leak to the plasma. Thus, the two enzymes, particularly ALT, rise in liver disease (such as cirrhosis and hepatitis) and serve as indices of the integrity of this vital organ.

The serum aminotransferase concentrations also rise when an athlete uses anabolic androgenic steroids that damage the liver. Finally, since aminotransferases are present in muscle, their serum concentrations increase

▶ **Table 16.1** Reference Intervals for Aminotransferase Concentrations in Human Serum at 37° C

Age (years)	Reference interval (U · L⁻¹)		Reference interval (µKat · L⁻¹)ᵃ	
	Male	Female	Male	Female
ALT				
1-60	10-40	7-35	0.17-0.67	0.12-0.58
60-90	13-40	10-28	0.22-0.67	0.17-0.47
AST				
2-60	15-40	13-35	0.25-0.67	0.22-0.58
60-90	19-48	9-36	0.32-0.80	0.15-0.60

ᵃTo convert U · L⁻¹ to µKat · L⁻¹, divide by 60.

Adapted from *Clinical Guide to Laboratory Tests*, N.W. Tietz, pgs. 20 and 76, Copyright 1995, with permission from Elsevier.

after hard exercise because of muscle fiber damage. AST increases more than ALT, as its muscle concentration is higher. However, the rise in either aminotransferase with exercise is lower than the rise in CK.

16.4 Steroid Hormones

In part III, we met hormones as signal-transducing compounds, through which the organism achieves the concerted response of different organs to situations such as exercise. Measuring the serum concentration of a hormone helps to estimate the strength of the signal transduction pathway(s) that it takes part in. Hormone concentrations are much lower than those of the metabolites we considered in the previous chapter, as will become evident from the reference intervals that follow. This makes hormone action even more amazing.

We will examine two hormones that are useful in the biochemical assessment of exercising persons: **cortisol** and **testosterone.** Both are **steroid hormones.** This family of compounds encompasses about 30 members, which are derivatives of cholesterol and are divided into five categories based on structure and biological function:

1. *Glucocorticoids,* acting primarily to increase the blood glucose concentration

2. *Mineralocorticoids,* having as their basic mission the maintenance of the water and salt balance in the body

3. *Estrogens,* the main female hormones

4. *Progestins,* or *progestagens,* which aid in the initiation and maintenance of pregnancy

5. *Androgens,* the main male hormones

Cortisol is the major glucocorticoid, and testosterone is the major androgen. Their molecules (figure 16.2) bear the characteristic four rings of cholesterol but have fewer carbons: Cortisol has 21 and testosterone has 19, whereas cholesterol has 27.

The two hormones share a number of other features as well. Their production is controlled by the hypothalamus, which synthesizes two peptide hormones, *corticotropin-releasing hormone* (CRH) and *gonadotropin-releasing hormone* (GnRH). The two hormones are secreted to the pituitary gland, which lies right underneath the hypothalamus in the brain (figure 16.3). There CRH stimulates the

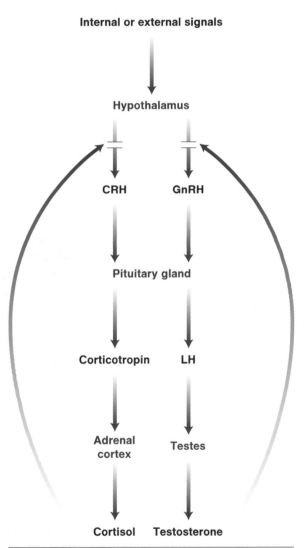

Internal or external signals

Hypothalamus

CRH **GnRH**

Pituitary gland

Corticotropin **LH**

Adrenal cortex **Testes**

Cortisol **Testosterone**

▶ **Figure 16.3** Control of steroid hormone production. The hypothalamus controls the development and function of many organs through the hormones it secretes to the pituitary gland in response to internal and external signals. Two of the hormones, corticotropin-releasing hormone (CRH) and gonadotropin-releasing hormone (GnRH), provoke the secretion of corticotropin and luteinizing hormone (LH, a gonadotropin) by the pituitary gland. These, in turn, stimulate the synthesis of cortisol by the adrenal cortex and testosterone by the testes. The plasma cortisol and testosterone concentrations are controlled by feedback inhibition (upward arrows): If they get too high, CRH and GnRH secretion is inhibited.

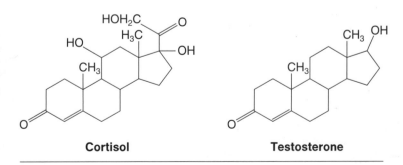

Cortisol **Testosterone**

▶ **Figure 16.2** Cortisol and testosterone. Compare the structure of the two steroid hormones to that of cholesterol, their parent compound (figure 5.15).

secretion of *corticotropin,* and GnRH stimulates the secretion of two *gonadotropins* to the plasma. All three *tropins* are proteins in nature.

Corticotropin then binds to receptors in the plasma membrane of target cells in the *adrenal cortex* (the outer layer of the adrenal glands) and stimulates cortisol synthesis. The gonadotropins, through their own receptors, control the development and operation of the gonads (the testes and ovaries). Their individual names are *follicle-stimulating hormone* (FSH) and *luteinizing hormone* (LH). Testosterone synthesis and secretion are controlled by LH, which also controls testis development. FSH, on the other hand, augments sperm production.

The plasma cortisol and testosterone concentrations are controlled by a reciprocal communication of the brain with the adrenals and testes. When the cortisol concentration drops, the hypothalamus elevates CRH secretion, resulting in increased corticotropin and cortisol secretion. If the blood cortisol rises excessively, the hypothalamus limits CRH secretion, resulting in decreased corticotropin and cortisol secretion. Likewise, when the blood testosterone concentration falls, the hypothalamus raises GnRH secretion, causing increased LH and testosterone secretion. If the blood testosterone rises excessively, the hypothalamus curbs GnRH secretion, causing decreased LH and testosterone secretion.

Steroid hormones differ from the hormones that we have encountered thus far in that they cross the plasma membrane of target cells thanks to the amphipathic character they share with cholesterol. Once inside the cells, they bind to receptors in the nucleus, which, in addition, are DNA-binding proteins (figure 16.4). Hormone attachment changes the conformation of the receptors, resulting in the induction or repression of certain genes and modification of gene expression.

Let's examine now the usefulness of measuring cortisol and testosterone in exercising persons.

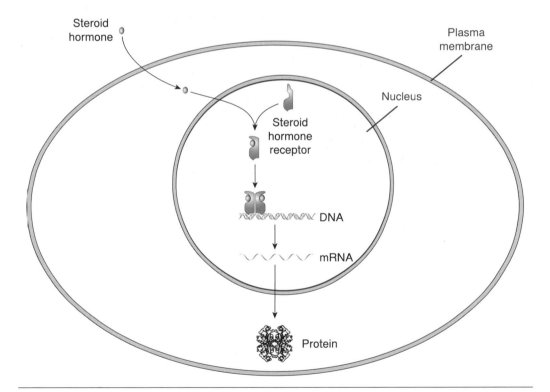

▶ **Figure 16.4** Steroid hormone action. Steroid hormones cross the plasma membrane of target cells to meet their receptors in the nucleus. Binding of the hormone to the receptor changes its shape. The steroid–receptor complex, usually as a dimer, binds to DNA and alters the expression of specific genes, resulting in modified mRNA and protein concentrations.

16.5   Cortisol

The serum cortisol concentration displays a distinctive fluctuation over the 24 h of the day (called *diurnal variation*): It is high in the morning and low in the evening. The reference intervals are

To convert μg · L⁻¹ to nmol · L⁻¹, multiply by 2.76.

50 to 230 μg · L⁻¹, or 138 to 635 nmol · L⁻¹, at 8 a.m. and

30 to 160 μg · L⁻¹, or 83 to 442 nmol · L⁻¹, at 4 p.m.

Cortisol and other glucocorticoids enhance gluconeogenesis, glycogen synthesis, and protein synthesis in the liver, and they raise the plasma glucose concentration. In muscle, they have an opposite effect on protein metabolism, causing proteolysis. At high concentrations, they suppress the immune system. This is why glucocorticoids such as *cortisone* are used as anti-inflammatory and antiallergic drugs. Athletes may use these drugs to counter the inflammation caused by injury and be able to compete. The WADA considers such practice unacceptable and prohibits the use of glucocorticoids and corticotropin.

Cortisol secretion by the adrenal cortex increases in hypoglycemia, obesity, physical stress, and mental stress. The plasma cortisol concentration increases in exercise exceeding 60% of $\dot{V}O_2$max in intensity, whereas it decreases if exercise intensity is below 50% of $\dot{V}O_2$max. The plasma cortisol concentration also increases after hard resistance exercise. Training mitigates the increase caused by a set exercise bout.

Training also affects the plasma cortisol concentration at rest. Athletes usually have higher resting concentrations than nonathletes. In addition, sprinters and weightlifters tend to have higher concentrations than endurance runners, probably because of the high intensities at which sprinters and strength athletes train. This finding raises the possibility that the resting concentration in an athlete depends on the phase of the training period he or she is in (high intensity vs. low intensity).

On the basis of all the information just presented, measuring cortisol at rest may aid in estimating physical or mental stress, while measuring cortisol after exercise may show how the organism receives a particular load. Because cortisol causes proteolysis in muscle, high concentrations of it are undesirable.

16.6   Testosterone

Testosterone is synthesized and secreted primarily by the *Leydig cells* in the testes. The adrenal cortex and ovaries also produce small amounts of testosterone. Its serum concentration is heavily influenced by sex and biological age. Table 16.2 shows the reference intervals.

Testosterone and other androgens exert a multitude of actions on the organism. They are responsible for the development and maintenance of the organs of the male reproductive system, including the testes, prostate, epididymides, and penis. They cause sperm production. They are necessary for the development and maintenance of the secondary sex characteristics of the male, such as deep voice and

▶ **Table 16.2**   Reference Intervals for Testosterone Concentration in Human Serum

	Reference interval (μg · L⁻¹)		Reference interval (nmol · L⁻¹)[a]	
	Male	**Female**	**Male**	**Female**
Children	0.03-0.3	0.02-0.2	0.10-1.04	0.07-0.69
Tanner stage 1[b]	0.02-0.23	0.02-0.1	0.07-0.80	0.07-0.35
Tanner stage 2	0.05-0.7	0.05-0.3	0.17-2.43	0.17-1.04
Tanner stage 3	0.15-2.8	0.1-0.3	0.52-9.72	0.35-1.04
Tanner stage 4	1.05-5.45	0.15-0.4	3.64-18.91	0.52-1.39
Tanner stage 5	2.65-8	0.1-0.4	9.20-27.76	0.35-1.39
Adults	2.8-11	0.15-0.7	9.72-38.17	0.52-2.43
After menopause		0.08-0.35		0.28-1.21

[a]To convert μg · L⁻¹ to nmol · L⁻¹, multiply by 3.47.

[b]Tanner stages are numbers on a scale rating the sexual maturation of boys and girls during adolescence from 1 (initial) to 5 (final) based on the development of secondary sex characteristics such as pubic hair, genital size, breast bud, and voice change.

Adapted from *Clinical Guide to Laboratory Tests*, N.W. Tietz, pgs. 278-579, Copyright 1995, with permission from Elsevier.

hairiness. They determine aggressiveness and libido in the male. They have anabolic effects on the skeleton, skeletal muscles, and skin, as they promote protein synthesis and curb proteolysis. Finally, they increase erythropoietin synthesis in the kidneys and the responsiveness of immature bone marrow cells to the hormone, thus promoting erythropoiesis.

Natural androgens and a wide variety of synthetic analogs termed **anabolic androgenic steroids (AAS)** are apparently the most widely used doping substances as a result of their positive effects on muscle mass and strength. The serum testosterone concentration of a man using AAS (except for testosterone itself) decreases because these drugs trick the hypothalamus into sensing an excessive plasma testosterone concentration and suppressing GnRH secretion (figure 16.3). This results in decreased gonadotropin secretion by the pituitary gland and thus decreased testosterone synthesis by the testes. Moreover, since the gonadotropins are needed for testis development and sperm production, their shortage results in testicular atrophy and the suppression or even disruption of sperm production. Gonadotropins are also banned as doping substances.

The serum testosterone concentration usually rises with exercise in men and, to a lesser extent, in women. The magnitude of the rise depends on exercise intensity and duration. Efforts that are shorter than 1 min, regardless of intensity, do not appear to modify the testosterone concentration significantly. If exercise lasts longer, the higher the intensity, the larger the increase in testosterone is. When exercise lasts longer than 1 h, the testosterone concentration begins to decrease and may drop below baseline after 3 h. A similar drop is seen during recovery from aerobic exercise.

The chronic effect of exercise on the serum testosterone concentration is less clear than the acute effect, but intensity and duration also appear to play a role. There are reports of endurance athletes having lower resting values than nonathletes, and of weightlifters having higher resting values than nonathletes. As with cortisol, the resting testosterone concentration may change with the phase of the training period. Therefore, one has to consider this when comparing athletes and nonathletes.

Because testosterone promotes protein synthesis and curbs proteolysis in muscle, high concentrations of it are desirable. There is still considerable uncertainty about what exercise is the most effective in eliciting high testosterone concentrations. This may well depend on the individual characteristics of an athlete, making it worthwhile to measure testosterone in training.

16.7   Overtraining

Hard-training athletes may sometimes exceed their natural capabilities because of overestimating them in their desire to achieve rapid adaptations and boost performance. Then a series of symptoms and signs may appear, including a drop in performance, chronic fatigue, heart function disturbances, susceptibility to infections, and slow healing of injuries. The terms *overtraining* and *overtraining syndrome* are used to describe this condition. A related term is *overreaching*, which refers to a milder condition that can be reversed within one or two weeks through reduction of the training load. By contrast, reversing overtraining requires rest for several weeks or even months.

For decades, many investigators have directed their efforts toward, on the one hand, diagnosing overtraining and, on the other hand, detecting the risk of overtraining early enough to prevent it. The efforts have met with only limited success because overtraining is a complex condition involving a multitude of factors. In the context of these efforts, biochemical assessment has also been recruited through the search for biochemical markers of overtraining.

One of the first markers to be proposed was the serum cortisol-to-testosterone ratio at rest. Several researchers observed an increased ratio in athletes when they became overtrained. An undesirable effect of this increase may be the stimulation of proteolysis and suppression of protein synthesis in muscle. In addition, the high cortisol concentration may suppress the immune system, which may explain the increased susceptibility to infections and the slow healing of injuries in overtraining.

Other researchers, however, have found no correlation between overtraining and the cortisol-to-testosterone ratio. Nevertheless, measuring the ratio regularly could indicate the balance between catabolic and anabolic processes and suggest probable required changes in the training program of the athlete. Several studies have shown a positive relationship between, on the one hand, the testosterone concentration, the change in testosterone concentration, or the change in the cortisol-to-testosterone ratio and, on the other hand, performance parameters such as maximal strength and speed.

An alternative marker of overtraining may be a diminished cortisol response to exercise because of impaired hypothalamic or pituitary function. Investigators have reported lower than normal increases in the serum cortisol concentration of overtrained athletes in response to exhaustive exercise.

16.8 Epilogue

In part IV, I brought together almost two dozen blood parameters that I consider the most useful in assessing persons who exercise. I left out a number of parameters that offer information of limited value regarding the response of the organism to exercise, either because exercise does not affect them significantly or because other parameters that I have presented provide the same information more accurately. Thus, I have omitted the following:

- Fatty acids, which could serve as indices of lipolysis. Glycerol is a better index of lipolysis.

- Proteins that abound in muscle, such as myoglobin, aldolase (the enzyme catalyzing the fourth reaction of glycolysis), and lactate dehydrogenase. Although the serum concentration of these proteins may serve as a marker of muscle fiber damage, CK is a more sensitive marker.

- Steroid hormones other than cortisol and testosterone, as well as other hormones such as IGF1, thyroid hormones, and leptin (a hormone secreted by the adipocytes). Although all of these hormones play pivotal roles in a number of bodily functions, exercise does not usually change their blood concentrations significantly.

- Minerals such as sodium, potassium, magnesium, and calcium, all of which are important for the neural and muscular processes of movement (see part II). Exercise does not affect their blood concentrations to a very great extent (except for a transient rise in [K^+]) because all are subject to powerful homeostatic mechanisms. Moreover, the serum Mg^{2+} and Ca^{2+} concentrations are not reliable markers of their adequacy in the body unless their intake through the food is very low.

In addition, I have not examined parameters whose measurement, although useful, is particularly demanding technically. Such parameters include the following:

- Catecholamines, which are useful in estimating the stimulation of the sympathoadrenergic system. The stress caused by piercing the skin for blood sampling increases the blood concentrations of the catecholamines; thus blood sampling has to be performed via insertion of a catheter into a vein followed by a half-hour wait.

- Growth hormone, an important anabolic hormone secreted by the pituitary gland (although less important than IGF1 when it comes to muscle growth). Its secretion is pulsatile (that is, in waves at intervals shorter than 1 h), which necessitates multiple blood samplings for reliable results to be obtained.

My purpose in this critical review of the value of measuring certain parameters was to give a clearer picture of what the biochemical assessment of exercising persons can offer. Compared to other ways of assessing exercising persons, for example physiologically, biomechanically, and psychologically, biochemical assessment is more expensive because of the high cost of chemical analysis. Moreover, to be as beneficial as possible, it must be regular and programmed. In particular, athletes subject to high training loads and frequent changes in training program should be assessed biochemically at least four times a year. It is therefore imperative to avoid meaningless expenses for analyses of limited utility. On the other hand, the cost of the biochemical assessment is a small fraction of the amounts usually spent in competitive sport. Therefore, sport organizations should not hesitate to bear such costs, realizing that by doing so they are investing in their own future.

Problems and Critical Thinking Questions

1. Which blood test could show the burden of a training session on the muscle system of an exercising person?

2. Which blood test could show the integrity of the liver?

3. Which blood test could show the presence of physical or mental stress?

4. Which serum parameters, of the ones discussed in all of part IV, could be affected by AAS use?

Part IV Summary

The biochemical assessment of persons who exercise can offer valuable information to the benefit of their health and performance. The biochemical assessment is usually performed via the measurement of substances in the blood, serum, or plasma, and through comparison of the concentrations of these substances to appropriate reference intervals or values from previous tests of the examinee. Of the hundreds of biochemical parameters that are measured, I presented those that I consider the most useful in assessing exercising persons. I divided them into four classes: parameters of the iron status, metabolites, enzymes, and hormones.

Iron Status

The iron status is intimately linked to health and aerobic capacity. Of the parameters in this class, blood hemoglobin shows the capacity to carry oxygen; serum iron shows the circulating amount of iron; the total iron-binding capacity shows the maximal capacity of the plasma to carry iron; the transferrin saturation shows how full the iron-carrying system is; and the soluble transferrin receptor and ferritin show the amount of stored iron. Hematological parameters such as the hematocrit and erythrocyte count complete the picture of the iron status, while the leukocyte and platelet counts inform us about the state of the immune and blood clotting systems, respectively. Iron deficiency may lead to anemia, which can be prevented if iron deficiency is diagnosed early enough and measures are taken to reverse it. Athletes do not seem to run a higher risk for iron deficiency or anemia than the general population.

Metabolites

Lactate is the metabolite displaying the most spectacular concentration changes in the muscles and blood with exercise. Measuring these changes affords an estimation of the anaerobic and aerobic capacities of an athlete, as well as a scientific design of training. To ensure safe conclusions one has to be careful about the testing protocol, sampling time, and nutrition. The serum glucose concentration shows the balance between the mechanisms of supply to and uptake from the blood of this important fuel during prolonged muscle work. Triacylglycerols, cholesterol, and the distribution of cholesterol in HDL and LDL relate to the risk for atherosclerosis. Monitoring these parameters warns against possible undesirable changes and contributes to assessing the effectiveness of aerobic training programs. Glycerol indicates the lipolytic rate, urea indicates the metabolism of amino acids, creatinine indicates the metabolism of creatine and CP, and ammonia indicates the deamination of adenine ribonucleotides and amino acids.

Enzymes and Hormones

The serum concentration of certain enzymes shows the integrity of the organs in which they abound. Thus, creatine kinase serves as a marker of muscle fiber damage; it is the parameter displaying the largest differences in resting values between athletes and nonathletes. In addition, two aminotransferases, ALT and AST, serve as markers of hepatocyte and, secondarily, muscle fiber damage. Finally, the serum concentrations of two steroid hormones, cortisol and testosterone, are useful in estimating the balance between catabolic and anabolic processes in muscle, as well as the possible diagnosis of overtraining or overreaching.

Answers to Problems and Critical Thinking Questions

Chapter 1

1. Detergents are by chemical structure both hydrophilic and hydrophobic. That way they remove the—usually hydrophobic—stains from fabrics or dishware into the water. Compounds that are hydrophilic and hydrophobic are called amphipathic (see sections 5.6, 5.8, and 5.9).

2. The evaporation of a fluid is a physical process: The fluid merely goes from the liquid to the gaseous phase. The rusting of a metal is a chemical reaction: The metal reacts with oxygen to form an oxide.

3. The following equation answers the question:

$$
\begin{array}{l}
H_2COCOR \\
| \\
HCOCOR + 3\,H_2O \longrightarrow \\
| \\
H_2COCOR
\end{array}
\qquad
\begin{array}{l}
H_2COH \\
| \\
HCOH + 3\,RCO_2{}^- + 3\,H^+ \\
| \\
H_2COH
\end{array}
$$

4. Based on equation 1.3, the $[H^+]$ changes from 10^{-7} to 10^{-6} mol $\cdot$ L^{-1}. Since $10^{-6}/10^{-7} = 10$, d is the correct answer.

Chapter 2

1. Small molecules produced in catabolism serve as raw materials for anabolism; ATP synthesized in catabolism fuels anabolism; and NADH or NADPH produced during oxidations of metabolites in catabolism is used as a reducing agent in anabolism.

2. d; see equation 2.4.

3. Since ATP hydrolysis releases 7.3 kcal $\cdot$ mol^{-1}, gluconeogenesis must be coupled to the hydrolysis of at least $35/7.3 = 4.8$ mol ATP. In reality, it is powered by the hydrolysis of $6 \sim P$ (section 9.19).

4. When a substance burns, oxygen is added to it; therefore, it is oxidized.

Chapter 3

1. a, 99; b, 1; c, 1; d, 99; e, 99.

2. Myoglobin has a higher affinity for O_2 than hemoglobin; muscle has a lower $[O_2]$ than the blood; and muscle has higher $[CO_2]$ and $[H^+]$ than the blood (the Bohr effect), especially during exercise.

3. a, increase until the enzyme is saturated; b, increase (see figure 3.18); c, decrease; d, decrease because of denaturation (see figure 3.17); e, increase.

Chapter 4

1. 23% G; 27% A; 27% T.

2. He or she will have to label a component of DNA that is missing from RNA. Such components are thymine and deoxyribose.

3. It belongs to RNA. Its complementary sequence is 3' UGAUCGCGAU 5'.

4. 5' . . . ACGACAUGUU . . . 3'.

5. One mRNA, four rRNA (in the ribosome), and as many tRNA as the number of amino acid residues of the protein.

6. Met-Arg-Ser-Tyr-Pro-Thr-Gln.

7. Insertion of U after the second base.

Chapter 5

1. Cotton is almost pure carbohydrate (cellulose), whereas wool and silk are proteins. Wool and silk are so different to the touch because they are made up of different proteins: Wool contains α-keratin (in which α helices dominate), whereas silk contains fibroin (in which β pleated sheets dominate).

2. Corn, like all grains, contains a lot of starch. Being hydrophilic, starch retains water. When we heat the corn kernels, the water they contain goes from liquid to gas. Because a substance takes up much more volume as a gas than as a liquid, and because the kernel casing is hard and airtight, pressure builds up inside the kernels until finally they explode. As the water vapor expands violently, it pushes on the starch and leaves it fluffy and dry: popcorn.

3. Hydrogenation, that is, addition of hydrogens to the double bonds of the unsaturated fatty acids, which turns them into saturated fatty acids.

4. Cholesterol ester = sphingomyelin (one acyl group) < phosphatidyl inositol (two) < triacylglycerol (three).

5. The egg yolk is rich in phosphatidyl choline, which, being amphipathic, lets oil and water mix just as detergents do (see question 1 in chapter 1). Detergents and phospholipids are *emulsifiers,* as they help emulsions (water-based liquids or creams in which lipid droplets remain suspended) form.

Chapter 6

1. Na^+-K^+ ATPase.

2. Na^+ channel in nerve cells; nicotinic acetylcholine receptor in muscle cells.

3. The resting potential is due to active transport, and the action potential is due to passive transport (an oxymoron).

4. Voltage-gated channels open when the membrane voltage changes, whereas ligand-gated channels open when ligands bind to them. The Na^+, K^+, and Ca^{2+} channels are voltage-gated channels. The nicotinic acetylcholine receptor is a ligand-gated channel.

Chapter 7

1. Muscle, muscle fiber, myofibril, sarcomere, thick and thin filaments. The reverse is also a logical order.

2. See the table below.

	Rest	Contraction
Sarcomere	2.5	2
A band	1.5	1.5
H zone	0.7	0.2
I band	1	0.5
Thick filaments	1.5	1.5
Thin filaments	0.9	0.9
Overlap	0.8	1.3

3. ATP; myosin.

4. Death stops ATP production. Thus, actin and myosin remain bound to each other (figure 7.9d).

5. It is. In fact, it is feasible regardless of the $[Ca^{2+}]$.

6. Caffeine acts as an *ionophore;* that is, it carries Ca^{2+} through the sarcoplasmic reticulum membrane. This may be part of the reason why coffee causes tremor to unaccustomed drinkers.

Chapter 8

1. Add the two equations and add their $\Delta G°'$.

2. Most of the ATP will be replenished by CP. Thus, probable concentrations of the two compounds are, respectively, 4 and 16 mmol · kg^{-1} and 5 and 15 mmol · kg^{-1}.

3. Fluoro-2,4-dinitrobenzene inhibits CK.

4. Creatine kinase replenishes ATP from CP as ATP decreases and ADP increases. Adenylate kinase produces one ATP from two ADP as ADP increases. Adenylate deaminase is activated as ATP decreases and shifts the adenylate kinase reaction further toward ATP production by eliminating AMP.

5. ATP and CP decrease; ADP, AMP, creatine, P_i, IMP, and ammonia increase.

Chapter 9

1. P_i, AMP, IMP, Ca^{2+}, and epinephrine.

2. Although it yields ATP, glycolysis cannot start without a small amount of it, as ATP is needed in the first and third reactions.

3. Glucose, glucose-6-phosphate, AMP, and ADP increase; ATP and CP decrease.

4. *S. cerevisiae* breaks down the starch in the dough to glucose; then it breaks down glucose to CO_2. As CO_2 accumulates, it forms bubbles that are trapped in the dough and make it rise. When the dough is heated during baking, the bubbles expand and make bread, cakes, and so on fluffy.

5. See the table below.

	Aerobic	Anaerobic
ATP/glucose	30	2
Maximal rate of ATP resynthesis ($mmol \cdot kg^{-1} \cdot s^{-1}$)	0.5	1.5
Exercise tasks	Maximal, hard, or moderate-intensity; >1 min	Maximal; 7 s to 1 min

6. The muscles and all the cells in the body do not burn oxygen. Instead, oxygen burns fuel molecules. Moreover, if the muscles and all the cells burn sugar, CO_2 rather than lactic acid is produced. An acceptable statement, then, would have been, "If the muscles and all the cells in the body do not have enough oxygen to burn sugar, lactic acid is produced."

7. Gluconeogenesis and glycogenolysis through increases in plasma epinephrine and glucagon, as well as a decrease in plasma insulin.

8. The increased uptake of glucose by the exercising muscles tends to decrease its blood concentration; the increased secretion of epinephrine and glucagon and the decreased secretion of insulin tend to increase it.

Chapter 10

1. Increases in epinephrine and norepinephrine; decrease in insulin.

2. Laurate, 78; myristate, 92; palmitate, 106; palmitoleate, 104; stearate, 120; oleate, 118; linoleate, 116; α-linolenate, 114; arachidonate, 126; eicosapentaenoate, 124; docosahexaenoate, 136.

3. An elongase can add two C to linoleate, and a desaturase can introduce two additional double bonds to produce arachidonate.

4. Hormone-sensitive lipase, pancreatic lipase, and lipoprotein lipase.

5. HDL cholesterol.

Chapter 11

1. Proteolysis: Both types of exercise increase it.

2. Protein synthesis: Resistance exercise increases it, whereas endurance exercise decreases it.

3. All three are essential amino acids. Valine is glucogenic, leucine is ketogenic, and isoleucine is partly glucogenic and partly ketogenic.

Chapter 12

1. Increase in cytosolic $[Ca^{2+}]$; increase in [ADP] and [AMP]; decrease in glycogen concentration; increase in radical concentration.

2. Increase in cytosolic $[Ca^{2+}]$; decrease in [ATP] and concomitant increase in [ADP] and [AMP]; hypoxia; increase in radical concentration; stretching; increase in [IGF1].

3. Increase in cytosolic $[Ca^{2+}]$; decrease in [ATP] and concomitant increase in [ADP] and [AMP]; hypoxia; increase in radical concentration; decrease in pH.

Chapter 13

1. Lactate system, 34%; oxygen system, 66%.

2. ATP-CP system, 15%; lactate system, 39%; oxygen system, 46%.

3. a, ↓; b, ↑; c, ↑; d, ↓; e, ↓.

4. Consult table 13.6.

Chapter 14

1. (Plasma volume post)/(plasma volume pre) = 0.96. Thus, the plasma volume has decreased by 4%.

2. Erythropoietin and blood doping.

3. Iron deficiency increases TIBC and sTfR; it decreases transferrin saturation and ferritin.

4. The transferrin saturation was originally 30%. The new iron and TIBC values are 81 and 330 $\mu g \cdot dL^{-1}$, respectively. Therefore, the new transferrin saturation is 24.5%, a relative decrease by 18%.

5. He or she probably has adequate iron stores but has reduced iron intake lately. We would advise this person to eat more iron-rich foods.

6. He or she probably has inadequate iron stores but has increased iron intake lately.

Chapter 15

1. A sprinter's plot will lie to the left of an endurance athlete's plot and will reach a higher lactate concentration because of genetic factors (the former usually has a higher percentage of type IIA and IIX muscle fibers, hence a higher lactate concentration in a given exercise) and training factors (endurance training shifts the plot to the right). See the following figure.

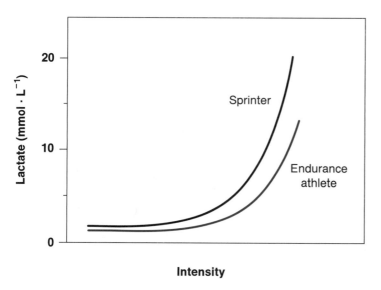

2. $220 - 50 - 100/5 = 150$ mg $\cdot$ dL^{-1}.

3. The triacylglycerol concentration is desirable; the total, HDL, and LDL cholesterol concentrations are intermediate; and the atherogenic indices ($220/50 = 4.4$ and $150/50 = 3$) are desirable. Overall, the athlete has a rather healthy lipidemic profile.

4. Glycerol.

5. Urea and creatinine; they rise.

6. High-protein diets or protein supplementation could raise serum urea. High-meat diets or creatine supplementation could raise serum creatinine.

Chapter 16

1. Serum CK.

2. Serum ALT.

3. Serum cortisol.

4. Serum testosterone may be too high if testosterone itself has been administered and too low if other AAS have been used; HDL cholesterol may be abnormally low; and aminotransferases may be abnormally high. Note, however, that none of these findings is acceptable as proof of AAS use in doping control. Rather, doping control is performed through the detection of AAS and their metabolites in the urine.

Glossary

β oxidation—The breakdown of fatty acids to acetyl CoA.

β-D-2-deoxyribose—A five-carbon monosaccharide, component of DNA.

β-D-ribose—A five-carbon monosaccharide, component of RNA.

ΔG (free-energy change)—A thermodynamic term indicating whether a reaction is favored. A reaction is favored if $\Delta G < 0$.

$\Delta G°$ (standard free-energy change)—The ΔG of a reaction when the concentration of every participating substance in solution is $1 \text{ mol} \cdot \text{L}^{-1}$.

$\Delta G°'$—The $\Delta G°$ of a reaction at pH 7.

AAS (anabolic androgenic steroids)—A group of pharmaceutical agents including natural androgens, such as testosterone, and synthetic analogs of these androgens.

acetyl coenzyme A—An activated carrier of the acetyl group that introduces it to the citric acid cycle, fatty acid synthesis, and ketone body formation.

acetyl group—A two-carbon unit having the formula (see page 153).

acetylcholine—The neurotransmitter at the neuromuscular junction.

actin—A muscle protein interacting with myosin during contraction.

action potential—The electrical potential difference across the plasma membrane of a neuron or muscle fiber when it is excited.

activator—A substance that increases the activity of an enzyme by binding to it noncovalently.

active site—The area on an enzyme where it catalyzes a reaction.

active transport—The passage of a substance from a compartment of low concentration to a compartment of high concentration at the expense of energy.

acyl group—A fatty acid without its terminal OH or O⁻.

adaptation to exercise—A change in a characteristic or function of the body (such as increased lipid combustion during exercise or muscle hypertrophy) because of exercise, usually chronic.

adenine—A nitrogenous base of DNA and RNA, a member of the purines.

adenosine diphosphate—See ADP.

adenosine monophosphate—See AMP.

adenosine triphosphate—See ATP.

adenylate—See AMP.

adenylate cyclase—The enzyme catalyzing the synthesis of cAMP from ATP.

adenylate deaminase—The enzyme catalyzing the deamination of AMP to IMP.

adenylate kinase—The enzyme catalyzing the interconversion of two ADP to AMP and ATP.

adipocyte—A fat cell.

ADP (adenosine diphosphate)—A compound produced by ATP hydrolysis.

adrenaline—See epinephrine.

adrenergic receptor—A receptor recognizing epinephrine and norepinephrine.

aerobic—Oxygen dependent.

aerobic system—See oxygen system.

affinity—The binding strength of one molecule for another.

albumin—A protein carrying fatty acids and other substances in the plasma.

allostery—The property of certain proteins to change their affinity for a molecule at one site when another molecule binds to a different site.

amino acid—A compound containing an amino group and a carboxyl group. α-Amino acids are the building blocks of proteins.

aminotransferase—An enzyme catalyzing a transamination.

AMP (adenosine monophosphate, or adenylate)—A compound produced by ATP hydrolysis.

amphipathic, or amphiphilic—Being partly hydrophilic and partly hydrophobic.

anabolic androgenic steroids—See AAS.

anabolism—The phase of metabolism that includes biosynthetic processes.

anaerobic—Oxygen independent.

anemia—Low hematocrit and blood hemoglobin concentration.

anticodon—A triplet of tRNA bases that base-pairs with a codon in mRNA.

atherogenic index—The serum concentration ratio of either total cholesterol or LDL cholesterol to HDL cholesterol, used to estimate the risk for atherosclerosis.

atherosclerosis—The presence of fatty deposits inside blood vessels.

ATP (adenosine triphosphate)—A compound releasing a high amount of energy when hydrolyzed, thus serving as the major energy currency of cells.

ATP-ADP cycle—The interconversion of ATP and ADP accompanying cellular functions such as metabolism.

ATP-CP system—An anaerobic energy system supplying a small amount of energy at high power during short maximal exercise.

ATP synthase—The enzyme synthesizing ATP from ADP and P_i in oxidative phosphorylation.

BCAA (branched-chain amino acids)—Three amino acids (leucine, isoleucine, and valine) that are preferentially degraded in muscle during exercise.

bicarbonate loading—See **soda loading.**

bioenergetics—A branch of the biological sciences dealing with the energy transactions in living organisms.

blood doping—The transfusion of a large volume of blood to an athlete before an event in order to increase his or her endurance performance.

branched-chain amino acids—See **BCAA.**

buffering capacity—The amount of strong acid or base needed to change the pH of a buffer system by one pH unit.

Ca²⁺ ATPase—See **Ca²⁺ pump.**

Ca²⁺ channel—A membrane protein that lets Ca^{2+} through.

Ca²⁺ pump—A membrane protein that accumulates Ca^{2+} inside the sarcoplasmic reticulum in a muscle fiber.

calmodulin—A Ca^{2+}-binding protein that mediates many of the ion's actions.

cAMP (cyclic AMP)—A compound mediating the actions of several hormones through the cAMP cascade.

cAMP cascade—A series of molecular interactions that begins with the binding of a hormone or other ligand to a receptor in the plasma membrane, involves the synthesis of cAMP, and ends in the modification of a metabolic process such as glycogenolysis and lipolysis.

capillary density—The number of blood capillaries per muscle fiber or muscle cross-sectional area examined under the light microscope.

carbohydrate—A biological compound having the molecular formula $C_nH_{2n}O_n$ or a substance derived from such a compound.

carbohydrate loading—A nutritional and training intervention involving a high-carbohydrate diet and light training before an endurance event in order to maximize glycogen stores and increase performance.

carnitine—A compound carrying acyl groups from the cytosol to the mitochondrial matrix.

catabolism—The phase of metabolism that includes degradation processes.

catecholamines—A group of compounds including epinephrine, norepinephrine, and dopamine.

cellulose—The structural polysaccharide of plants.

chemical kinetics—A branch of chemistry and physics dealing with the rates and mechanisms of chemical reactions.

chemical thermodynamics—A branch of chemistry and physics that deals with energy changes in chemical systems.

chemiosmotic hypothesis—A hypothesis linking the electron transport chain with oxidative phosphorylation through protons that are expelled from the mitochondrion as the electron transport chain operates and that then return through ATP synthase, thus powering ATP synthesis.

cholesterol—The parent compound of steroids, a component of cell membranes, and a lipid increasing the risk for atherosclerosis.

cholesterol ester—An ester of cholesterol with a fatty acid.

chromatography—A set of laboratory methods of separating the components of a mixture based on their differing migration along a solid or liquid medium, through which they are forced to pass by pressure or capillary action.

chromosome—A natural DNA molecule, with associated proteins, containing part or all of an organism's genome.

chylomicron—A lipoprotein carrying dietary triacylglycerols to extrahepatic tissues.

citrate—A compound of the citric acid cycle produced from acetyl CoA and oxaloacetate.

citric acid cycle—A cyclic metabolic pathway converting the acetyl group to two CO_2.

codon—A triplet of mRNA bases that encodes an amino acid.

coenzyme—Any of a number of nonprotein organic compounds present in the active sites of many enzymes and participating in catalysis.

committed step—The first irreversible reaction in a metabolic pathway.

compound of high phosphoryl transfer potential—A compound having a phosphoryl group and releasing a high amount of energy when hydrolyzed.

concentration—The amount of a substance dissolved in a certain amount of solution or solvent.

concentric contraction—Contraction in which the muscle is shortened.

controlled study—A study including a control group or condition as well as the study group or condition.

Cori cycle—The transport to muscle of glucose synthesized in the liver from lactate that has been produced from glucose in muscle.

cortisol—A hormone secreted by the adrenal cortex.

covalent modification—The addition of a chemical group to a molecule. The term is usually used with large molecules such as proteins.

CP (creatine phosphate, or phosphocreatine)—A compound of high phosphoryl transfer potential, which replenishes ATP during maximal exercise.

creatine—A compound serving as the precursor of creatine phosphate.

creatine kinase—The enzyme catalyzing the interconversion of ATP and CP.

creatine phosphate—See **CP.**

creatinine—A compound produced by dehydration of creatine.

cross-bridge—The myosin head in contact with actin in a myofibril.

cross-sectional study—A study in which researchers compare population groups at a certain point in time.

cyclic adenylate—See **cyclic AMP.**

cyclic AMP (cyclic adenylate)—See **cAMP.**

cytoplasm—The interior of a cell.

cytosine—A nitrogenous base of DNA and RNA, a member of the pyrimidines.

cytosol—The fluid of a eukaryotic cell outside the intracellular organelles.

deamination—Removal of an amino group.

decarboxylation—Removal of a carboxyl group.

denaturation—Disruption of the tertiary and quaternary structure of a protein or nucleic acid.

deoxyribonucleic acid—See **DNA.**

deoxyribonucleotide—A compound consisting of a nitrogenous base, a deoxyribose unit, and one to three phosphoryl groups. Deoxyribonucleotides are the building blocks of DNA.

dephosphorylation—Removal of a phosphoryl group.

diabetes mellitus—A disease caused by inadequate insulin production or by insulin resistance.

dihydropyridine receptor—A voltage-sensing protein of the transverse tubule membrane in muscle fibers that mediates the release of Ca^{2+} from the sarcoplasmic reticulum in response to action potentials.

DNA (deoxyribonucleic acid)—A large biological compound serving as the repository of an organism's genetic material.

DNA polymerase—The main enzyme catalyzing replication.

DNA replication—Duplication of DNA.

eccentric contraction—Contraction in which a muscle developing contractile force is lengthened by the application of a higher opposing force.

electron transport chain—The transport of electrons from NADH and $FADH_2$ to O_2, accompanied by the release of a large amount of energy.

electrophoresis—A laboratory method of separating ions in a mixture along a gel to which an electrical field is applied.

endergonic—A reaction in which $\Delta G > 0$. Such a reaction is not favored.

endocytosis—The uptake of extracellular material by a cell through invagination of the plasma membrane and formation of intracellular vesicles.

endoplasmic reticulum—An extensive network of tubules and flattened sacs inside a eukaryotic cell.

endurance exercise—Exercise characterized by prolonged continuous or intermittent periods of contractile activity against low resistance, for example, a marathon run.

enzyme—A protein that catalyzes (speeds up) a reaction.

enzyme activity—The increase in the rate of an enzyme reaction caused by the enzyme.

epinephrine (or adrenaline)—A hormone produced mainly by the adrenal medulla.

EPOC (excess postexercise oxygen consumption, or oxygen debt)—The difference between the oxygen consumed during recovery from exercise and that normally consumed at rest.

equilibrium constant—See K_{eq}.

ergogenic aid—A substance or method used to increase sport performance.

erythrocyte—A red blood cell.

erythrocyte count—The number of erythrocytes per volume (usually 1 μL) of blood.

erythropoietin—A hormone produced by the kidneys, causing erythrocyte formation in the bone marrow.

essential amino acid—An amino acid used in protein synthesis that has to be obtained from the diet, since we are unable to synthesize it.

essential fatty acid—A fatty acid that is necessary for cell function and has to be obtained from the diet, since we are unable to synthesize it.

ester linkage—The linkage joining a carboxylic acid and an alcohol.

euglycemia—Normal plasma glucose concentration.

eukaryotic cell—A cell having internal compartments.

excess postexercise oxygen consumption—See EPOC.

exergonic—A reaction in which $\Delta G < 0$. Such a reaction is favored.

exocytosis—The release of cellular material to the extracellular space by fusion of intracellular vesicles with the plasma membrane.

FAD/FADH$_2$—The oxidized and reduced forms, respectively, of flavine adenine dinucleotide, a compound participating in oxidation–reduction reactions.

fat loading—High fat intake for several days or weeks in order to maximize fat utilization during prolonged moderate-intensity exercise.

fatigue—Inability to maintain a prescribed exercise intensity.

fatty acid—A lipid consisting of a long hydrocarbon chain with a carboxyl group attached to the end.

feedback inhibition—The inhibition of an early step in a metabolic pathway by the product of the pathway.

ferritin—The iron-storing protein in the spleen, bone marrow, liver, and so on.

flavine adenine dinucleotide—See FAD/FADH$_2$.

fluorometry—A laboratory method of determining the concentration of a substance in solution by measuring how much light it emits after being excited by light of a higher energy.

free-energy change—See ΔG.

fructose—A six-carbon monosaccharide, similar in structure and function to glucose.

gene—A part of DNA that is transcribed to functional RNA.

gene expression—The presence in a cell, organ, or tissue of RNA or protein synthesized according to the genetic information contained in a gene.

gene product—A protein or RNA that are synthesized based on the information encoded in a gene.

genetic code—The set of instructions on how the base sequence of any mRNA is translated to amino acid sequence during translation.

genome—The genetic material of an organism.

genotype—The genetic makeup of an organism.

glucagon—A hormone produced by the α cells in the pancreas.

glucogenic amino acid—An amino acid that can serve as a precursor of glucose.

gluconeogenesis—The synthesis of glucose from compounds that are not carbohydrates.

glucose—A six-carbon monosaccharide serving primarily as an energy source.

glucose-alanine cycle—The transport to muscle of glucose synthesized in the liver from alanine that has been produced from glucose in muscle.

GLUT4—A protein importing glucose to muscle fibers, adipocytes, and other cells.

glycerol—A three-carbon compound, component of triacylglycerols and glycerophospholipids.

glycerophospholipid—A phospholipid containing a glycerol unit.

glycogen—The storage polysaccharide of animals.

glycogen phosphorylase—See **phosphorylase**.

glycogenolysis—The breakdown of glycogen to glucose and glucose 1-phosphate.

glycolysis—The breakdown of glucose to two pyruvates.

glycosidic linkage—The linkage joining two monosaccharide units.

guanine—A nitrogenous base of DNA and RNA, a member of the purines.

HDL (high-density lipoprotein)—A lipoprotein carrying cholesterol from extrahepatic tissues to the liver, chylomicrons, and VLDL.

hematocrit—The volume of the erythrocytes as a percentage of the total blood volume.

heme—The nonprotein part (prosthetic group) of hemoglobin and myoglobin that binds O_2.

hemoconcentration—Excessive loss of water from the blood vessels.

hemodilution—Excessive entry of water into the blood vessels.

hemoglobin—The oxygen-carrying protein in the blood.

hemopoiesis—The formation of erythrocytes in the bone marrow.

hepatocyte—A liver cell.

high-density lipoprotein—See HDL.

homeostasis—The maintenance of a biological parameter in an organism at a relatively stable level despite temporary fluctuations.

hormone—A compound secreted in the blood by an endocrine gland and serving as a messenger to other organs.

hormone-sensitive lipase—The enzyme that hydrolyzes triacylglycerols in adipose tissue and muscle. Its activity is affected positively by catecholamines and negatively by insulin.

hydrogen bond—An attraction between a hydrogen atom with partial positive charge and an atom with partial negative charge.

hydrolysis—Breakdown by water.

hydrophilic—Mixing readily with water.

hydrophobic—Not mixing readily with water.

hyperplasia—Referring to muscle, increase in the number of muscle fibers.

hypertrophy—Referring to muscle, increase in the cross-sectional area of muscle fibers.

hypoglycemia—Low plasma glucose concentration.

hypoxia—Decrease in the oxygen concentration in a biological fluid such as the blood or cytosol.

IGF1 (insulin-like growth factor 1)—A hormone involved in muscle hypertrophy.

in situ—In an organ exposed for manipulation in an anesthetized animal.

in vitro—In the test tube.

in vivo—In the living organism.

inhibitor—A substance that decreases the activity of an enzyme by binding to it noncovalently.

inorganic phosphate—See P_i.

inorganic pyrophosphate—See PP_i.

insulin—A hormone produced by the β cells in the pancreas.

insulin-like growth factor 1—See IGF1.

insulin resistance—Partial or total inability of insulin to exert its actions in the body due to malfunctioning of its signal transduction pathway.

iron status—The concentration of iron in the various compartments of the body.

isoform—One of two or more similar forms of a protein.

isometric contraction—Contraction in which the length of a muscle developing contractile force does not change, as an opposite force is applied to the muscle.

K^+ channel—A protein of the plasma membrane that lets K^+ out of the cell.

K_{eq} (equilibrium constant)—The ratio of the molar concentrations of the products to the molar concentrations of the reactants in a reaction at equilibrium.

ketogenic amino acid—An amino acid that cannot serve as a precursor of glucose.

ketone body—A compound resulting from the joining of two acetyl groups in the liver.

kinase—An enzyme catalyzing phosphorylation at the expense of ATP.

K_M (Michaelis constant)—The substrate concentration causing half-maximal rate in an enzyme reaction.

Krebs cycle—See citric acid cycle.

lactate—The product of anaerobic carbohydrate catabolism.

lactate dehydrogenase—The enzyme catalyzing the interconversion of pyruvate and lactate.

lactate system—An anaerobic energy system supplying a moderate amount of energy at moderate power during hard or maximal exercise.

LDL (low-density lipoprotein)—A lipoprotein carrying cholesterol to extrahepatic tissues.

leukocyte count—The number of leukocytes per volume (usually 1 μL) of blood.

ligand-gated channel—A transmembrane protein that lets substances through when a ligand binds to it.

lipase—An enzyme hydrolyzing tri-, di-, or monoacylglycerols.

lipid—A biological compound that is poorly soluble in water.

lipidemic profile—The serum concentrations of lipids such as triacylglycerols, total cholesterol, LDL cholesterol, and HDL cholesterol that are related to the risk for atherosclerosis.

lipolysis—The hydrolysis of triacylglycerols to three fatty acids and glycerol.

lipoprotein—A spherical aggregate of lipids and proteins serving the transport of lipids in the blood.

lipoprotein lipase—The enzyme that hydrolyzes lipoprotein-borne triacylglycerols in the capillaries.

longitudinal study—A study in which researchers perform an experimental intervention on a population group for a period of time and compare the final to the initial state.

low-density lipoprotein—See LDL.

maximal oxygen uptake—See $\dot{V}O_2$max.

mechano growth factor—See MGF.

messenger RNA—See mRNA.

metabolic control—The coordinated change in the rates of reactions taking place in a living system in response to changing circumstances.

metabolic pathway—A series of reactions in which the product of one is a reactant in the next.

metabolic syndrome—A morbid state characterized by visceral obesity, undesirable plasma lipid concentrations, high blood pressure, and hyperglycemia.

metabolism—The sum of the chemical reactions occurring in a living organism.

metabolite—A compound participating in metabolism.

MGF (mechano growth factor)—A variant of IGF1 involved in the muscle hypertrophic response to mechanical load or electrical stimulation.

Michaelis constant—See K_M.

mitochondrial biogenesis—Generation of new mitochondria.

mitochondrial (volume) density—The volume of mitochondria as a percentage of the total tissue volume.

mitochondrion—An intracellular organelle in which most of the energy of a eukaryotic cell is produced by aerobic processes.

mitosis—Division of a cell into two daughter cells.

monosaccharide—A carbohydrate containing three to seven carbons.

monounsaturated fatty acid—A fatty acid that has a double bond between two of its carbons.

motor neuron—A neuron in contact with muscle fibers that makes them contract.

motor unit—A motor neuron and the muscle fibers it innervates.

mRNA (messenger RNA)—A kind of RNA that contains information on the amino acid sequence of proteins.

muscle fiber—A muscle cell.

mutagen—A substance causing mutations.

mutation—A change in the DNA base sequence during replication.

myelin—The wrapping of the axons of many vertebrate neurons.

myofibril—A rod of contractile proteins inside a muscle fiber.

myoglobin—A protein storing O_2 in vertebrate skeletal muscles.

myokinase—See adenylate kinase.

myosin—The major muscle protein, moving the muscle at the expense of ATP.

Na$^+$ channel—A protein of the plasma membrane that lets Na$^+$ into the cell.

Na$^+$-K$^+$ ATPase—See Na$^+$-K$^+$ pump.

Na$^+$-K$^+$ pump—A protein of the plasma membrane that pumps Na$^+$ out of the cell and K$^+$ into the cell at the expense of ATP.

NAD$^+$/NADH—The oxidized and reduced forms, respectively, of nicotinamide adenine dinucleotide, a compound participating in oxidation–reduction reactions.

NADP$^+$/NADPH—The oxidized and reduced forms, respectively, of nicotinamide adenine dinucleotide phosphate, a compound participating in oxidation–reduction reactions.

neuromuscular junction—The interface between a motor neuron and a muscle fiber.

neuron—A cell of the nerve tissue.

neurotransmitter—A compound transmitting a signal from one neuron to another or from a neuron to a muscle fiber.

nicotinamide adenine dinucleotide—See NAD$^+$/NADH.

nicotinamide adenine dinucleotide phosphate—See NADP$^+$/NADPH.

nonpolar—Having even distribution of charges.

noradrenaline—See norepinephrine.

norepinephrine (or noradrenaline)—A neurotransmitter produced mainly in the sympathetic system.

nucleic acid—A large biological compound consisting of many nucleotide units joined in a row by phosphodiester linkages.

nucleotide—A compound consisting of a nitrogenous base, a monosaccharide, and one to three phosphoryl groups. Nucleotides are the building blocks of nucleic acids.

nucleus—An intracellular compartment containing the main genetic material of the cell.

oligosaccharide—A carbohydrate consisting of 2 to 10 monosaccharide units.

oxaloacetate—An intermediate of the citric acid cycle, also serving as a precursor of glucose.

oxidation—The loss of electrons from a compound.

oxidative phosphorylation—The synthesis of ATP from ADP and P_i, compliments of the electron transport chain.

oxygen debt—See EPOC.

oxygen system—The aerobic energy system supplying a large amount of energy at low power during exercise exceeding 1 min in duration.

pancreatic lipase—The enzyme that hydrolyzes triacylglycerols in the small intestine.

passive transport—The spontaneous passage of a substance from a compartment of high concentration to a compartment of low concentration.

peptide—A compound made up of amino acid residues linked in a row by peptide bonds. Typically, the term is used for small numbers of amino acid residues.

peptide bond—The bond linking amino acids in a protein.

perilipin—A protein at the surface of lipid droplets in the adipocytes.

phenotype—The visible or measurable characteristics of an organism.

phosphagen—See compound of high phosphoryl transfer potential.

phosphatase—An enzyme catalyzing dephosphorylation.

phosphoanhydride linkage—The linkage joining two phosphoryl groups.

phosphocreatine—See CP.

phosphodiester linkage—The linkage joining two nucleotides in a nucleic acid.

phosphofructokinase—An enzyme of glycolysis that is a control point of the pathway.

phosphoglyceride—See glycerophospholipid.

phospholipid—A lipid having an amphipathic character, a major component of cell membranes.

phosphorylase—The main enzyme catalyzing glycogenolysis.

phosphorylation—Addition of a phosphoryl group.

photosynthesis—The synthesis of glucose from CO_2 and H_2O, compliments of solar energy.

P_i (inorganic phosphate)—A phosphate ion, mainly of the HPO_4^{2-} form in biological fluids.

plasma—The fluid of the blood outside its cells. Plasma is obtained after centrifugation of a blood sample in which coagulation has been inhibited.

plasma membrane—The membrane surrounding a cell.

platelet count—The number of platelets per volume (usually 1 μL) of blood.

polar—Having uneven distribution of charges.

polynucleotide chain—A chain of nucleotide units (deoxyribonucleotide in the case of DNA, ribonucleotide in the case of RNA) linked in a row by phosphodiester linkages. Typically, the term is used for large numbers of nucleotide units and is synonymous with DNA or RNA.

polypeptide chain—A chain of amino acid residues linked in a row by peptide bonds. Typically, the term is used for large numbers of amino acid residues and is synonymous with protein.

polysaccharide—A carbohydrate consisting of over 10 monosaccharide units.

polyunsaturated fatty acid—A fatty acid that has more than one double bond between its carbons.

potassium channel—See **K⁺ channel**.

PP$_i$ (inorganic pyrophosphate)—An ion, mainly of the $HP_2O_7^{3-}$ form in biological fluids, produced from the joining of two P_i ions.

prokaryotic cell—A primitive cell having no internal compartments.

protease—An enzyme catalyzing the hydrolysis of a protein.

protein—A large biological compound consisting of many amino acid residues linked in a row by peptide bonds.

proteolysis—The hydrolysis of a protein to amino acids or peptides.

purine—A nitrogenous base of DNA and RNA consisting of a six-membered and a five-membered ring.

pyrimidine—A nitrogenous base of DNA and RNA consisting of a six-membered ring.

pyruvate—The product of glycolysis.

pyruvate dehydrogenase complex—An assembly of enzymes catalyzing the conversion of pyruvate to acetyl CoA.

receptor—A protein that recognizes and binds another biological compound.

reduction—The gain of electrons by a compound.

reference interval—The range of values of a biochemical or hematological parameter in the vast majority (usually the central 95%) of a population.

reference limit—The lower or upper end of the reference interval.

repeated-bout effect—The occurrence of less muscle fiber damage when an exercise, particularly eccentric exercise, is repeated several days or even weeks after the initial bout.

RER (respiratory exchange ratio)—The ratio of the volume of expired CO_2 to the volume of O_2 consumed.

resistance exercise—Exercise involving short periods of contractile activity against high resistance, for example, weightlifting.

respiratory chain—See **electron transport chain**.

respiratory exchange ratio—See **RER**.

respiratory quotient—See **RQ**.

resting potential—The electrical potential difference across the plasma membrane of a neuron or muscle fiber when it is not excited.

ribonucleic acid—See **RNA**.

ribonucleotide—A compound consisting of a nitrogenous base, a ribose unit, and one to three phosphoryl groups. Ribonucleotides are the building blocks of RNA.

ribosomal RNA—See **rRNA**.

ribosome—A complex of rRNA and proteins that performs protein synthesis.

RNA (ribonucleic acid)—A large biological compound serving in transmitting the genetic information from DNA to proteins.

RNA polymerase—The main enzyme catalyzing transcription.

RQ (respiratory quotient)—The ratio of the volume of CO_2 released to the volume of O_2 taken up at the cellular level.

rRNA (ribosomal RNA)—A kind of RNA that is part of the ribosomes.

ryanodine receptor—A protein of the sarcoplasmic reticulum membrane acting as the Ca^{2+} channel through which Ca^{2+} is released in response to action potentials.

sarcolemma—The plasma membrane of a muscle fiber.

sarcomere—The segment of a myofibril between two Z lines, the minimal complete functional unit in a muscle fiber.

sarcoplasm—The cytoplasm of a muscle fiber.

sarcoplasmic reticulum—A membranous system of sacs surrounding the myofibrils in a muscle fiber and containing Ca^{2+} at high concentration.

satellite cell—A cell lying adjacent to a muscle fiber, containing little more than a nucleus, and offering its genetic material by fusion to the muscle fiber for growth or repair.

saturated fatty acid—A fatty acid that has only single bonds between its carbons.

serum—The fluid obtained as supernatant after centrifugation of a blood sample that has been allowed to coagulate.

siderophilin—See **transferrin**.

signal transduction pathway—A series of molecular interactions through which a signal (for example, the secretion of a hormone) causes a response (for example, the acceleration of a reaction).

sliding-filament model—A model explaining that muscle contraction is due to the active sliding of thick and thin filaments past each other.

soda loading—Intake of $NaHCO_3$ before maximal exercise in order to counter the drop in blood and muscle pH and to delay fatigue.

sodium bicarbonate loading—See **soda loading**.

sodium channel—See **Na⁺ channel**.

sodium-potassium pump—See **Na⁺-K⁺ pump**.

soluble transferrin receptor—See **sTfR**.

spectrophotometry—A laboratory method of determining the concentration of a substance in solution by measuring how much light it absorbs.

sprint exercise—Exercise consisting of short periods of maximal contractile activity against low resistance, for example, a competitive 100 m run.

standard free-energy change—See **ΔG°**.

starch—The storage polysaccharide of plants.

steroid—A lipid containing three six-membered and one five-membered carbon ring, as in cholesterol.

steroid hormone—A member of a family of hormones, all derivatives of cholesterol.

sTfR (soluble transferrin receptor)—A plasma-borne form of the transferrin receptor, the protein that binds transferrin on the surface of cells.

substrate—A reactant in an enzyme reaction.

subunit—A polypeptide chain that is part of a protein consisting of more than one polypeptide chain. Also, one of the two parts of a ribosome.

synapse—The point of contact between neurons or between a neuron and a muscle fiber.

synaptic cleft—The space between the presynaptic and postsynaptic membranes in a synapse.

synaptic vesicle—A vesicle near a synapse that gathers neurotransmitter molecules.

T system—See transverse tubules.

testosterone—The major male sex hormone, mainly secreted by the testes.

thick filament—A filament containing myosin inside a muscle fiber.

thin filament—A filament containing actin, tropomyosin, and troponin inside a muscle fiber.

thymine—A nitrogenous base of DNA, a member of the pyrimidines.

TIBC (total iron-binding capacity)—The serum iron concentration in the hypothetical case that the iron-binding proteins in the plasma are fully saturated.

total iron-binding capacity—See TIBC.

transaminase—See aminotransferase.

transamination—The transfer of an amino group from an α-amino acid to an α-keto acid.

transcription—The synthesis of RNA based on the genetic information contained in DNA.

transcription factor—A protein controlling transcription.

transfer RNA—See tRNA.

transferrin—The major iron-carrying protein in the plasma.

transferrin saturation—The serum iron concentration as a percentage of the TIBC.

translation—Protein synthesis based on the information contained in mRNA.

transverse tubules—A system of tubules conducting action potentials to the interior of a muscle fiber.

triacylglycerol (or triglyceride)—A lipid consisting of three acyl groups and a glycerol unit; the major energy depot in the body.

tricarboxylic acid cycle—See citric acid cycle.

triglyceride—See triacylglycerol.

tRNA (transfer RNA)—A kind of RNA that transfers amino acids for protein synthesis.

tropomyosin—A muscle protein controlling the interaction of actin and myosin.

troponin—A muscle protein mediating the control of Ca^{2+} on contraction.

turnover number—The number of substrate molecules converted to product by an enzyme molecule in a specified time when the enzyme is fully saturated with substrate.

unsaturated fatty acid—A fatty acid that has one or more double bonds between some of its carbons.

uracil—A nitrogenous base of RNA, a member of the pyrimidines.

urea cycle—A cyclic metabolic pathway producing urea in the liver.

very low-density lipoprotein—See VLDL.

VLDL (very low-density lipoprotein)—A lipoprotein carrying hepatic triacylglycerols to extrahepatic tissues.

$\dot{V}O_2max$— The maximal amount of oxygen taken up by the tissues in the body per unit of time (usually minute) during maximal exertion.

voltage-gated channel—A transmembrane protein that lets substances through when the membrane voltage changes.

Suggested Readings

Suggested readings are organized by part of this book, further categorized as books or review articles, and further broken down by topic.

Part I

A number of excellent recent textbooks of general biochemistry are available for the reader who wishes to delve into the topics presented in part I.

Berg JM, Tymoczko JL, Stryer L (2002). *Biochemistry* (Freeman, New York).

Horton HR, Moran LA, Ochs RS, Rawn DJ, Scrimgeour KG (2001). *Principles of Biochemistry* (Prentice Hall, Boston).

Nelson DL, Cox MM (2004). *Lehninger Principles of Biochemistry* (Freeman, New York).

Voet D, Voet JG, Pratt CW (2001). *Fundamentals of Biochemistry* (Wiley, New York).

I also recommend the following books devoted to particular areas of biochemistry: proteins, nucleic acids and gene expression, carbohydrates, and lipids.

Bloomfield VA, Crothers DM, Tinoco I (2000). *Nucleic Acids: Structures, Properties, and Functions* (University Science Books, Sausalito, California).

Branden C-I, Tooze J (1999). *Introduction to Protein Structure* (Garland, New York).

Fersht A (1998). *Structure and Mechanism in Protein Science: A Guide to Enzyme Catalysis and Protein Folding* (Freeman, New York).

Vance DE, Vance JE, eds. (2002). *Biochemistry of Lipids, Lipoproteins and Membranes* (Elsevier, Amsterdam).

Varki A, Cummings R, Esko J, Freeze H, Hart G, Marth J (1999). *Essentials of Glycobiology* (Cold Spring Harbor Laboratory, Cold Spring Harbor, New York).

Part II

Books

Hille B (2001). *Ion Channels of Excitable Membranes* (Sinauer, Sunderland, Massachussetts).

Solaro RJ, Moss RL, eds. (2002). *Molecular Control Mechanisms in Striated Muscle Contraction* (Kluwer, Boston).

Sugi H, ed. (2004). *Molecular and Cellular Aspects of Muscle Contraction* (Kluwer/Plenum, New York).

Review Articles

Davis JP, Wahr PA, Rall JA (2004). Molecular aspects of muscular contraction. In Poortmans JR, ed., *Principles of Exercise Biochemistry* (Karger, Basel), pp. 62-86.

Dulhunty AF, Haarmann CS, Green D, Laver DR, Board PG, Casarotto MG (2002). Interactions between dihydropyridine receptors and ryanodine receptors in striated muscle. *Progress in Biophysics and Molecular Biology* 79: 45-75.

Gordon AM, Regnier M, Homsher E (2001). Skeletal and cardiac muscle contractile activation: Tropomyosin "rocks and rolls." *News in Physiological Science* 16: 49-55.

Kaplan JH (2002). Biochemistry of Na,K-ATPase. *Annual Review of Biochemistry* 71: 511-535.

McElhinny AS, Kazmierski ST, Labeit S, Gregorio CC (2003). Nebulin: The nebulous, multifunctional giant of striated muscle. *Trends in Cardiovascular Medicine* 13: 195-201.

Toyoshima C, Inesi G (2004). Structural basis of ion pumping by Ca^{2+}-ATPase of the sarcoplasmic reticulum. *Annual Review of Biochemistry* 73: 269-292.

Unwin N (2003). Structure and action of the nicotinic acetylcholine receptor explored by electron microscopy. *FEBS Letters* 555: 91-95.

Part III

Review Articles

Compounds of High Phosphoryl Transfer Potential

Sahlin K (2004). High-energy phosphates and muscle energetics. In Poortmans JR, ed., *Principles of Exercise Biochemistry* (Karger, Basel), pp. 87-107.

Carbohydrate Metabolism in Exercise

Bonen A (2000). Lactate transporters (MCT proteins) in heart and skeletal muscle. *Medicine and Science in Sports and Exercise* 32: 778-789.

Coker RH, Kjaer M (2005). Glucoregulation during exercise: The role of the neuroendocrine system. *Sports Medicine* 35: 575-583.

Gladden LB (2004). Lactate metabolism during exercise. In Poortmans JR, ed., *Principles of Exercise Biochemistry* (Karger, Basel), pp. 152-196.

Greenhaff PL, Hultman E, Harris RC (2004). Carbohydrate metabolism. In Poortmans JR, ed., *Principles of Exercise Biochemistry* (Karger, Basel), pp. 108-151.

Hargreaves M (2004). Muscle glycogen and metabolic regulation. *Proceedings of the Nutrition Society* 63: 217-220.

Henriksen EJ (2002). Effects of acute exercise and exercise training on insulin resistance. *Journal of Applied Physiology* 93: 788-796.

Rose AJ, Richter EA (2005). Skeletal muscle glucose uptake during exercise: How is it regulated? *Physiology* 20: 260-270.

Lipid Metabolism in Exercise

Aschenbach WG, Sakamoto K, Goodyear LJ (2004). 5' Adenosine monophosphate-activated protein kinase, metabolism and exercise. *Sports Medicine* 34: 91-103.

Bülow J (2004). Lipid mobilization and utilization. In Poortmans JR, ed., *Principles of Exercise Biochemistry* (Karger, Basel), pp. 197-226.

Donsmark M, Langfort J, Holm C, Ploug T, Galbo H (2005). Hormone-sensitive lipase as a mediator of lipolysis in contracting skeletal muscle. *Exercise and Sport Sciences Reviews* 33: 127-133.

Durstine JL, Grandjean PW, Davis PG, Ferguson MA, Alderson NL, DuBose KD (2001). Blood lipid and lipoprotein adaptations to exercise: A quantitative analysis. *Sports Medicine* 31: 1033-1062.

Luiken JJ, Miskovic D, Arumugam Y, Glatz JF, Bonen A (2001). Skeletal muscle fatty acid transport and transporters. *International Journal of Sport Nutrition and Exercise Metabolism* 11: S92-S96.

Ramsay RR, Gandour RD, van der Leij FR (2001). Molecular enzymology of carnitine transfer and transport. *Biochimica et Biophysica Acta* 1546: 21-43.

Ranallo RF, Rhodes EC (1998). Lipid metabolism during exercise. *Sports Medicine* 26: 29-42.

van Loon LJC (2004). Use of intramuscular triacylglycerol as a substrate source during exercise in humans. *Journal of Applied Physiology* 97: 1170-1187.

Protein Metabolism in Exercise

Gibala MJ (2001). Regulation of skeletal muscle amino acid metabolism during exercise. *International Journal of Sport Nutrition and Exercise Metabolism* 11: 87-108.

Poortmans JR (2004). Protein metabolism. In Poortmans JR, ed., *Principles of Exercise Biochemistry* (Karger, Basel), pp. 227-278.

Tipton KD, Wolfe RR (2001). Exercise, protein metabolism, and muscle growth. *International Journal of Sport Nutrition and Exercise Metabolism* 11: 109-132.

Effects of Exercise on Gene Expression

Baar K (2004). Involvement of PPARγ co-activator-1, nuclear respiratory factors 1 and 2, and PPARα in the adaptive response to endurance exercise. *Proceedings of the Nutrition Society* 63: 269-273.

Booth FW, Baldwin KM (1996). Muscle plasticity: Energy demand and supply processes. In Rowell LB, Shepherd JT, eds., *Handbook of Physiology, Section 12: Regulation and Integration of Multiple Systems* (Oxford University Press, New York), pp. 1075-1123.

Flück M, Hoppeler H (2003). Molecular basis of skeletal muscle plasticity—from gene to form and function. *Reviews of Physiology, Biochemistry and Pharmacology* 146: 159-216.

Irrcher I, Adhihetty PJ, Joseph A-M, Ljubicic V, Hood DA (2003). Regulation of mitochondrial biogenesis in muscle by endurance exercise. *Sports Medicine* 33: 783-793.

Machida S, Booth FW (2004). Insulin-like growth factor-1 and muscle growth: Implication for satellite cell proliferation. *Proceedings of the Nutrition Society* 63: 337-340.

Michel RN, Dunn SE, Chin ER (2004). Calcineurin and skeletal muscle growth. *Proceedings of the Nutrition Society* 63: 341-349.

Stoughton RB (2005). Applications of DNA microarrays in biology. *Annual Review of Biochemistry* 74: 53-82.

Williams RS, Neufer PD (1996). Regulation of gene expression in skeletal muscle by contractile activity. In Rowell LB, Shepherd JT, eds., *Handbook of Physiology, Section 12: Regulation and Integration of Multiple Systems* (Oxford University Press, New York), pp. 1124-1150.

Integration of Exercise Metabolism

Bemben MG, Lamont HS (2003). Creatine supplementation and exercise performance: Recent findings. *Sports Medicine* 35: 107-125.

Boisseau N, Delamarche P (2000). Metabolic and hormonal responses to exercise in children and adolescents. *Sports Medicine* 30: 405-422.

Claing A, Laporte SA, Caron MG, Lefkowitz RJ (2002). Endocytosis of G protein-coupled receptors: Roles of G protein-coupled receptor kinases and β-arrestin proteins. *Progress in Neurobiology* 66: 61-79.

Fitts RH (2004). Mechanisms of muscular fatigue. In Poortmans JR, ed., *Principles of Exercise Biochemistry* (Karger, Basel), pp. 279-300.

Greenhaff PL, Timmons JA (1998). Interaction between aerobic and anaerobic metabolism during intense muscle contraction. *Exercise and Sport Science Reviews* 26: 1-30.

Holloszy JO (2005). Exercise-induced increase in muscle insulin sensitivity. *Journal of Applied Physiology* 99: 338-343.

McMahon S, Jenkins D (2002). Factors affecting the rate of phosphocreatine resynthesis following intense exercise. *Sports Medicine* 32: 761-784.

Mujika I, Padilla S (2000). Detraining: Loss of training-induced physiological and performance adaptations. Part I: Short term insufficient training stimulus. *Sports Medicine* 30: 79-87.

Mujika I, Padilla S (2000). Detraining: Loss of training-induced physiological and performance adaptations. Part II: Long term insufficient training stimulus. *Sports Medicine* 30: 145-154.

Newsholme EA (2004). Enzymes, energy and endurance. In Poortmans JR, ed., *Principles of Exercise Biochemistry* (Karger, Basel), pp. 1-35.

Noble EG, Rice CL, Thayler RE, Taylor AW (2004). Evolving concepts of skeletal muscle fibers. In Poortmans JR, ed., *Principles of Exercise Biochemistry* (Karger, Basel), pp. 36-61.

Pette D, Staron RS (1997). The molecular diversity of mammalian muscle fibers. *International Review of Cytology* 170: 143-223.

Tomlin DL, Wenger HA (2001). The relationship between aerobic fitness and recovery from high intensity intermittent exercise. *Sports Medicine* 31: 1-11.

Part IV

Books

Bishop ML, Fody EP, Schoeff LE, eds. (2005). *Clinical Chemistry: Principles, Procedures, Correlations* (Lippincott Williams & Wilkins, Philadelphia).

Viru A, Viru M (2001). *Biochemical Monitoring of Sport Training* (Human Kinetics, Champaign).

Weltman A (1995). *The Blood Lactate Response to Exercise* (Human Kinetics, Champaign).

Review Articles

Bassett DR, Howley ET (2000). Limiting factors for maximum oxygen uptake and determinants of endurance performance. *Medicine and Science in Sports and Exercise* 32: 70-84.

Billat LV (1996). Use of blood lactate measurements for prediction of exercise performance and for control of training.

Recommendations for long-distance running. *Sports Medicine* 22: 157-175.

Bosquet L, Léger L, Legros P (2002). Methods to determine aerobic endurance. *Sports Medicine* 32: 675-700.

Consitt LA, Copeland JL, Tremblay MS (2002). Endogenous anabolic hormone responses to endurance versus resistance exercise and training in women. *Sports Medicine* 32: 1-22.

Green S, Dawson B (1993). Measurement of anaerobic capacities in humans. Definitions, limitations and unsolved problems. *Sports Medicine* 15: 312-327.

Halson SL, Jeukendrup AE (2004). Does overtraining exist?: An analysis of overreaching and overtraining research. *Sports Medicine* 34: 967-981.

McHugh MP, Connolly DA, Eston RG, Gleim GW (1999). Exercise-induced muscle damage and potential mechanisms for the repeated bout effect. *Sports Medicine* 27: 157-170.

Shaskey DJ, Green GA (2000). Sports haematology. *Sports Medicine* 29: 27-38.

Urhausen A, Kindermann W (2002). Diagnosis of overtraining. What tools do we have? *Sports Medicine* 32: 95-102.

Weight LM (1993). "Sports anaemia." Does it exist? *Sports Medicine* 16: 1-4.

References

American College of Sports Medicine (1998). Position stand on the recommended quantity and quality of exercise for developing and maintaining cardiorespiratory and muscular fitness, and flexibility in adults. *Medicine and Science in Sports and Exercise* 30: 975-991.

Blomstrand E, Celsing F, Newsholme EA (1988). Changes in plasma concentrations of aromatic and branched-chain amino acids during sustained exercise in man and their possible role in fatigue. *Acta Physiologica Scandinavica* 133: 115-121.

Bogdanis GC, Nevill ME, Boobis LH, Lakomy HK, Nevill AM (1995). Recovery of power output and muscle metabolites following 30 s of maximal sprint cycling in man. *Journal of Physiology* 482 (part 2): 467-480.

Bogdanis GC, Nevill ME, Lakomy HK, Boobis LH (1998). Power output and muscle metabolism during and following recovery from 10 and 20 s of maximal sprint exercise in humans. *Acta Physiologica Scandinavica* 163: 261-272.

Booth FW, Thomason DB (1991). Molecular and cellular adaptation of muscle in response to exercise: Perspectives of various models. *Physiology Reviews* 71: 541-585.

Gastin PB (2001). Energy system interaction and relative contribution during maximal exercise. *Sports Medicine* 31: 725-741.

Goldspink G (2005). Mechanical signals, IGF-I gene splicing, and muscle adaptation. *Physiology* 20: 232-238.

Hisatome I, Morisaki T, Kamma H, Sugama T, Morisaki H, Ohtahara A, Holmes EW (1998). Control of AMP deaminase 1 binding to myosin heavy chain. *American Journal of Physiology* 275: C870-C881.

Ploug T, van Deurs B, Ai H, Cushman SW, Ralston E (1998). Analysis of GLUT4 distribution in whole skeletal muscle fibers: Identification of distinct storage compartments that are recruited by insulin and muscle contractions. *Journal of Cell Biology* 142: 1429-1446.

Ross A, Leveritt M (2001). Long-term metabolic and skeletal muscle adaptations to short-sprint training: Implications for sprint training and tapering. *Sports Medicine* 31: 1063-1082.

Tietz NW, ed. (1995). *Clinical Guide to Laboratory Tests* (Saunders, Philadelphia).

Index

Note: The italicized *f* and *t* following page numbers refer to figures and tables, respectively.

A

α-amylase 139
α-bungarotoxin 103
acid-base interconversions
    buffering capacity 10
    buffer system 10
    description of 9
    example with alanine 5*f,* 9-10
    Henderson-Hasselbalch equation 9-10
    K_{eq} (dissociation constant, K) 9
actin
    actomyosin 110
    ATP 16*f,* 110
    F-actin 109-110, 110*f*
    G-actin 109
    polymerization 110
action potential
    depolarization 91, 92*f*
    description of 93
    electrical nerve signal 91, 92*f*
    excitation threshold 91
    potassium channel (K$^+$ channel) 93
    propagation of 93-94, 93*f,* 94*f*
    simple diffusion, passive transport and active
        transport 91-92
    sodium channel (Na$^+$ channel) 91, 92*f,* 93
    voltage-gated channel 91
active site
    description of 39
    induced-fit and lock-and-key models 40
    mutual recognition of enzyme and substrate 40,
        40*f*
adaptations to exercise 122
adenine 16
α-dextrin 139
adrenal medulla 144
aerobic endurance, estimating 288, 288*f*
aerobic exercise 122
affinity 37
α-ketoglutarate dehydrogenase complex 157
allosteric regulation
    description of 124
    enzyme activity, modification of 124
    feedback inhibition 125, 125*f*
    inhibitor 124

Altman, Sidney 65
amino acid metabolism
    acetyl CoA (acetoacetyl CoA) 219
    amino acid catabolism 215, 216*f*
    branched-chain amino acids (BCAA) 216, 217-218
    carbon skeletons of amino acids, fates of 216,
        217*f*
    classification of amino acids 219, 219*t*
    glucose-alanine cycle 220, 220*f*
    glutamate 216, 218-219
    glutamate dehydrogenase 216
    a-keto acids 215
    a-ketoisocaproate 217-218
    in liver during exercise 219-220, 219*t,* 220*f*
    in muscle during exercise 215-219, 216*f,* 217*f*
    oxidative deamination 216
amino acid(s). *See also* amino acid metabolism
    α-amino acids 25, 25*f*
    description of 25-26
    enantiomers 26
    general formula of 25, 25*f*
    names and abbreviations 26, 26*t*
    side chains 26, 27*f,* 28
    structure of 26, 27*f,* 28
    synthesis 222, 222*t*
amino group 5
amino terminal (N-terminal residue) 28
aminotransferases
    alanine aminotransferase (ALT) 297
    amino acid metabolism 215
    aspartate aminotransferase (AST) 297
    reference intervals for 297, 297*t*
ammonia 293
AMP deamination
    adenylate deaminase 133
    ATP controlled adenylate deaminase 133
    ATP depletion, positive effect of 134
    conversions 134, 134*f*
    deamination 133
    inosine monophosphate (IMP) 133, 134
    significance of 133-134, 134*f*
    xanthine oxidase 134
anabolic androgenic steroids (AAS) 301
anabolism
    ATP-ADP cycling 18, 18*f*

biosynthetic processes of 18
central concept of metabolism 18, 18f
metabolites 18
anaerobic carbohydrate catabolism, features of 164
anaerobic exercise 122
anaerobic lactic capacity, estimating 286-287
angstrom (Aod) 4
anions 6
anticodon 62
antioxidant 78
ascorbate (vitamin C) 78
a-tocopherol (vitamin E) 78
atomic orbitals 3
ATP (adenosine triphosphate)
    ADP (adenosine diphosphate) 16, 16f
    AMP (adenosine monophosphate) 16, 16f, 17
    description of 16, 16f, 17
    discrete units of 16
    high-energy compound 17
    phosphoanhydride linkages 16, 17
    P_i (inorganic phosphate) and PPi (inorganic
        pyrophosphate) 16, 17, 17f
    replenishment 260-261
    synthase 159
ATP-ADP cycle
    adenylate kinases 129
    ATP breakdown 128-129
    ATP content of skeletal muscle at rest 129, 129f
    Biochemistry of Exercise Conference 129, 129f
    description of 128
    energy exchange 128, 128f
    in exercise 128-129, 129f
    exercise processes and ATP 128
    functions degrading ATP 128
ATP-CP system. See energy systems

B
balanced chemical equation 8
β-ᴅ-2-deoxyribose 47, 47f
β-ᴅ-ribose 16
beriberi 78
biochemical assessment of exercising persons
    aims and scope of 274-275
    biochemical parameters, classes of 276, 276t
    blood 273, 273f, 274, 274f
    reference interval 275-276, 275f
    suitable tissue for 273, 273f
biochemistry 1
biological substances, classes of 10-11
biological warfare 101-103, 102f
biotin 78
blood and biochemical assessment
    anticoagulants 274
    blood components, separations of 274, 274f
    coagulation 274
    description of 273, 273f
    ethylenedinitrilotetraacetate (EDTA) 274
    serum or plasma 274, 274f

blood lactate accumulation
    exercise and 177
    lactate-time plot 177, 177f
    peak concentration 177
    steady rate 176-177
blood lactate decline 177, 177f, 178

C
calcitriol 78
calcium 78
carbohydrate metabolism in exercise
    anaerobic carbohydrate catabolism 164
    blood lactate 176-178, 177f
    carbohydrate oxidation 160-161, 161f
    carbohydrates, major energy source 137
    citric acid cycle 155-157, 156f-157f
    cyclic-AMP cascade 144-147, 144f-147f
    electron transport chain 157-158, 158f, 159-160,
        159f
    gluconeogenesis 165-169, 170-173
    glycogen 137, 137f
    glycogen metabolism 138-142, 138f-139f, 141f-142f
    glycogenolysis 142-144, 143f
    glycolysis 148-152, 149f, 151f, 152f
    lactate 164-165, 165f
    lactate production 162-164, 163f
    muscle glycogen metabolism 147-148, 148f
    oxidative phosphorylation 158-159, 161-162
    plasma glucose concentration control 173-176,
        174f-176f
    pyruvate oxidation 152-155, 152f-154f
    thresholds 178-179, 178f, 179f
carbohydrate(s). See also carbohydrate metabolism in
        exercise
    from depots to consumption 137, 137f
    description of 67, 83
    digestion 139, 139f, 140
    as energy sources during exercise 267-268, 268f
    glycogen 137
    glycogen metabolism 138-140, 138f-142f
    loading 249
    oxidation, energy yield of 160, 161, 161f
carbohydrates and lipids
    carbohydrates, description of 67, 83
    cell membranes 77-78, 77f
    fatty acids 73-74, 73f, 74t
    interconversion of 199, 199f
    lipids 72, 83
    in living organisms 67
    monosaccharides 68-69, 68f, 69f
    oligosaccharides 69-70, 70f
    phospholipids 75, 75f, 76f
    polysaccharides 70-72, 71f, 72f
    steroids 77, 77f
    triacylglycerols 74-75, 74f, 75f
    vitamins and minerals 78
carboxyl terminal (C-terminal residue) 28
carnitine 190

catabolism
    acetyl coenzyme A 22
    acetyl group 22, 23
    amino acids 22
    β oxidation 22
    catabolic processes 81
    citric acid cycle 22, 23
    degradation of 17-18
    electron transport chain 23
    function of 81
    glucose 22
    glycerol 22
    glycolysis 22
    lipid fatty acids 22
    oxidative phosphorylation 23
    phase of metabolism 17-18
    stages of 21-23, 22f
cations 6
cDNA 231
Cech, Thomas 65
cell membranes
    description of 77, 77f
    integral proteins and peripheral proteins 78
cell structure
    animal cell 11, 12f
    common features of cells 11, 12f
    cytoplasm and plasma membrane 11
    endoplasmic reticulum and eukaryotic cells 11
    mitochondria and cytosol 11
    prokaryotic cells 11
    size, diversity, and common features 11, 12f
cellulose 70, 71f
central fatigue
    BCAA 258, 259
    causes of 258
    during prolonged exercise 258
    tryptophan and tyrosine 258
chemical bonds
    covalent bonds 3t, 4-5, 16f
    description of 4
chemical elements
    atomic mass (atomic weight) 4, 4t
    atomic number 3t, 4
    atoms and electrons 3
    description of 3
    in living organisms 3, 3t
    prefixes of units 4, 4t
chemical reactions and equilibrium
    balanced chemical equation 8
    chemical equation, example 7
    chemical kinetics 40
    chemical reactions, description of 7
    enzyme activity 40
    equilibrium constant (K_{eq}) 8
    free energy of activation (êG‡) 41, 41f
    hydrolysis 7-8
    principle of charge conservation 8
    principle of mass conservation 8

rate 40-41
    reactants and products 7
    transition state 41
    vs. physical processes 7
chloride 78
cholecalciferol 78
cholesterol
    cholesterol ester 77
    concentrations in serum 290, 290t
    HDL 290-291, 291t
    LDL 291, 291t
    measuring, reasons for 292, 292t
    steroids and 77, 77f
Chondodendron tomentosum 103
chromatography 23
chromosomes 46
chylomicrons. See lipoprotein(s)
citrate 155
citric acid cycle 155-157, 156f, 157f
cobratoxin 103
codons 60
coenzymes 78
compound 5
concentration of a solute 7
constitutional formulas 3t, 5, 5f
copper 78
Cori, Carl 170
Cori, Gerty 170
cori cycle 170, 170f
corticotropin 299
corticotropin-releasing hormone (CRH) 298-299, 298f
cortisol
    benefits of measuring 300
    in biochemical assessment 298-299, 298f
covalent modification 125
CP replenishment 261-262, 261f
creatine kinase
    athletes and 296
    description of 295, 296
    repeated-bout effect 296-297
    serum CK after eccentric exercise 296, 296f
creatine phosphate
    ATP regeneration by CP 130-131, 131f
    CK-MM 130
    creatine 129, 130f
    creatine kinase (CK) 130
    description of 129, 130, 130f
creatinine 293-294, 294t
Crick, Francis 48
curare 103
cyclic-AMP cascade
    adrenal medulla 144
    adrenergic 145
    β-adrenergic receptor 145
    cAMP cascade, beginning of 145-146, 145f, 146f
    cAMP cascade and glycogenolysis control 146, 147f
    catecholamines 144

cyclic adenylate (cyclic AMP or cAMP) 145, 145f
    drawbacks 128f, 146
    epinephrine (adrenaline) and norepinephrine (noradrenaline) 144, 144f
    G protein 145
    PKA (protein kinase) 146, 147
    protein kinase, activation of 146, 146f
    sympathetic system 145
cytoplasm and plasma membrane 11
cytosine 47

D
dADP 47
dalton (Da) 4
dAMP 47, 47F
Darwin, Charles 64
dATP 48
denaturation 32-33
dendrites 88
detraining, metabolic changes 264-265, 265t
dinoflagellates 101
discs 97
diurnal variation 300
DNA microarray (DNA chip) 231
dNMP, dNDP, and dNTP 48
double helix (B DNA) 48, 48f, 49
d-tubocurarine 103

E
Ebashi, Setsuro 114
electric eel (Electrophorus electricus) 101
electric organs 101, 102
electrocytes 101, 102f
electrolytes 78
electron clouds 3
electron microscope vs. light microscope 89
electron transport chain (respiratory chain)
    cells, gaining from 158
    chemiosmotic hypothesis 159, 159f, 160
    description of 157-158
    diagram of 158, 158f
    energy yield of 159, 159f, 160
    mobile carriers 157
    protein complexes 157
electrophoresis 23
electroplaxes 101
electrostatic bonds 31, 31f
energy metabolism and anaerobic training
    muscle adaptations to training 253, 253t
    sprint training 252-253
    strength training 248t, 252
    training program parameters 253
energy sources, choice of
    age, effect of 253
    carbohydrates 241, 243
    during exercise 241
    exercise duration 243-244, 243f, 244f
    exercise intensity 241-243, 242f
    exercise program, effects of 246, 246f

    fiber types 247-248, 248t. See also muscle fiber types, conversions
    heredity, effects of 247-248, 248t
    lipids, reduced contribution of 242-243
    during maximal exercise 243, 243f, 244
    during moderate-intensity exercise 244, 244f
    sex and ambient temperature 254
energy sources during exercise
    carbohydrates 267-268, 268f
    classes of 126
    compounds of high phosphoryl transfer potential 267
    detecting adaptations of proportions 250
    effects on proportion of energy sources 250-251, 251f
    gene expression 269
    integration of exercise metabolism 269, 269f, 270
    lipids 268
    modifying proportions during exercise 251-252
    other energy sources 240
    proteins 268-269
    ratio of lipids to carbohydrates, benefits of 251
    summary of 239, 240, 240t, 241
energy state after exercise, restoring
    ATP replenishment 260-261
    CP replenishment 261-262, 261f
    EPOC 263-264, 263f
    glycogen replenishment 262-263, 262f
    myocellular triacylglycerol replenishment 263, 263f
    sources needing replenishment 260
energy systems 239
Engelhardt, Vladimir 108
enzyme reaction rate
    enzyme concentration 42
    ionic strength 43
    Michaelis constant 42
    pH 42, 43f
    substrate concentration 42, 42f
    temperature 42, 43f
enzymes
    aminotransferases 295, 297, 297t
    concentration, changing 125
    creatine kinase 295, 296-297, 296f
    hormones and 304
    measuring 295
    names of 32f, 33f, 39
    reaction rate 40-43, 41f, 42f
epinephrine and exercise metabolism 254-256, 255f
EPOC 263-264, 263f
equilibrium constant (K_{eq}) 8
essential and nonessential amino acids for humans 222, 222t
eukaryotic cells 11
excitation-contraction coupling
    acetylcholine receptor, opening of 115
    Ca^{2+} gradient 116, 117, 117f
    calsequestrin 116

Ca²⁺ pump 115, 115f, 116
dihydropyridine receptor 116
from nervous excitation to muscle contraction 116, 116f
ryanodine receptor 116, 116f
sarcoplasmic reticulum 115
transverse tubules 115
excitation threshold 91
exercise
duration 122
energy sources during 267-270
intensity 122
types of 122
exercise metabolism
age and choice of energy sources 253
carbohydrates 237
central fatigue 258-259
detraining 264-265, 265t
duration and intensity 244-246, 245f, 245t, 246t
endurance training 251-252
energy metabolism and anaerobic training 252-253, 253t
energy sources 126, 239-244, 240t, 242f-244f
energy state restoration 260-264, 261f-263f
energy systems 239
exercise parameters 122
exercise program and choice of energy sources 246, 246f
experimental models 122-124, 123f-124f
fatigue 257-258
features of 120
heredity and choice of energy sources 247, 248t
hormonal effects on 254-257, 255f, 256f
integration of 269, 269f, 270
lipids 238
metabolic control in exercise 124-125
metabolic pathway interconnections 237, 238, 238f
muscle fiber types 248, 248f
nutrition and choice of energy sources 249-250, 249f
peripheral fatigue 259-260, 260t
principles of 121-122
proportion of energy sources 250-251, 251f
proportion of fuels 254, 254f
proteins 238-239
pyruvates 237, 238
sex and ambient temperature 254

F
FAD (flavin adenine dinucleotide)
FADH₂ 20, 21
FMN 19, 20
as hydrogen acceptor 20, 20f
isoalloxazine 19, 20f
oxidizing and reducing agents 20, 21, 21f
reducing agent 20
fat. See triacylglycerol metabolism
fatigue 257-258, 270. See also central fatigue; peripheral fatigue

fatty acid(s). See also lipolytic products during exercise
acetyl CoA 194
AMP-dependent protein kinase 197
β oxidation 191, 192f, 193
carnitine acyltransferase I and II 190, 191, 191f
degradation of 190-193, 191f, 192f
description of 73, 73f
energy demand 194, 195f, 196, 196f
essential fatty acids 196
exercise speeds up oxidation of 196-197
in humans and animals 74, 74t
monounsaturated and polyunsaturated 73
oxidation, energy yield of 193
palmitate degradation 191, 192f
palmitate synthesis 194, 195f
plasma fatty acid concentration and exercise 197-199, 198f
saturated and unsaturated 73, 73f
substrates, preparing 194
synthesis 193-196, 195f, 196f
synthesis of other fatty acids 196
ferritin
description of 281
serum ferritin 281-282
tissue ferritin 281
fluorescence 23
fluorine 78
fluorometry 23
folate 78
follicle-stimulating hormones (FSH) 299
force generation
ADP and ATP 112f, 113
excitation of muscle 111-113, 112f
force of contraction 111, 112f
lever arm and power stroke 112f, 113
free-energy change(s)
bioenergetics 14
chemical thermodynamics 14
determinants of 15
enthalpy change (ΔH) 14
entropy change (ΔS) 14
in exergonic reactions 14-15
joule (J) 15
kilocalorie (kcal) 15
of a reaction 14
standard free-energy change 15
free radicals. See radicals
fructose (d-fructose) 68-69, 69f. See also hexoses
fuels, measuring
fuel breath test 254, 254f
respiratory exchange ratio (RER) 254

G
gene chip 231
gene expression and exercise
changes modifying 231-232
control points affected by exercise 229
doping 235
energy sources during exercise 269

gene expression, definition of 57, 82, 83, 227
genes 227
    kinetics of gene product 230-231, 230f
    mitochondrial biogenesis 235-236, 235f, 236f
    muscle hypertrophy 232-235, 232f-234f
    stages in control of 228-229, 228f
    terra incognita 227-228, 227f
    tools in study of 231
genetic code 60, 61f
genome of living organisms 49-50, 49f
glucagon and exercise metabolism 256, 256f
glucogenic amino acids 219
gluconeogenesis
    description of 165, 166f-167f
    glucose 6-phosphatase 167, 168
    glucose synthesis from pyruvate 165, 167
    input of energy 166
    muscle glycogen from lactate 168, 168f
gluconeogenesis in liver during exercise
    epinephrine and glucagon 145f, 146f, 169, 169f
    fructose 1, 6-bisphosphatase 168
    fructose 2, 6-bisphosphatase 168-169, 169f
glucose 289
glycerol 292. See also lipolytic products during
        exercise
glycogen 71f, 72, 72f
glycogen metabolism
    branching enzyme 140
    carbohydrate digestion 139, 139f, 140
    debranching enzyme 140, 142f
    glucogen synthase 140
    glucose 1-phosphate 140
    glycogen 138, 138f
    glycogenolysis 140, 142f
    glycogen phosphorylase (phosphorylase) 140
    glycogen synthesis 140, 141f
    hepatocytes 140
    monosaccharides 139, 140
    sucrose hydrolyzed to glucose and fructose 139,
        139f
    uridine diphosphate glucose (UDP-glucose) 140
glycogenolysis 140, 142f
glycogenolysis in liver during exercise
    α_1 -adrenergic receptor 171, 171f
    control of 170, 171, 171f, 172f
    diacylglycerol 172
    epinephrine and glucagon 170, 171, 171f, 173
    glycogen in hepatocytes 170, 170f
    inositol 1,4,5-trisphosphate (IP3) 171f, 172
    phosphatidyl inositol 4,5-bisphosphate (PIP) 76f,
        171, 172
    phosphoinositide cascade 171, 171f
    phospholipase C 76f, 171
glycogenolysis in muscle during exercise
    AMP and IMP increase; ATP decrease 143f,
        144
    Ca^{2+} increase 117f, 144
    calmodulin 144
    dephosphorylation 143

epinephrine increase 144
hormone epinephrine 144
phosphatases 143
phosphoserine 143
P_i increase 143, 143f
protein phosphatase 1 (PP1) 143
glycogen replenishment 262-263, 262f
glycolipids and glycoproteins 67
glycolysis
    1,3-biphosphoglycerate 150
    description of 148, 149f
    dihydroxyacetone phosphate 150
    energy produced by 150
    fructose 1, 6-biphosphate 148, 150
    glucose 1-phosphate 149f, 150
    glyceraldehyde 3-phosphate 150
    pyruvate 148
    reactions of 148-150, 149f
glycolysis in muscle and exercise
    GLUT4 150, 151, 151f
    increased substrates 150-151, 151f
    phosphofructokinase activation 132f, 151-152
    pyruvate kinase activation 152, 152f
glycolytic 247
Golgi apparatus (Golgi complex) 11
gonadotropin-releasing hormone (GnRH) 298-299,
        298f
gonadotropins 299
G protein 145
guanidinum 102
guanine 47

H
Hanson, Jean 107
HDL cholesterol 290-291, 291t
HDL lipoproteins. See lipoprotein(s)
hematologic parameters
    erythrocyte count 278-279, 279t
    hematocrit 278
    leukocyte count 278, 279
    platelet count 278, 279
hemodilution 280
hemoglobin
    aerobic capacity and 38
    allostery 38, 39
    Bohr effect 37
    cooperativity 37-38
    description of 36-37, 36f
    erythrocytes 36
    globins 36-37
    iron status and 278, 278t
    oxyhemoglobin 36
    properties of 37-38, 39
    $\dot{V}O_2max$ 38
Henseleit, Kurt 220
heparin 274
hepatocytes 140
heredity and choice of energy sources 247-248, 248t,
        270, 286

hexoses 68-69, 69f. *See also* monosaccharides
high phosphoryl transfer potential
    AMP deamination 133-134, 134f
    ATP-ADP cycle 127-129, 128f, 129f
    creatine phosphate 129-131, 130f, 131f
    description of 127
    energy sources and 267
    equations 127
    phosphagens 127
    sarcoplasm 131-133, 132f
homeostasis 122
hormonal effects on exercise metabolism 254-257, 255f, 256f
hormones 125, 304
    description of 125, 304
    overtraining 301-302
    steroid hormones 297-299, 297-301, 298f, 299f
Huxley, Andrew 107
Huxley, Hugh 107
hybridization 231
hydrolysis 7-8
hydrophilic 7
hydrophobic 7
hyperpolarization 97
hyperventilation 38
hypoxanthine 134

I
inosine 134
inosine monophosphate (IMP) 133
insulin. *See also* plasma glucose concentration
    exercise metabolism 151f, 256-257
    resistance 174
iodine 78
ions
    description of 6
    example of 5f, 6
    hydrogen ions (H$^+$) 6
    molecules 6
    positively and negatively charged 6
    radicals (free radicals) 6
iron deficiency
    description of 282, 282f
    iron deficiency anemia 282, 282f
    risks for athletes 282-283
iron status
    in biochemical assessment 304
    ferritin 281-282
    hematologic parameters 278-279, 2798t
    hemoglobin 278, 278t
    iron 277, 280
    iron deficiency 282-283, 282f
    parameters for assessing 277
    soluble transferrin receptor (sTfR) 281
    sports anemia 279-280
    total iron-binding capacity (TIBC) 280
    transferrin saturation 280-281
isomeric 5
isotopes 4

K
katal 295
ketogenic 219
ketone body formation 156f, 168f, 209, 210f, 211f
ketones 68
kinases 129
kinetics of gene product after exercise
    corresponding proteins 230, 230f
    mRNA and proteins, stabilities of 231
    postexercise mRNA kinetics 230, 230f
Krebs, Hans 155, 220
Krebs cycle 155

L
lactate. *See also* blood lactate
    anaerobic lactic capacity, estimating 286-287
    in blood, measuring 285
    blood lactate response, factors 286
    concentration during prolonged exercise 286, 286f
    description of 285
    utilizing 164-165, 165f
lactate production during exercise
    anaerobic glycolysis 164
    carbohydrates converted to lactate 163-164
    gain from lactate production 163, 163f
    lactate concentration, increase 163
    lactate dehydrogenase 162
    muscle lactate concentration 162
    proton production in muscle 163
    regenerating cytosolic NAD$^+$ 162
lactate system 239
LDL cholesterol 291, 291t
LDL lipoproteins. *See* lipoprotein(s)
lipid metabolism
    controlled study 186f, 187
    exercise and lipolysis 185-188, 185f-188f
    exercise and plasma triacylglycerols 205-207, 205f-206f
    fatty acids 181, 190-199, 191f-192f, 195f-196f, 198f
    ketone body formation and exercise 209, 210f, 211f
    lipid categories 181
    lipids and carbohydrates 199, 199f
    lipolytic products during exercise 188-190, 189f
    lipoproteins 200-205, 200f, 201t, 202f-204f, 205t
    plasma cholesterol and exercise 207-209, 208f
    triacylglycerols 181-185, 182f
lipids. *See also* carbohydrates and lipids
    description of 72, 83
    as energy sources during exercise 268
lipolysis
    β-arrestin 187, 188
    cAMP cascade 186
    controlled study 186f, 187
    description of 184-185
    desensitization of β-adrenergic receptor 187, 187f, 188
    exercise and 185-188, 185f-188f

hormonal control of 185, 185*f*
hormone-sensitive lipase 186
insulin 186
phosphodiesterase 186
PKA 185, 186
triacylglycerols in muscle fiber 188, 188*f*
lipolytic products during exercise 188-190, 189*f*
lipoprotein(s)
chylomicrons 200, 201-203, 202*f*
classes, roles of 200, 201-204, 202*f*-204*f*, 205*t*
HDL 200, 203-204, 204*f*, 204*t*
LDL 200, 203, 203*f*, 205*t*
lipase 202
main plasma lipoproteins 200, 201*t*
plasma lipoproteins 200, 200*f*, 201*t*
separation 200, 200*f*
structure 200, 200*f*
VLDL 200, 203, 205*t*
luteinizing hormone (LH) 299
lysosomes 11
Lyubimova, Militsa 108

M
macromolecules 11
magnesium 78
maltotriose 139
manganese 78
mass spectrometers 23
Menten, Maud 42
messenger RNA (mRNA) 58, 58*f*, 59
metabolic control in exercise
allosteric regulation 124, 125, 125*f*
covalent modification 125
enzyme concentration 125
nervous and hormonal control 125
substrate concentration 125
metabolism
ATP 16-17
complexity and order of 13, 13*f*
description of 13, 81
features of 13, 14*f*
free-energy change (ΔG) 14-15
metabolic pathways 13, 14*f*
oxidation-reduction reactions 18-21
phases of 17-18
tools in study of 23
metabolites
aerobic endurance 288, 288*f*
ammonia 293
anaerobic lactic capacity 286-287
in biochemical assessment 285, 285*t*, 304
cholesterol 290, 290*t*, 292, 292*t*
creatinine 293-294, 294*t*
glucose 289
glycerol 292
HDL cholesterol 290-291, 291*t*
lactate 285-286, 286*f*, 304
LDL cholesterol 291, 291*t*
programming training 287

triacylglycerols 289-290, 289*t*
urea 292-293
methyl group 5
Michaelis, Leonor 42
microvilli 139, 139*f*
miscibility 6
Mitchell, Peter 159
mitochondrial biogenesis, exercise-induced
Ca^{2+} 236, 236*f*
mitochondrion divides 235, 235*f*
nuclear respiratory factors 1 and 2 (NRF1 and NRF2) 235-236
peroxisome proliferator-activated receptors (PPAR) 236
signal molecules, hierarchy of 236, 236*f*
transcription coactivators 236
transcription factors controlling 235-236
molar concentration 7
molecular formulas 5
molecular mass 5
molecular orbitals 4
molecules
carboxyl group 5, 5*f*
constitutional formula 3*t*, 5, 5*f*
description of 5
ions 6
molecular formula 5
polar compound 6, 6*f*, 7, 7*f*
polarity and miscibility 6
mole (mol) 5
monomers and polymers 11
monosaccharides 68-69, 68*f*
movement
action potential 91-94, 92*f*-94*f*
birth of nerve impulse 97, 98*f*
description of 85, 87
involuntary and reflexive 87
lethal arsenal 101-103, 102*f*
motor cortex, brain stem, and cerebellum 87
motor unit 85-86
muscle 85
nerve impulse transmission 94-97, 95*f*, 96*f*
nerve signals 88, 88*f*, 89*f*
nervous motorways 87, 87*f*
neuromuscular junction 98-101, 98*f*, 99*f*, 100*f*
resting potential 89-91, 90*f*, 90*t*, 91*f*
voluntary movement 87
Mullis, Kary 231
multiple sclerosis 94
muscarine 100
muscarinic receptor 99
muscle cell structure
components of 105
features of 105
muscle fibers 105-106, 106*f*
myofibrils 105-106
sarcomere 106, 106*f*
thick filaments 106-107, 106*f*, 107*f*
thin filaments 106*f*, 107, 107*f*

muscle contraction
  actin 109-110, 110f
  control of 114, 114f, 115f
  excitation-contraction coupling 114-117, 115f-117f
  force generation 111-113, 112f
  muscle cell structure 105-107, 106f-107f
  myosin isoforms and muscle fiber types 113
  myosin properties and structure 108-109, 109f
  sarcomere 110-111, 111f
  sliding-filament theory 107-108, 107f
muscle contraction control
  Ca^{2+} 114, 115, 115f, 116
  muscle at rest 114
  thin-filament proteins 114, 114f
  tropomyosin 114, 114f
  troponin (Tn) 114
muscle fiber types, conversions 248, 248f
muscle glycogen metabolism in exercise 147-148, 148f
muscle hypertrophy, exercise-induced
  basal lamina 232, 233
  cytosolic [Ca^{2+}] 235
  gene doping 234-235
  IGFI 233-235, 233f, 234f
  MGF 234
  myostatin 235
  PKB 233-234, 233f, 234f
  satellite cells 232, 232f, 233, 233f
mutations
  defective protein 53
  kinds of 53
  understanding 53
myasthenia gravis 101
myocellular triacylglycerol replenishment 263, 263f
myoglobin
  description of 35-36, 36f
  oxymyoglobin 35
  prosthetic groups and heme 35
myokinases 129
myosin
  light and heavy meromyosin 108-109, 109f
  papain 108, 109f
  properties of 108, 109f
  proteases 108
  proteolysis 108
  S1 109, 109f
  structure of 108-109, 109f
  trypsin 108, 109f
myosin isoforms and muscle fiber types
  fast-twitch and slow-twitch 113
  fiber types, differences between 113
  force of contraction, controlling 113
  gel electrophoresis 113
  isoforms 113
  motor units 113
  myosin 113

N
NAD (nicotinamide adenine dinucleotide)
  forms of 19

H$^-$ (hydride ion) 19
NAD$^+$ 19, 19f
NADP (nicotinamide adenine dinucleotide phosphate)
  NADP$^+$ 19, 20
  NADPH 19, 20f
nerve impulse
  birth of 96, 97f
  Ca^{2+} channel 95, 96, 96f
  cones 96
  exocytosis 95
  neurotransmitters 95, 96-97
  retinal 96
  rhodopsin 96
  rods 96, 97f
  Schwann cells 88
  synapse 94, 95, 94f
  synaptic cleft 95
  synaptic vesicles 95
  transmission 94-96
  visual stimulus 96
nerve signals
  axon 88
  cell body 88
  myelin 88, 89f
  neurons 88, 88f
  nodes of Ranvier 88, 88f
  oligodendroglial cells (oligodendrocytes) 88
  transmission of 88, 89, 89f
nervous and hormonal control
  hormones 125
  signal transduction pathways 125
neuromuscular junction
  acetylcholine receptor 99
  acetylcholinesterase 100
  acetylicholine 98-100, 99f, 100f
  choline acetyltransferase 99
  description of 98, 98f
  free acetylcholine 100, 101
  ligand-gated channel 100, 100f
  motor unit 98
  nicotinic acetylcholine receptor 99-101, 99f
  quaternary amino group 99
  signal transmission across 100, 100f
neutrons 3
niacin 19
nicotinamide 19
nicotine 99
Niedergerke, Ralph 107
nonpolar 6, 7
Northern blotting 231
nuclear magnetic resonance (NMR) spectroscopy 131-133, 132f
nuclear run-on transcription assay 231
nucleic acids and gene expression
  biological organic compounds 10-11
  chromosomes 46
  deoxyribonucleic acid (DNA) 45, 48-53, 48f, 51f, 52f

deoxyribonucleotides 45-48, 46*f*, 47*f*
DNA, primary structure of 48, 48*f*
DNA replication 50-53, 51*f*, 52*f*
double helix of DNA 48, 48*f*, 49
genes and gene expression 54-55, 57
genetic code 60, 61*f*
genetic information flow 46, 46*f*
genome of living organisms 49-50, 49*f*
kinds of 45
mutations 53
phosphodiester linkage 48
ribonucleotides (RNA) 45
transcription 56-57, 56*f*
translation 59-60, 62-64
nucleus 3
nutrition and choice of energy sources
    blood lactate response 286
    carbohydrate loading 249, 249*f*
    fat intake after exercise 250
    fat loading 249, 250
    proportion of energy sources 270
    sport drinks 249

O
oligosaccharides
    disaccharides 69-70, 70*f*
    glycosidic linkage 69, 70*f*
    maltose and sucrose 69-70, 70*f*
overreaching. *See* overtraining
overtraining (overtraining syndrome) 301-302
oxidation-reduction reactions
    dehydrogenations 18
    description of 18
    FAD (flavin adenine dinucleotide) 19-21, 20*f*, 21*f*
    H (hydrogenations) 19
    (NAD) nicotinamide adenine dinucleotide 19, 19*f*
    NADP (nicotinamide adenine dinucleotide phosphate) 19, 20*f*
    relay 21, 21*f*
oxidative phosphorylation
    ATP synthase 159
    description of 158-159
    during exercise 161-162
oxygen carriers 35
oxygen system (aerobic system). *See* energy systems

P
pancreatic lipase 201
pantothenate 78
partial negative charge (δ⁻) (delta minus) 6
partial positive charge (δ⁺) 6
pentoses 68, 68*f*. *See also* monosaccharides
peptide bond
    description of 28, 28*f*
    dipeptide Asp-Phe 28, 28*f*
    peptides 28
    Phe-Asp (phenylalanyl aspartate) 28, 28*f*
    polypeptide chains 28
perilipins 181

peripheral fatigue, causes of 259-260, 260*t*
peroxisomes 11
pH
    acidic and alkaline (basic) 9
    description of 8
    neutral 8, 8*f*
    pH scale 8, 8*f*, 9
phosphoanhydride linkages 16, 17
phosphofructokinase activation 151-152
phospholipids
    alcohols 75, 76*f*
    glycerophospholipids (phosphoglycerides) 75, 76*f*
    phosphatidate 75, 75*f*
    sphingomyelin 75, 76*f*
phosphorolysis 140
phosphorus 78
phosphorylation 23
photoreceptor. *See* nerve impulse
photosynthesis 35
phylloquinone (vitamin K) 78
plasma amino acid, ammonia, and urea concentrations 223
plasma cholesterol and exercise
    acute effects of 208, 208*f*
    chronic effects of 208-209
    LDL and HDL 207
    studies of 207
plasma glucose concentration in exercise
    cytoplasmic proteins 174
    diabetes mellitus 174, 176
    euglycemia 176
    exercise 176
    homeostasis of 176, 176*f*
    homeostatic mechanisms 173
    hypoglycemia 173
    insulin 173-176, 174*f*, 175*f*
    maintenance of 173
plasma glycerol. *See* glycerol
plasma lipoproteins. *See* lipoproteins
plasma triacylglycerols and exercise
    acute effects of 205-207, 206*f*
    atherosclerosis 205, 205*f*
    chronic effects of 207
    postprandial lipemia 206, 206*f*
polar 6, 6*f*, 7
polarity and miscibility 6, 6*f*
    miscibility, description of 6
    polar and nonpolar 6, 6*f*
    polarity, description of 6
polymerase chain reaction (PCR) 231
polymorphism 54
polynucleotide chain 48
polysaccharides
    cellulose 70, 71*f*
    glycogen 71*f*, 72, 72*f*
    heteropolysaccharides and homopolysaccharides 70
    starch 71, 71*f*
potassium 78

principle of charge conservation 8
principle of mass conservation 8
products. *See* chemical reactions and equilibrium
prokaryotic cells 11
propagation of action potential
    along neuron 93, 93*f*
    description of 93
    myelin 93, 94
    node of Ranvier 93, 94, 94*f*
    salvatory conduction 93, 94*f*
protein metabolism
    amino acids 214
    degrading dietary proteins and proteases 214
    effects of exercise on 214, 214*f*, 215
    protein, life cycle of 213, 213*f*
    proteolysis 213-214
    ubiquitin 213-214
protein(s)
    active site 39-40, 40*f*
    amino acids 25-28, 81
    biological organic compounds 10-11
    β pleated sheet 30-31, 31*f*, 81
    contributions to energy expenditure 223
    denaturation 32-33
    description of 25, 81
    as energy sources during exercise 268-269
    enzymes 39-43, 82
    α helix 30, 31*f*, 81
    hemoglobin 36-39, 81-82
    hydrogen bond 30, 30*f*
    myoglobin 35-36, 36*f*, 81-82
    oxygen carriers 35
    peptide bond 28, 28*f*
    primary structure of 29-30, 29*f*, 30*f*
    protein function 34-35, 81
    purposes of 81
    quaternary structure 33, 33*f*
    secondary structure of 30-31, 31*f*
    stability 229
    tertiary structure 31-32, 31*f*, 32*f*
protons 3
pyridoxal phosphate 78
pyridoxine (vitamin B_6) 78
pyrimidines 47
pyruvate kinase activation 152, 152*f*
pyruvate oxidation
    acetyl CoA 153
    CoA (coenzyme A) 152, 153, 153*f*
    decarboxylation 153
    mitochondria 152, 153*f*
    pyruvate 152
    pyruvate oxidation 154
    sulfhydryl group (-SH) 153
pyruvate oxidation during exercise
    [NADH]/[NAD$^+$] ratio 155
    pyruvate dehydrogenase control 154, 154*f*
    pyruvate dehydrogenase kinase 154-155

Q

quaternary structure 33, 33*f*

R

radicals (free radicals) 6
radioactivity counters 23
reactants. *See* chemical reactions and equilibrium
reference interval 275-276, 275*f*
repeated-bout effect 296-297
residue 28
resistance exercise 122
respiratory exchange ratio (RER) 254
resting potential
    active transport 89-90, 90*t*
    concentration gradients 89, 90*t*
    Na^+ and K^+ 89-90, 90*t*
    polarized membrane 91, 91*f*
    resting potential 91
    sodium-potassium adenosine triphosphatase (Na^+-K^+ ATPase) 90-91
    sodium-potassium pump (Na^+-K^+ pump) 90-91, 90*f*
retinal. *See* nerve impulse
retinoic acid 78
retinol (vitamin A) 78
reverse transcriptase PCR 231
ribonucleic acid (RNA) 45, 55-56, 58-59, 61-62, 64-66
    alternative splicing 58, 58*f*
    messenger RNA 58, 58*f*, 59
    RNA editing 59
ribosomal RNA (rRNA) 59
ribosomes 59, 59*f*
RNA processing 229
running and swimming, energy sources in
    energy sources in freestyle swimming events 246, 246*t*
    energy sources in running events 245, 245*t*
    graph differences 243*f*, 244, 245*f*
    percentage contribution of energy systems 245, 245*f*
    proportion of energy systems 244, 245*f*

S

Sanger, Frederick 49, 231
sarcomere
    concentric contraction 111
    eccentric contraction 111
    isometric contraction 111
    nebulin 110
    thin and thick filaments 110-111, 111*f*
    titin 110
sarcoplasm
    drawbacks of NMR 132
    exercise metabolism, NMR study of 132-133, 132*f*
    nuclear magnetic resonance (NMR) spectroscopy 131
    powerful cylindrical magnet 131
    resonation of nucleus in compound, frequency of 131
sarcoplasmic reticulum 115
saxitoxin 101, 102
scurvy 78
selenium 78
serum 274, 274*f*

serum urea concentration. *See* urea
sex, ambient temperature, and choice of energy sources 254
siderophilin 280
signal transduction pathways 125
sliding-filament theory
    sarcomere shortening 107, 107*f*
    sliding-filament model 107-108
sodium 78
soluble transferrin receptor (sTfR) 281
solute(s) 7
solutions
    concentration 7
    description of 7
    molar concentration 7
    water 7, 7*f*
solvent 7
spectrophotometer 23
spectrophotometry 23
sports anemia 279-280
sprint 122
starch 71, 71*f*
steroid hormones 298-299, 299*f*
steroids
    cholesterol 77, 77*f*
    cholesterol ester 77
substrate concentration 42, 42*f*, 125
subunits 33, 59, 59*f*, 60
swimming. *See* running and swimming, energy sources in
synaptic body 97

T
tertiary structure of proteins
    disulfide bond 32, 32*f*
    electrostatic bond 31, 31*f*
    hydrophobic interactions 32, 32*f*
    interactions of 31-32
    Van der Waals interactions 32
testosterone
    anabolic androgenic steroids (AAS) 301
    in biochemical assessment 298-299, 298*f*
    doping 301
    reference intervals for 300, 300*t*
    role of 300-301
    serum testosterone concentration 301
tetrahydrofolate 78
tetrodotoxin 101, 102
thiamine pyrophosphate 78
thiamine (vitamin B$_1$) 78
thresholds
    lactate-intensity plot 178, 178*f*, 179
    proposed thresholds 178
    pyruvate's dilemma 179, 179*f*
thymine 47
thyroid hormones 78
total iron-binding capacity (TIBC) 280
training
    endurance, and energy sources 250-251, 251*f*
    state and blood lactate response 286

training and protein metabolism
    endurance training and muscle proteins 224, 224*f*
    hypertrophy 224
    mitochondrial biogenesis 224
    resistance training and muscle proteins 224, 224*f*
transcription 56-57, 56*f*, 57*f*, 229
transferrin and transferrin saturation 280-281
transferrin receptor 281
transfer RNA (tRNA) 61, 62, 62*f*
translation
    ribosomal RNA (rRNA) 59
    ribosomes 59, 59*f*, 229
    steps for 62, 63*f*, 64
    subunits 59, 59*f*, 60
transmination 215
transverse tubules 115
triacylglycerol metabolism
    adipocytes (fat cells) 181, 182, 182*f*
    adipose tissue 181, 182*f*
    lipid droplets 181, 182*f*
    lipolysis 184-185
    triacylglycerols (fat) 181
    triacylglycerol synthesis 149*f*, 183-184
triacylglycerols (triglycerides)
    acyl group 75
    ester linkage 74
    glycerol (glycerin) 74, 74*f*
    as insulator 75, 75*f*
    melting point, saturated *vs.* unsaturated fatty acids 75
    reference intervals 289, 289*t*
    serum triacylglycerol concentrations 289, 290
tricarboxylic acid cycle 155
trioses (glyceraldehyde and dihydroxyacetone) 68, 68*f*. *See also* monosaccharides
T system 115

U
units. *See* enzymes
urea 292-293
urea cycle
    aspartate as amino group acceptor 221, 222, 222*f*
    carbamoyl phosphate 220-221
    description of 220, 221*f*
    urea production, rate of 222

V
villi 139, 139*f*
vitamins and minerals 78
VLDL. *See* lipoprotein(s)

W
water-soluble vitamins 78
Watson, James 48

X
xanthine oxidase 134
x-ray diffraction 108

Z
zwitterions 6

About the Author

Vassilis Mougios, PhD, is an associate professor of exercise biochemistry at the University of Thessaloniki in Greece. A teacher of exercise biochemistry, sport nutrition, and ergogenic aspects in sport for 17 years, Mougios served on the Scientific Committee of the 2004 Pre-Olympic Congress. He has coauthored many articles in international scientific journals and has done research on muscle contraction, exercise metabolism, biochemical assessment of athletes, and sport nutrition. Mougios is a member of the American College of Sports Medicine, the American Physiological Society, the European College of Sport Science, the New York Academy of Sciences, and Index Copernicus Scientists. He serves as a reviewer for the *Journal of Applied Physiology,* the *British Journal of Sports Medicine,* the *European Journal of Clinical Nutrition, Acta Physiologica, Annals of Nutrition and Metabolism, and Obesity Research.* In his leisure time, he enjoys bicycle riding, hiking, and photography.